The Gymnosperms

THE GYMNOSPERMS

Chhaya Biswas □ **B.M. Johri**

Springer-Verlag

Narosa Publishing House

Cover photograph: *Taxus baccata*. Courtesy Prof. B.D. Sharma,
JNV University, Jodhpur, India

Dr. Chhaya Biswas
Formerly Principal, Gargi College
University of Delhi South Campus
New Delhi 110 049, India

Prof. (Retired) B.M. Johri
Central Reference Library
University of Delhi,
Delhi 110 007, India

ISBN 3-540-61283-1 Berlin Heidelberg New York
ISBN 0-387-61283-1 New York Berlin Heidelberg
ISBN 81-7319-080-1 Narosa Publishing House, New Delhi

Printed in India.

To
Our Teacher

Professor Panchanan Maheshwari
(1904-1966)

Preface

The Gymnosperms is a well-illustrated comprehensive account of living and fossil plants of this group. Chapters 1 and 2 give a general account, and describe similarities and dissimilarities with pteridophytes and angiosperms. Chapter 3 deals with classification. The next 18 chapters (4-21) deal sequentially with fossil and living taxa. Phylogenetic relationships are considered for each order. Chapter 22 discusses the in vitro experimental studies on the growth, development and differentiation of vegetative and reproductive organs and tissues. Chapter 23 summarizes the economic importance of gymnosperms. Chapter 24 gives the concluding remarks.

Thus, there is a complete coverage of significant findings concerning morphology, anatomy, reproduction, development of embryo and seed, cytology, and evolutionary trends and phylogeny. Ultrastructural and histochemical details are given wherever considered necessary.

There is a comprehensive list of literature citations, and a plant index. This book is essentially meant for the postgraduate students in India and abroad. Undergraduate students can also use it profitably. The entire course should be taught in 25–30 lectures/hours and about 75 hours of field and laboratory work.

CHHAYA BISWAS
BRIJ MOHAN JOHRI

Acknowledgements

For the preparation of this book we are grateful to:

Dr. B.S. Venkatachala, former Director, Birbal Sahni Institute of Palaeobotany, Lucknow, for allowing Dr. Chhaya Biswas to work in the Library of the Institute, and to Dr. H.K. Maheshwari, Dr. Sheila Chandra, Dr. A.K. Srivastava, Dr. Usha Bajpai, Dr. Sukh Devi, Dr. S.C. Srivastava and Dr. Jayashree Bannerjee for helping her to update the information, fruitful discussions and suggestions concerning fossil members of the gymnosperms, to the Librarian of the Institute for providing the literature.

(The Late) Professor P.N. Mehra, Department of Botany, Punjab University, Chandigarh, for scrutinizing the outlines of this book and giving valuable advice, for providing his monograph on the *Indian Conifers, Gnetophytes and Phylogeny of Gymnosperms*; Professor B.D. Sharma, Department of Botany, JNV University, Jodhpur, for a critical review of several chapters of the manuscript; to Professor S.P. Bhatnagar, and Dr. S.S. Bhojwani, Department of Botany, University of Delhi, for providing relevant literature and fruitful discussions; Dr. P.S. Srivastava, Department of Botany, Hamdard (deemed) University, New Delhi, for important literature and valuable suggestions; our colleagues in Gargi College, University of Delhi, Dr. Bharati Bhattacharyya for useful suggestions, Dr. Asha Juneja and Dr. Lalita Sehgal for making available several publications from their personal collections, and Dr. Gita Mathur for scrutinizing some of the chapters; our students Arindam Bhattacharyya and Somdutta Sinha Roy for technical assistance in the preparation of the manuscript.

Professor Upendra Baxi (former Vice-Chancellor), Professor A.L. Nagar (former Pro-Vice-Chancellor), Professor A.P. Srivastava (former Librarian) and Mr. Sher Singh (Deputy Librarian), University of Delhi, for providing an office in the Library where this work was completed.

Mr. Kishen Lal and Mr. S.K. Das for preparing the photographs, Mr. R.K. Gupta for typing the manuscript, and Professor N.N. Bhandari (then Head of the Department of Botany, University of Delhi) for allowing us to use the departmental facilities during the preparation of this book.

Our family members, especially Mr. Alok Biswas and Mrs. Raj Johri, for their encouragement and interest in this work.

We wish to express our gratitude to all the persons and institutions mentioned above.

CHHAYA BISWAS
B.M. JOHRI

Contents

Gymnospermopsida—Cycadophytes
(Chapters 6-8)

Gymnospermopsida—Gymnosperms of Uncertain Relationship (Chapters 9-12)

12. Ginkgoales 98-126

Gymnospermopsida—Coniferophytes
(Chapters 13-16)

13. Cordaitales 127-136

19. Welwitschiales   343-365

20. Gnetales   366-404

21. Phylogenetic Considerations: *Ephedra*, *Welwitschia* and *Gnetum*

22. *In Vitro* Experimental Studies

23. Economic Importance

The Gymnosperms

1. Introduction

Indubitable seeds first appeared in the Devonian (395–359 my B.P. = million years Before Present). These seeds are not enclosed in ovaries, as in flowering plants, and are designated naked. This character is the basis for grouping a large number of varied plants as a natural group, the Gymnospermae (gymnos = naked, sperm = seed). Palaeobotanical evidence suggests that the seeds evolved independently in more than one group of Palaeozoic plants and diversified rapidly during the Lower Carboniferous (345 my B.P.). In the course of their evolution, the seed plants have tended to evolve a variety of structures to protect the ovules. Some of these evolutionary "experiments" were successful, and provide insight into the origin of the carpel, while others ended in extinction (see Stewart 1983).

Presently, with the advanced knowledge of fossil plants, the gymnosperms are regarded as a heterogeneous group of seed plants. These plants constituted most of the world's dominant vegetation throughout the late Paleozoic and Mesozoic, and steadily declined thereafter. A modern approach is to interpret the phylogeny of gymnosperms (as already in flowering plants) on the basis of phytogeographical and ecological conditions in which the taxa survived.

Antiquity and Fossil History

The gymnosperms are an ancient group which dates back to the Devonian (Figs. 1.1, 1.2). They began ca. 395 my B.P., lasted for ca. 50 my, and gave rise to aneurophytes, progymnosperms, archaeopterids and pteridosperms. During the Carboniferous (ca. 345–280 my B.P.), a large variety of cordaitels and seed ferns existed. In the Permian and Triassic (ca. 280 and 225 my B.P.) the Carboniferous pteridosperms became extinct. The early conifers (Voltziales) diversified and the cycads and cycadeoids became evident for the first time. Glossopteridales formed a conspicuous flora of the Southern Hemisphere during the Permian. The ginkgophytes appeared some time during the Permian and became more widely spread in the Triassic. In the Jurassic (ca. 195–141 my B.P.), the cycads, cycadeoids, conifers and ginkgophytes reached their peak of diversification, and the glossopterids became extinct. During the Upper Cretaceous (ca. 141–65 my B.P.), the

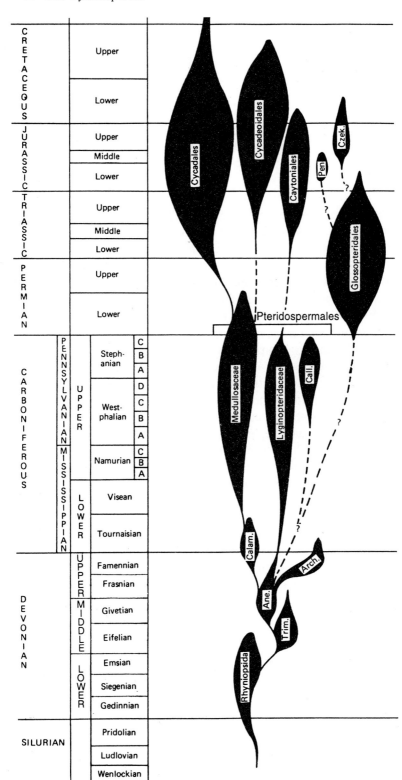

Fig. 1.1   The distribution in geological time of major groups of Cycadophytes and their presumed relatives; their origin and relationship is also suggested. *Ane* Aneurophytales. *Arch* Archaeopteridales. *Calam* Calamopityaceae. *Call* Callistophytaceae. *Czek* Czekanowskiales. *Pen* Pentoxylales. *Trim* Trimerophytopsida. (After Stewart 1983)

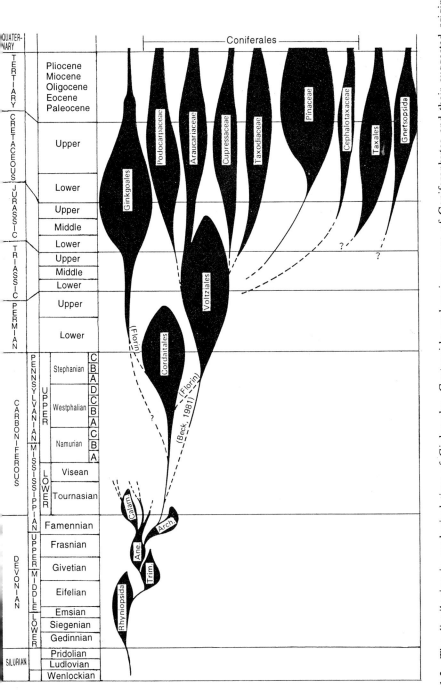

Fig. 1.2 The distribution in geological time of Ginkgoales, Gnetopsida, and major groups of Coniferophytes and their suggested origin and relationships. *Ane* Aneurophytales, *Arch* Archaeopteridales, *Calam* Calamopityaceae, *Trim* Trimerophytopsida. (After Stewart 1983)

angiosperms appeared and diversified rapidly. They began to replace the already declining cycads, cycadeoids, conifers and ginkgophytes. Mesozoic pteridosperms and other smaller groups became extinct. However, the majority of the conifer families have continued up to the present. In the Tertiary (65 my B.P.) the angiosperms evolved steadily while the conifers declined in diversity. Thereafter, the angiosperms became dominant and the gymnosperms occupied a secondary position, although they still dominated landscapes that angiosperms have yet to conquer. The long evolutionary history of gymnosperms provides many examples of taxa which flourished, and finally became completely or nearly extinct. Presently, there are ca. 13 000 genera and 240 000 spp. of angiosperms (Takhtajan 1980), while the gymnosperms comprise only 69 genera and 750–760 spp. These are grouped in seven orders: Cycadales, Ginkgoales, Coniferales, Taxales, Ephedrales, Welwitschiales, and Gnetales.

Geographical Distribution

The Cycadales and Ginkgoales represent the surviving members of extremely ancient groups. The cycads were once a large and dominant group with widest distribution in the Mesozoic (Fig. 1.1), which is sometimes referred to as the age of cycads (Scott 1923). Now there are only 11 living taxa (including *Chigua,* a new genus; see Stevenson 1990a, b) confined to limited areas of the tropics and subtropics. They form neither extensive nor conspicuous features of the vegetation. Six taxa occur in the Eastern Hemisphere, and five in the Western Hemisphere (Fig. 1.3A); no single genus is represented in both hemispheres. Among the eastern cycads, *Macrozamia, Lepidozamia* and *Bowenia* are confined to Australia, and *Encephalartos* and *Stangeria* exclusively to South Africa. The genus *Cycas* occurs from Australia to Japan, touching India and China. Of the western genera, *Dioon* and *Ceratozamia* are confined entirely to Mexico, *Microcycas* to western Cuba, *Zamia* to both areas, and *Chigua* in Colombia (the only endemic cycad in south America). *Stangeria* resembles the ferns so much that it was placed, for a long time, in the family Polypodiaceae. According to Arnold (1953), these isolated genera are "leftovers" from the widely distributed cycads of the Mesozoic.

The order Ginkgoales included many genera and species, reported from the Permian; from the Triassic onwards and during the Mesozoic they had worldwide distribution (Fig. 1.2), but, before the end of the Jurassic, they began to disappear, and all the members of this order, except *Ginkgo*, became extinct by the Cretaceous. The sole surviving member, *G. biloba,* is restricted in geographical distribution. At present its natural occurrence is confined to a small, inaccessible region in southeastern China. There is, however, some uncertainty whether or not such specimens are escapes from gardens (see Andrews 1961). It is doubtful whether it actually exists in the wild/natural state today. One of the reasons for its survival from extinction is

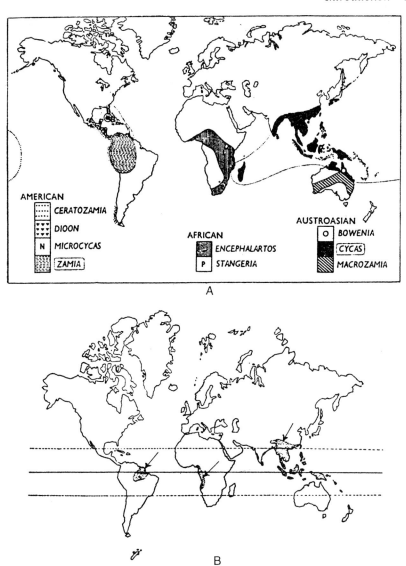

Fig. 1.3 A, B. World distribution. **A** Living cycad genera. **B** *Gnetum* . (**A** After Schuster 1932, **B** after P. Maheshwari and V. Vasil 1961a)

that for centuries priests in China and Japan cultivated and worshipped it. The modern *Ginkgo* is remarkably resistant to attacks of insects and fungi. The reason probably is its immense vigour, which enabled it to survive for millions of years. *Ginkgo* is to be regarded as one of the wonders of the world because it has persisted with very little change through a long succession of ages inhabited by plants and animals quite different from those present in the modern age.

The cycads and *Ginkgo* are designated living fossils because they still exhibit many of their ancestral features without much change. One such character is the presence of motile or swimming sperms, displayed only in these plants among the existing seed plants.

The Coniferales form 75% of the modern gymnosperms, and are an important constituent of the flora today. There are ca. 52 genera and 560 spp. (Mehra 1988), usually grouped into six families. They have a markedly disjunct geographical distribution and several taxa are endemic. Each family is widely distributed, although individual taxa often show extremely limited distribution. Such a distribution pattern indicates antiquity. This is evident in the evolutionary history of each of the six families, which extend back to the Mesozoic (Fig. 1.2). Most modern taxa of the Coniferales were already present by the onset of the Tertiary (Stewart 1983).

The conifers are plants of the more temperate regions of the world, and only a few taxa are strictly tropical. Western North America, Eastern and Central China, and parts of Australia and New Zealand have abundant conifers, which show exceptional diversity.

Pinaceae. The Pinaceae have ten genera, which form prominent coniferous forests of the Northern Hemisphere (Fig. 1.4). This family is totally unrepresented in the Southern Hemisphere (Coulter and Chamberlain 1917).

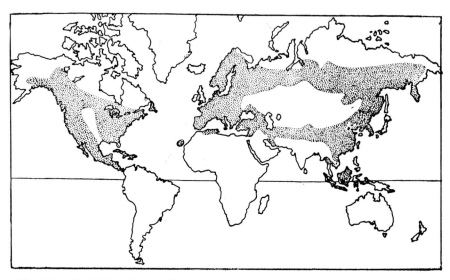

Fig. 1.4. World distribution of pines. (After P. Maheshwari and Konar 1971)

Taxodiaceae. There are ten genera, seven monotypic. In the past, these taxa were an important constituent of the forests of the Northern Hemisphere. At present, they show a relict distribution. Of the ten taxa, only three occur in the USA; *Sequoia sempervirens* (Californian redwood) restricted to a narrow coastal belt in California, *Sequoiadendron giganteum* (big tree) in central California, and *Taxodium* spp. which grow in lowland swamps of the Southeastern USA and Mexico. Of the remaining seven taxa, *Cryptomeria japonica* (Japanese cedar) and *Cunninghamia* spp. are distributed in Japan and parts of China, *Sciadopitys verticillata* (Japanese umbrella pine) in Japan, *Glyptostrobus* (Chinese deciduous cypress) in China, and *Taiwania* in Formosa. *Metasequoia glyptostroboides* (Dawn redwood) was known only as a fossil from Pliocene deposits until 1948, when living specimens were discovered from Szechuan Province, China (Hu and Cheng 1948). *Athrotaxis* spp. (Tasmanian cedars) is the only southern taxon confined to Tasmania.

Cupressaceae. The largest family of the conifers includes ca. 19 genera, 8 monotypic. Of these 19 genera 9 are distributed in the Northern and 10 in the Southern Hemisphere. Of the northern taxa, *Cupressus* (cypress), *Chamaecyparis* (false cypress) and *Thuja* (arbor vitae) have a disjunct distribution. *Juniperus* (Juniper) is distributed in a broad belt round the Northern Hemisphere. This may be due to its female cones which are berry-like and distributed by birds. Some of the southern taxa are *Callitris* (Cypress pines) confined to Australia, Tasmania, and New Caledonia, *Libocedrus* to New Zealand and New Caledonia, and *Papuacedrus* to New Guinea, extending somewhat across the equator into the Northern Hemisphere.

Podocarpaceae. Podocarpaceae is the most important family of the Southern Hemisphere, where it originated. It comprises seven taxa, two monotypic. Some of the taxa are represented in the Northern Hemisphere also, and are assumed to be recent arrivals. The largest genus, *Podocarpus* (yellow woods), occurs in the mountain forests of warm temperate and subtropical regions of the Southern Hemisphere. Some species occur in Japan, China, India, Malaya and The Philippines. *Dacrydium* and *Phyllocladus* are distributed chiefly in New Zealand and Tasmania, *Dacrydium* also occurs in Malaysia, Burma, Southern China, Patagonia and Chile, while *Phyllocladus* is found in New Guinea, North Borneo and the Philippines. They have representatives north of the Equator. *Saxegothaea* (Prince Alberts' yew) and *Microstrobus* (= *Pherosphera*) are monotypic. The former is confined to Chile and the latter to Tasmania. *Acmopyle* is confined to New Caledonia and Fiji.

Araucariaceae. An extremely old family with fossil record extending back to the Triassic of both North and South Hemispheres. Both the present-day taxa are restricted to the Southern Hemisphere. *Araucaria* occurs in South America, Australia, New Guinea and New Caledonia, and *Agathis* (kauri pines) is exclusively eastern, extending from The Philippines to New Zealand and Malaya to Fiji.

Cephalotaxacae. A single taxon, *Cephalotaxus*, is restricted to eastern Asia. It is distributed from the Eastern Himalayas to Japan in subtropical forests.

Taxales. The order Taxales includes a single family, Taxaceae, with five genera. *Taxus* (Yew) occurs in North America, Europe and Asia, and extends up to Malaya. This wide geographical distribution is partly due to birds. *Pseudotaxus* (=*Nothotaxus*) grows in a small part of east China. *Amentotaxus* presently occurs in east Asia, but fossil remains have been reported from Europe and western America. *Austrotaxus* is restricted to New Caledonia. *Torreya* occurs only in California, Florida and eastern Asia.

Ephedrales. The Ephedrales comprise a single monogeneric family, Ephedraceae, and *Ephedra* with ca. 42 spp. has a worldwide distribution. None of the species is common to the Old and the New Worlds or Eastern and Western Hemispheres. Most of the species inhabit arid or desert regions including saline tracts where they act as a sand-binder. The plants grow up to 5000 m above sea level.

Welwitschiales. *Welwitschia* is a monotypic genus of limited distribution. It is restricted to a strip ca. 1200 km long along the *west coast* of southern Africa from Angola (Nicolau River) to South-West Africa, Namibia (Kuiseb River). In the Welwitschia Flats (ca. 45 km east of Swakopmund) 5000–6000 specimens grow (von Willart 1985). *Welwitschia* never reaches the coast. It occupies only the northern and central parts of the Namib desert in its south-north extension and stretches over subtropical grassland, entering the Mopane Savanna in its west-east extension. Thus, it occupies an area with a wide ecological range. It grows mainly where the annual rainfall is between 0.0 and 100 mm and precipitation (except dew and fog) is only in the summer months, January to March. In the Mopane Savanna, annual rainfall can exceed 200 mm, and fog is absent.

Gnetales. The single family, Gnetaceae, includes only one genus, *Gnetum.* It inhabits moist tropical forests in parts of Asia, Africa, northern South America, and certain islands between Asia and Australia (Fig. 1.3 B). Most species, ca. 30, are endemic within the areas of their distribution. There is no species common to both the hemisphere, and none of the Asian species occurs in Africa or America. The center of the present diversification appears to be Eastern Malaysia (see P. Maheshwari and V. Vasil 1961a).

Characteristic Features

The seed plants are usually classified into two major groups: gymonsperms and angiosperms, based on the protection to the ovule (at the time of pollination). The gymnosperms bear naked ovules, i.e. the ovules are borne directly on the sporophyll or an equivalent structure, and are exposed. In some of the gymnosperms, overlapping scales and sporophylls may protect the ovules but they are freely exposed at pollination. In the angiosperms, the ovules develop within an ovary. This distinction between naked and enclosed ovules and later seeds is of considerable importance.

Some of the other features typical of gymnosperms are: There are no herbaceous gymnosperms, and the plants whether trees, shrubs or lianas are woody and evergreen. They have a tap root which usually persists for a long time. The xylem consists of tracheids, parenchyma and rays. Vessels are present in *Ephedra, Welwitschia* and *Gnetum*. They have evolved from pitted tracheids, as shown by intermediate stages between pits and perforations. In phloem only sieve cells differentiate; sieve areas commonly occur on the radial walls as well, and are numerous where the end of one sieve element overlaps that of another. The companion cells are absent (see Chaps. 8, 12, 15, 18, 19, 20 for details.). Secondary growth takes place in all gymnosperms; mature metaxylem shows bordered pits of various types; *Stangeria* and *Zamia* show scalariform thickenings (Pant 1973). The anther has an exothecium. There are numerous light pollen grains which land directly on the surface of the nucellus during pollination. Prothallial cells are formed in the male gametophyte. The ovule is unitegmic and orthotropous. There is a prolonged free-nuclear phase in the development of the female gametophyte, a long interval between pollination and fertilization, a free-nuclear phase in the development of the proembryo, and haploid nutritive and storage tissue (the female gametophyte).

Gymnosperms and Pteridophytes

The gymnosperms and pteridophytes share some common features: (a) Both the groups have independent sporophyte, mostly differentiated into root, stem and leaf. (b) A well-developed vascular tissue is present, the xylem lacks true vessels, and phloem is without companion cells. (c) The leaves of cycads are compound with circinate vernation, while in *Ginkgo* the venation is dichotomous (cf. ferns). (d) Motile male gametes are present in cycads and in *Ginkgo*, and (e) Archegonia are present in the female gametophyte of gymnosperms except *Welwitschia* and *Gnetum*.

The gymnosperms show a definite advance over pteridophytes: (a) There is a continuously growing, long-lasting tap root system in gymnosperms which provides better anchorage to a perennial plant. In pteridophytes the roots are mostly adventitious. (b) The majority of gymnosperms are monostelic. The pteridophytes have a wide range of primary vascular system— protostele, actinostele, siphonostele, solenostele, dictyostele, polycyclic stele and polystele. (c) All gymnosperms show secondary growth, but it is absent in pteridophytes. In *Isoetes* and *Botrychium,* secondary growth occurs, but is not as extensive as in gymnosperms. The mature metaxylem, in most gymnosperms, shows bordered pits. In most pteridophytes, the xylem is typically scalariform. The phloem of gymnosperms differs from the pteridophytes in the type of conducting cells, their cellular composition, and the degree and nature of functional interrelationships between the conducting and parenchyma cells (see Paliwal 1992). A typical pteridophytic sieve element is identical to the elongate parenchyma cells but is longer;

the gymnospermous sieve cells are shorter. The sieve elements have a smaller diameter, their endwalls are horizontal or slightly oblique. The sieve areas are small and their outline varies. The sieve pores occur in the sieve areas as well as scattered on the vertical walls. Callose deposition has been observed in *Psilotum*, lycopods and the ferns (see Paliwal 1992). (d) All gymnosperms are heterosporous. At pollination the microspores are usually carried to the megasporangia by wind; entomophily is reported in *Ephedra aphylla* Forsk. (Bino et al. 1984), *Gnetum* sp. (Kato et al. 1995), and *Welwitschia* (Kubitzki 1990). The need for water to bring about fertilization has been eliminated in gymnosperms. A major difference between the motile spermatozoids of cryptogams and those of cycads and *Ginkgo* is that the former swim freely in water, whereas the latter swim in a fluid medium enclosed within the developing ovule (see Chaps. 8, 12). The majority of the pteridophytes are homosporous with monoecious gametophytes. With this combination there is a greater possibility of self-fertilization; this is disadvantageous since the sporophytes produced are mostly homozygous for all characters and are, therefore, relatively slow in evolutionary progress (see Sporne 1965). A few pteridophytes, e.g. *Selaginella, Isoetes, Stylites, Marsilea, Pilularia, Regnellidium, Salvinia* and *Azolla* are heterosporous. (e) The female gametophyte in gymnosperms is retained inside the megaspore, and is dependent on the sporophyte for nutrition. There is progressive reduction in the archegonia which is without neck canal cells and, occasionally, ventral canal cell. Gymnosperms are seed-bearing plants. Seed formation is absent in pteridophytes.

Gymnosperms and Angiosperms

Both gymno- and angiosperms are seeded plants, and there are many differences in vegetative and reproductive structures: (a) The gymnosperms are slow-growing perennial plants with limited vegetative reproduction. In angiosperms, the plants are annual, biennial or perennial, and have varied means of vegetative multiplication. (b) In angiosperms, both tracheids and vessels are present. The sieve-tube elements with the sieve plates are confined to the end walls, and companion cells are present.

The wood in gymnosperms is : (i) Manoxylic—relatively meager xylem with wide parenchymatous rays, and of no economic use. (ii) Pycnoxylic—xylem relatively dense with small and narrow wood rays, making up most of the trunk and branches. The wood is commercially very useful. The angiosperms do not show such a distinction. (iii) Pollination and dispersal of seeds in gymnosperms is mostly anemophilous; in angiosperms entomophilous, hydrophilous and zoophilous. (iv) The ovules are naked and unitegmic in gymnosperms. During pollination, the pollen grains enter the ovules directly through the micropylar canal and rest on the nucellus. However, in *Gnetum* (P. Maheshwari and V. Vasil 1961a), *Tsuga, Pseudotsuga, Abies* (Doyle 1945), *Torreya* (Favre-Duchartre 1967) and the extinct Caytoniales (Thomas 1925, Harris 1951), the pollen grains germinate (a short distance) away from the nucellus (an angiospermic tendency) or inside the micropyle. In angiosperms the carpel is closed and its upper part is differentiated into style and stigma. Pollen grains land on the stigma and the pollen tubes grow all the way down through the style before they can reach the ovules, which are uni- or bitegmic. Johri (1936), in a few carpels of *Butomopsis lanceolata,* reported, germinated pollen grains and bits of pollen tubes in the stylar canal and ovary. In one case, a pollen grain had germinated on an ovule. Sahni (see Sahni and Johri 1936) considered these features to be very significant, since a typically gymnospermous character has been discovered in a confirmed angiosperm. Sahni also pointed out that the pollen grains must have been sucked in (into the style) as in the micropyles of ovules in gymnosperms. The pollen grains were three-celled and positively of *Butomopsis*. Sahni recalls the inter-micropylar pollen grains in *Caytonia* (see Harris 1951; Chap. 9).

It may be added that (entire) intercarpellary pollen grains (but no pollen tubes comparable to the condition in *Butomopsis*) have been reported in several angiosperms (see Johri and Ambegaokar 1984). (v) The development of female gametophyte in gymnosperms is monosporic (tetrasporic in *Gnetum* and *Welwitschia*). It undergoes a variable number of free-nuclear divisions followed by wall formation, and remains haploid. It bears archegonia at the micropylar end. In angiosperms the development of the female gametophyte may be mono-, bi- or tetrasporic, and a 4–16–nucleate embryo sac is formed. (vi) The haploid female gametophyte in gymnosperms functions as endosperm, after fertilization. In angiosperms the endosperm is triploid, formed by the fusion of the two polar nuclei and the second male gamete (double fertilization). (vii) In gymnosperms, the zygote undergoes free nuclear divisions (except in *Sequoia, Gnetum* and *Welwitschia*) followed by wall formation. Later, only the apical cells give rise to the embryo. Cleavage polyembryony is known in several taxa. In angiosperms, the division of zygote is invariably followed by wall formation (except in *Paeonia*). Cleavage polyembryony is absent.

The angiosperms are much more advanced than gymnosperms, and have diverse habit and vegetative form. They grow successfully in all habitats and have an efficient means of vegetative propagation which ensures rapid multiplication. There is a greater reduction in the gametophytes—prothallial cells are absent in the male gametophyte, and the female gametophyte lacks archegonia. The ovules are enclosed in an ovary which protects the ovules. This protection gives biological advantages to the angiosperms over the gymnosperms. The seeds develop within the ovary; when mature, these are effectively and efficiently dispersed by insects, birds, wind and water.

The antiquity and the fascination of a group of plants which stand between cryptogams and flowering plants (in their reproductive development) make the gymnosperms a very interesting and fruitful area for investigations.

2. Seed Development

While the plant form is quite diverse, most gymnosperms show a more or less similar mode of reproduction. The developmental aspects and important variations are dealt with under each order. *Gnetum* and *Welwitschia* show remarkable deviations from the rest of the gymnosperms.

Microsporangium

There are two main patterns of initiation of microsporangium (Fagerlind 1961, 1971): (a) In most taxa it is initiated from a meristem which is differentiated into the epidermis and the hypodermal archesporium, e.g. *Cephalotaxus* (Fig. 15.6.5 A, B). (b) It is initiated from a meristem with exposed (surface) initial cells which divide to form an outer epidermis and an inner archesporium, e.g. *Cedrus*, *Larix*, *Picea*, *Pinus* and *Pseudotsuga* (see H. Singh 1978). The archesporial cells show a higher protein content, RNA and histone.

Microsporogenesis

The archesporial cells divide periclinally to form the primary parietal and primary sporogenous layers. The primary parietal layer undergoes periclinal divisions and gives rise to a multilayered microsporangial wall, the innermost layer functions as tapetum. The cells of the epidermal layer develop thick walls (except along the line of dehiscence) and is designated exothecium. The middle layers become crushed during enlargement and maturation of the microsporangium. The tapetum encloses the sporogenous tissue, and comprises a single (occasionally double) layer of large, richly cytoplasmic multinucleate cells. The tapetum is mostly of the secretory type. The walls of tapetal and sporogenous cells are mostly pectinaceous, those of the epidermis and wall layers predominantly cellulosic. The tapetal cells show maximal activity during meiosis in microspore mother cells, and degenerate after the spores are released from the tetrads. The main function of the tapetum is to nourish the sporogenous cells and young microspores. In addition the tapetum is concerned in the production and release of callase, transmits PAS-positive material into the locule, forms an acetolysis-resistant membrane (see Pacini et al. 1985), and finally the exine of the microspores.

In most gymnosperms, cytokinesis following meiosis is simultaneous and centripetal callose walls are laid down between the four microspore nuclei. The Golgi bodies become active and produce a variety of vesicles during anaphase II and telophase II. These vescicles become arranged in the region of the future wall. A two-layered membrane of ER origin separates the four microspores. Eventually, the callose spreads between the two layers. Plasmodesmatal connections do not exist either between the dividing mother cells or between the microspores in a tetrad. Callose with a low permeability isolates the mother cells and their products from the influence of adjoining tissues. This helps the microspore mother cells to follow an independent course of development during meiosis.

Male Gametophyte

The development of the male gametophyte mostly follows a uniform pattern, except in *Welwitschia* and *Gnetum* (see Chaps. 19, 20).

The initial step in the pollen grain is the formation of one (cycads) or two prothallial cells which are usually inconspicuous and ephemeral. The prothallial cells are absent in Cupressaceae, Taxodiaceae, Cephalotaxaceae and Taxales. The microspore functions directly as the antheridial initial.

After the formation of prothallial cells, the antheridial initial divides to form a small antheridial cell and a large tube cell. The latter is usually vacuolate and shows a large nucleus, while the small antheridial cell remains attached to the intine at the site of prothallial cell(s). The antheridial cell generally divides periclinally to form the stalk cell towards the pollen wall, and body cell. Initially, the stalk cell is surrounded by a distinct wall which eventually breaks down, and its cytoplasm merges with that of the tube cell. In later stages the stalk and tube nuclei are indistinguishable from each other. The body cell enlarges, has dense cytoplasm, and a large nucleus. It divides and gives rise to male gametes.

The formation of male gametes is schematized below:

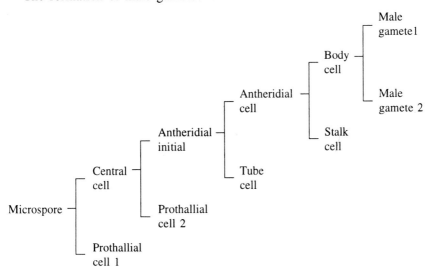

The male gametes may be flagellate spermatozoids, as in cycads (Fig. 8.14 E, F) and *Ginkgo* (Fig. 12.15 E), or unflagellate male cells (equal/unequal), as in all the other taxa. The male cells are common in the members of Cupressaceae, Taxodiaceae, Araucariaceae and *Gnetum* spp., while male nuclei (needs confirmation) have been reported in taxa where the archegonium is placed individually, as in Pinaeae. Reports of male nuclei are erroneous, and all male gametes have a cytoplasmic sheath even if there is no cellulosic wall. In a few species of *Cupressus*, and occasionally in *Juniperus*, the body cell divides several times to produce multiple male gametes.

Winged pollen grains occur in several taxa, e.g. *Podocarpus* (Fig. 15.4.7 K), *Microcachrys*, *Pherosphera*, and members of Pinaceae. The shape and structure of the wings vary.

The number of cells in mature pollen, at shedding, differs in various taxa. The pollen grain is one-celled in *Taxus, Chaemaecyparis, Callitris, Cryptomeria* and *Cupressus*; two-celled in *Cephalotaxus, Athrotaxis, Torreya* and *Taxodium*; three-celled in most cycads, *Welwitschia* and *Gnetum*; four-celled in *Pinus*; five-celled in *Cedrus* and *Ephedra*; muticelled (due to a large number of prothallial cells) in podocarps and araucarias.

Ovule

The young ovule has a central nucellus covered by a single integument. *Ephedra* has two coverings and *Welwitschia* and *Gnetum* three. All along the central region of integument up to the nucellus is a narrow passage, the micropyle. There is a conspicuous chalaza, the funicule is not recognizable. The ovule is mostly orthotropous in gymnosperms, both extinct and extant. In the family Podocarpaceae the ovule is anatropous (Konar and Oberoi 1969 b).

Megasporogenesis

The young nucellus has one to several hypodermal archesporial cells. They divide periclinally to form the primary parietal and primary sporogenous cells. The parietal cell and its derivatives divide periclinally and give rise to a massive parietal tissue capping the primary sporogenous cells. The latter may also divide once or twice, or one or more cells may function directly as megaspore mother cell. The latter are elongated, have prominent nuclei, dense cytoplasm, and a thick wall. They undergo meiosis and produce triads (due to undivided upper dyad cell) or linear tetrads, generally the lowermost megaspore functions. The division takes place in a layer of callose, indicating that meiosis takes place under controlled conditions, comparable to the division of the microspore mother cell giving rise to microspores (see Konar and Moitra 1980). One to several layers of densely cytoplasmic cells differentiate around the sporogenous cells, and become progressively conspicuous during subsequent stages. This is designated the spongy tissue or tapetum. Eventually, it degenerates and becomes compressed between the female gametophyte and outer tissues of the ovule.

Megaspore Wall. Pettitt (1966) has reinterpreted the structure of the megaspore wall in gymnosperms and the following generalizations have been put forward: (a) The megaspore wall is usually a multiple structure, and various parts are laid down in an orderly and controlled manner. They can be linked to other formative events in the morphogenesis of the ovule. There is considerable variation between the species. (b) For the construction of wall, the material is supplied by both the sporophyte and gametophyte. From a very early stage of development, polymer sporopollenin begins to accrue in the wall.

Pettitt (1977) investigated the ontogeny of megaspore wall (preferring to call it wall instead of membrane), and demonstrated features comparable with that of pollen wall.

The foundation for the megaspore wall, laid on the free surface of the coenocytic megaspore, consists of fibrillar matrix with radially aligned hexagonal chambers. The fibrillar coating is interpreted as primexine, and serves as the template of the sculptured sexine of the mature megaspore wall (this role of sexine is similar in both pollen and megaspore wall). In the beginning, a finely granular material (the probacula) accumulates in the chambers in the proximal region, and later extends radially to form a series of interconnecting columns (homologous to the bacula of sexine of pollen wall).

Simultaneously, a continuous granular basement (homologous to nexine I or foot layer) is deposited within the undersurface of the fibrillar matrix. As development proceeds, the columns and basement layer become electron-opaque and amorphous. At the same time, they become resistant to acetolysis. These are regarded as evidence for the introduction of the polymer sporopollenin into the wall, and the changes in appearance are visible consequences thereof. Three separate and distinct substrata are subsequently formed below the patterned layer. The first layer (which appears to be absent in pollen wall) is rich in polysaccharide, and the material for its development possibly originates in the dictyosomes of the megaspore cytoplasm. The second substratum (equivalent to nexine II of pollen wall) is acetolysis-resistant. It is formed when electron-opaque globules are extruded across the plasma membrane from large, membrane-bound vesicles of unknown origin. The third layer (interpreted as equivalent to intine of the pollen wall), as the first, contains polysaccharides in vesicles, possibly of dictyosomal origin. This layer appears when the female gametophyte enters the cellular phase, and is structurally continuous with the radial walls of the alveoli.

The principle features of construction and structure of the megaspore wall in a number of cycad genera are comparable but differ from each other, and from *Ginkgo* in several details (Pettitt 1977).

The nucellar tissue degenerates progressively as the megaspore and its wall develop. This results in the formation of a thick covering of cellular detritus (containing sporopollenin and lipids) which accumulates against the surface of the female gametophyte.

Various investigations have revealed that a megaspore wall occurs in all extant gymnosperms. However, it has a more elaborate structure in the most primitive members (cycads and *Ginkgo*) where fertilization is affected by motile male gametes (Pettitt 1977).

Female Gametophyte

In monosporic development, the chalazal megaspore undergoes several free-nuclear divisions; the number of nuclei varies from ca. 256 in *Taxus baccata* and *Torreya nucifera*, to ca. 8000 in *Ginkgo biloba*. The number of free-nuclear divisions is usually constant for a particular species. These nuclei become arranged in a thin peripheral layer of cytoplasm enclosing a central large vacuole. The free-nuclear gametophyte enters the cellular phase through the formation of alveoli followed by anticlinal walls laid down centripetally. In the beginning, the alveoli have no walls on the inner side facing the central vacuole, and the persistent spindles appear to guide the laying down of wall material. Different alveoli close at various distances from the centre of the gametophyte. A large number of alveoli are present at the micropylar and the chalazal pole of the usually oval gametophyte. Repeated periclinal divisions occur in the closed alveoli, except in those which function as archegonial initials. Initially, the gametophyte shows rows of radiating cells, but in older gametophytes this arrangement is lost due to the laying down of irregular walls. Generally, the nuclei of the cells adjacent to the megaspore membrane are larger than those in the middle of the gametophyte.

In *Welwitschia* and *Gnetum*, instead of one large central vacuole in the female gametophyte, there are several vacuoles and the nuclei are dispersed throughout the cytoplasm. Wall formation takes place by free-cell formation.

In all gymnosperms, the female gametophyte develops a central cavity, the corrosion cavity, to contain and nourish the developing embryos. Reserve food materials like fat, protein and starch accumulate in the gametophyte during maturation of the seed. This is utilised at the germination of the seed. Thus, the female gametophyte (usually termed endosperm after fertilization) serves a dual function of bearing the gametes as well as nourishing the embryo.

Archegonia.   The archegonia develop in all gymnosperms, except *Gnetum* and *Welwitschia*, mostly at the micropylar pole; lateral and chalazal archegonia are rare. Apical archegonial complex with a varying number of archegonia are reported in members of the Cupressaceae and Taxodiaceae (except *Sciadopitys* where four or five archegonia occur singly).

The closed and undivided alveoli function as archegonial initials. The nucleus divides, and a periclinal wall is laid down, forming a large central cell and a smaller neck initial. The latter undergoes several divisions to form a short neck. It helps in the entry of the male gamete into the egg cell, at the time of fertilization. The central cell enlarges considerably, its nucleus lies just below the neck, and the cytoplasm becomes highly vacuolate, foam stage of the archegonium. The nucleus of the central cell divides to form an ephemeral ventral canal cell/nucleus and egg nucleus. The latter enlarges considerably, and the cytoplasm of the egg accumulates numerous proteid vacuoles. A closer look with an electron microscope has shown that the proteid vacuoles are only cytoplasmic formations of several types; large inclusions, small inclusions, microbodies and vesicular bodies (Fig. 15.1.14). Besides these inclusions, the cytoplasm of the egg also shows a conspicuous ER, plastids, mitochondria, Golgi bodies and ribosomes. The egg nucleus is usually large and filled with nucleoplasm (see Konar and Moitra 1980).

EM studies of egg of various species (*Pinus nigra*—Camefort 1962; *Biota orientalis*—Chesnoy 1967) show that a large number of mitochondria, present in the cytoplasm, organize in a circle around the egg nucleus and form the perinuclear zone. The latter increases in width due to continued migration and multiplication of mitochondria from the general cytoplasm. The egg nucleus is Feulgen-negative and the perinuclear zone is Feulgen-positive.

One to three layers of densely cytoplasmic cells, the jacket layers, surround the central cell/egg/proembryo. There is a thick wall between the jacket cells and the central cell of the archegonium, due to the secondary deposition of the wall material. This thick wall with simple pits forms the egg membrane during post-fertilization stages. The jacket layer is concerned in the nutrition of egg/proembryo. The food reserve (starch or protein) in the gametophytic cells becomes available to the egg/proembryo through plasmodesmata on the inner tangential walls of jacket cells. These cells also secrete the enzyme.

The gametophytic tissue circumjacent to the archegonia undergoes divisions and the cells enlarge, resulting in an upward growth forming a depression, the archegonial chamber; the archegonia appear sunken.

Pollination and Fertilization

Pollination. In different gymnosperms the ovules receive pollen at various stages of development: (a) sporogenous cells or megaspore mother cell in *Ginkgo*, conifers and taxads; (b) free-nuclear gametophyte in many cycads and *Gnetum*; (c) young archegonia in *Macrozamia*, and (d) mature archegonia in *Ephedra*.

The pollen is produced in large quantities, is dispersed by wind, and the adjoining areas become covered by yellow dust, the sulphur shower. In *Ephedra aphylla* (Bino et al. 1984), *Gnetum* sp. (Kato et al. 1995) and *Welwitschia* (Kubitzki 1990) there is effective insect pollination (see Chaps. 18, 19, 20).

At the time of pollination, the gymnosperms investigated so far show a sugary exudate at the micropyle. The exceptions are: (a) *Abies, Cedrus, Larix, Pseudotsuga* and *Tsuga* have a stigmatic micropyle; (b) *Araucaria, Agathis* and *Tsuga dumosa*, where the pollen grains do not land on the micropyle.

The pollen is received on a pollination drop secreted at the micropyle of the ovule, a short distance away from the nucellus. The pollination drop is sucked in into the micropyle after the pollen grains are caught in it. Finally, the pollen lands on the nucellus, where it germinates.

The pollination drop, its secretion, chemical composition and the mode of transport of pollen grains require fresh study using modern tools and techniques.

The pollen tubes grow towards the female gametophyte. At its tip the tube contains the body cell, and later the male gametes. The short and unbranched tube is quite prominent. In cycads and *Ginkgo* (see Chaps. 8, 12) the pollen tube is non-siphonogamous. It arises from the distal (upper) end of the pollen grain, elongates and grows laterally and intercellularly into the apical region of nucellus. The tube branches in *Cycas* and *Ginkgo* but remains unbranched in *Zamia*. These pollen tube branches are enucleate and are not concerned with the transport of male gametes for fertilization. Their main function is to absorb nutrition from the circumjacent tissue. From the proximal end also a small tube is formed through which the motile spermatozoids pass into the pollen chamber (see Friedman 1987, Choi and Friedman 1991, Johri 1992, Norstog pers. comm).

Fertilization.   The release of male gametes varies in different taxa of gymnosperms. (a) In cycads and *Ginkgo*, the gametes are released in the archegonial chamber which may contain a fluid. (b) In gymnosperms where the archegonia occur singly (*Pinus, Podocarpus, Cephalotaxus, Taxus, Ephedra*), the neck cells degenerate and the pollen tube penetrates the egg cell and releases the male gametes. (c) In taxa where the archegonial complexes are laterally placed as in *Athrotaxis* and *Callitris*, the pollen tubes grow adpressed to the neck of several archegonia. The male gametes, while still inside the pollen tube, become closely attached to the neck cells (which form a passage), which eventually degenerate. The male gametes then pass into the egg cell leaving behind its cytoplasmic sheath. Two male gametes can enter two archegonia (Brennan and Doyle 1956, Baird 1953). On entering the egg cytoplasm, the male gamete (along with some cytoplasm) comes in contact with the egg nucleus. The non-functional male nuclei, once inside the egg cytoplasm, usually persist for some time.

Studies with the light microscope, as well as EM, on the characteristics of the male and female cytoplasm during fusion and subsequent development (in a few conifers), have led to the concept of neocytoplasm. The nucleoplasm of both male and female nuclei form the ground substance of neocytoplasm. The neocytoplasm alone takes part in the formation of embryonal cytoplasm; the remaining cytoplasm of zygote degenerates. Further studies are called for.

Embryo

The development of embryo can be divided into three phases.

(a) Proembryogeny. Development beginning with the division of zygote up to the stage prior to elongation of the suspensor. There is heterogeneity in the development and four types can be distinguished: (i) Cycad and Ginkgo type, (ii) Conifer type, (iii) Ephedra and Sequoia type, and (iv) Gnetum and Welwitschia type.

(b) Early embryogeny. Varies in different taxa; it includes elongation and proliferation of suspensors and formation of young embryonal mass.

(c) Late embryogeny. Further development of the embryo, and establishment of polar meristems, i.e. root and shoot.

Seed

Seed Coat. The seed coat may develop (a) mainly from the tissue derived from the chalazal portion of the ovule, e.g. cycads, members of Pinaceae, and *Cephalotaxus*; (b) both from chalaza and integument, e.g. *Ginkgo*, podocarps, taxads and *Gnetum*; (c) mainly from integument, e.g. araucarias, taxodioids etc. Whatever the mode of origin, the initial changes are (i) an increase in the number of cell layers in the particular region followed by (ii) differentiation of mucilage canals (cycads) and resin ducts (in conifers), (iii) differentiation of xylem and phloem in the provascular strands (cycads, *Cephalotaxus*, taxads, *Ephedra*, *Gnetum* and *Welwitschia*), (iv) deposition of tannin in several cells distributed all over the young seed coat, and finally (v) the differentiation of three specific layers in seed coat—outer parenchymatous sarcotesta, middle sclerenchymatous sclerotesta, and innermost thin-walled endotesta. In members of the Pinaceae, a few layers of cells of the ovuliferous scale, adjacent to the ovule, give rise to a wing. In Araucariaceae, the wing arises from the entire bract scale, and in *Welwitschia* from the outer envelope, the perianth (see Chap. 19).

Temporal Considerations

The time lag between ovule initiation and seed maturation in angiosperms may usually be only a month or so, while in gymnosperms it takes much longer. There are 1-year, 2-year and 3-year types of reproductive cycle. The year is counted by the number of winter rests, are followed by growth in the spring, through which the ovule passes. In most temperate gymnosperms, the course of development of the reproductive structures is seasonal, i.e. not continuous, while in the tropical and subtropical gymnosperms like cycad, *Gnetum* and *Ephedra*, the development of the ovule is a continuous process. In *Ephedra* the reproductive cycle takes about 3 months to complete. In other gymnosperms, the development of the male cone and the ovule on a plant or a stand is usually synchronous. The details of pollination and fertilization in different ovules are more or less the same.

3. Classification

One of the objectives of systematists is to discover a phylogenetic system of classification. The evolutionary history of gymnosperms extends far back in geological time (Figs. 1.1, 1.2); its fossil record begins in the Upper Devonian and spreads over nearly 350 my B.P. (million years Before Present) (Beck 1985). Most branches of the phylogenetic tree are extinct or have left only fragmentary palaeobotanical records. Several presumed gymnospermous remains have been described from the Upper Devonian and Lower Carboniferous. It is, therefore, natural that there will be diversity of opinion on the phylogenetic relationships and classification of plants. However, it is felt that classification schemes should be flexible, and should be continuously revised in the light of growing knowledge (Arnold 1948). The evolutionary relationships of fossil plants are, in most instances, not completely known (Miller 1985). Until reproductive organs are discovered in organic connection with the vegetative parts, their taxonomic placement is entirely optional (Sporne 1965).

In classification, both vegetative and reproductive organs are important. A knowledge of anatomy, embryology, cytology, palynology and phytochemistry should be used in determining the validity of the system of classification (see H. Singh 1978).

Sahni (1920) recognied two main phyletic lines in gymnosperms:

1 **Phyllosperms.** Leaves large and much divided, ovules borne on leaves or organs regarded as such: (a) Pteridospermales, (b) Cycadales, and (c) Bennettitales.

2 **Stachysperms.** Leaves simple, ovules borne on stems:
(a) Cordaitales, (b) Ginkgoales, (c) Coniferales, and (d) Taxales including *Taxus, Torreya and Cephalotaxus*; separated from Coniferales.

Florin (1948) upheld the separation of Taxales as an Order, coordinate in rank with the Cordaitales, Ginkgoales and Coniferales (restricted). However, he included only *Taxus, Torreya, Nothotaxus, Amentotaxus, Austrotaxus* and their fossil relatives, and retained *Cephalotaxus* within Coniferales.

Chamberlain (1920, 1935)—almost simultaneously with Sahni—recognized two main groups amongst gymnosperms:

1 **Cycadophytes.** (a) Cycadofilicales, (b) Bennettitales, and (c) Cycadales.

2 **Coniferophytes.** (a) Cordaitales, (b) Ginkgoales, (c) Coniferales, and (d) Gnetales.

The main difference between Sahni's and Chamberlain's classification is the emphasis on the primary characters. Sahni's main basis is the morphological nature of the ovule-bearing organs. This, according to Scott (1923), is mainly a theoretical distinction for groups like Cordaites, Bennettitales and Coniferales. Chamberlain emphasizes mainly the differences in habit, stem anatomy and leaves.

Arnold (1948) gave a palaeobotanical foundation to Chamberlain's (1935) system. He interpreted the class Gymnospermae as an artificial polyphyletic group where the seed habit arose independently in cycadophytes and coniferophytes. Arnold, therefore, recommended that the class Gymnospermae should be dropped from the systems of classification. He divided Gymnosperms into three phyla.

1 **Cycadophyta.** (a) Pteridospermales, (b) Cycadeoidales, and (c) Cycadales.

2 **Coniferophyta.** (a) Cordaitales, (b) Ginkgoales, (c) Taxales, and (d) Coniferales.

3 **Chlamydospermophyta.** (a) Ephedrales, and (b) Gnetales.

According to Arnold, the Cycadophytes and Coniferophytes represented natural groups and had separate origin.

Pant (1957) modified Arnold's classification:

Division 1. Cycadophyta: Class Pteridospermopsida: (a) Lyginopteridales, (b) Medullosales, (c) Glossopteridales, (d) Peltaspermales, (e) Corystospermales, and (f) Caytoniales. Class Cycadopsida: (g) Cycadales. Class Pentoxylopsida: (h) Pentoxylales. Class Bennettitopsida (Cycadeoideopsida): (i) Bennettitales (Cycadeoidales).

Division 2. Chlamydospermophyta: Class Gnetopsida: (a) Gnetales, (b) Welwitschiales.

Division 3. Coniferophyta: Class Coniferopsida: (a) Cordaitales, (b) Coniferales, (c) Ginkgoales. Class Ephedropsida: (d) Ephedrales. Class Czekanowskiopsida: (e) Czekanowskiales. Class Taxopsida: (f) Taxales.

Bierhorst (1971) presented a slightly different classification:

1 **Cycadopsida:** (a) Pteridospermales, (b) Cycadales, (c) Cycadeoidales, and (d) Caytoniales.

2 **Coniferopsida:** (a) Cordaitales, (b) Coniferales, (c) Taxales, and (d) Ginkgoales.

3 **Gnetopsida:** (a) Ephedrales, (b) Gnetales, and (c) Welwitschiales.

Most classifications of gymnosperms are based on criteria proposed by Chamberlain (1935) and Arnold (1948). Of the two major lineages recognized by them, the Cycadophyta was presumed to have evolved from ferns, while the origin of the Coniferophyta was unknown. *Pitys* and *Callixylon*, petrified tree trunks with coniferophytic secondary xylem, were regarded to be the oldest representatives. It was later discovered that *Pitys* is a lyginopterid pteridosperm (Long 1963, 1979), *Callixylon* is a progymnosperm (Beck 1960), and the Mesozoic seed ferns are much more diverse and abundant than was formerly realized (Miller 1985). These discoveries conflict with Arnold's scheme of classification.

Beck (1981) and Rothwell (1982) proposed schemes to account for the role of progymnosperms. The progymnosperms show a combination of pteridophytic (spore-producing reproductive structure) and gymnospermic (arborescent, abundant secondary xylem, bordered pits on xylem tracheids, and leaf traces without leaf gaps) characters (see Beck 1970). This group is considered to be the progenitor of gymnosperms, i.e. the Cycadophyte and Coniferophyte lines converging at the base.

Stewart (1983) proposed a classification which is a synthesis of several different natural systems. He used evidence from the fossil record which has contributed extensively to establishing the natural relationships within the major groups of vascular plants:

1 **Progymnospermopsida** (ancestors of gymnosperms): (a) Aneurophytales, (b) Archaeopteridales, and (c) Protopityales.
2 **Gymnospermopsida** "Cycadophytes": (a) Pteridospermales, (b) Cycadales, (c) Cycadeoidales; "Gymnosperms of uncertain relationship": (d) Caytoniales, (e) Glossopteridales, (f) Pentoxylales, (g) Czekanowskiales, (h) Ginkgoales; "Coniferophytes": (i) Cordaitales, (j) Voltziales, (k) Coniferales, (l) Taxales.
3 **Gnetopsida** *Gnetum, Ephedra, Welwitschia.*

Meyen (1984, 1986) proposed a classification for all gymnosperms, past and present. He erects three classes, namely, Ginkgoopsida, Cycadopsida and Pinopsida, which agree with the currently available palaeobotanical data. The classes are based on the analysis of the structural variation in gymnosperms and interpretations of homologies of various organs. The Ginkgoopsida is established as a distinct Class and a major phylogenetic line coordinate with the Cycadopsida and Pinopsida. This is a significant departure from the preceding classification. It comprises Calamopityaceae, Callistophytales, Glossopteridales (Arberiales), Peltaspermales, Caytoniales, and Ginkgoales, among others. The Cycadopsida and Pinopsida are relatively natural groups. Meyen treats Gnetales and Welwitschiales in the Cycadopsida (the reason for this treatment is not stated), and Ephedrales in the Ginkgoopsida (because of its platyspermic seed and striated pollen). According to Meyen (1986), the gymnosperms are a monophyletic taxon originating from one of the progymnospermic Orders, probably Archaeopteridales.

During the past (nearly) 50 years, significant progress has been made in understanding the characteristic features of the established taxa. However, new taxa have not been described, and the interrelationships between the existing taxa have also not changed significantly (Rothwell 1985).

In this book the classification (given below) is based on Stewart's (1983) suggestions:

1 **Progymnospermopsida:** (a) Aneurophytales, (b) Archaeopteridales, and (c) Protopityales.

2 **Gymnospermopsida:** (Cycadophytes): (a) Pteridospermales, (b) Cycadeoidales, (c) Cycadales; (Gymnosperms of uncertain relationships): (d) Caytoniales, (e) Glossopteridales, (f) Pentoxylales, (g) Czekanowskiales, (h) Ginkgoales; (Coniferophytes): (i) Cordaitales, (j) Voltziales, (k) Coniferales, (l) Taxales.

3 **Gnetopsida:** (a) Ephedrales, (b) Welwitschiales, (c) Gnetales.

For a more convincing scheme of classification of gymnosperms, data from all disciplines (mentioned earlier) should be critically analysed to frame a phylogenetic tree — as "natural" as possible.

4. Progymnospermopsida

Beck's (1960) discovery of the organic connection between the leaf of a free-sporing fern, *Archaeopteris*, and a stem, *Callixylon,* with gymnospermous characters led to the establishment of the Class Progymnospermopsida, the probable progenitors of gymnosperms (Namboodiri and Beck 1968). This finding of "plants having gymnospermic secondary wood and pteridophytic reproduction" had a profound effect on the interpretation of the origin of gymnosperms and their systematics.

The above specimen, from the Upper Devonian (ca. 395 my B.P.) beds in New York State, is 80 cm long. It had compressed fertile and sterile foliage of *Archaeopteris macilenta* type attached to partially permineralized stem and rachis axes of *Callixylon zalesskyi.* Beck (1962) reconstructed the plant as a tree up to 18 m tall with a crown of spirally arranged bipinnate fronds (Fig. 4.1). He used the name *Archaeopteris* for the entire plant, according to nomenclatural priority.

Before 1960, *Archaeopteris* was presumed to be the foliage of a Late Devonian fern-like plant with large compound leaves. Nearly 15 spp. are now known. The bipinnate frond with opposite or subopposite fertile and sterile pinnae are the penultimate branches of the frond. The fertile pinnules bear one or two rows of adaxial fusiform sporangia which dehisced longitudinally. *Callixylon* is also a Late Devonian/Lower Carboniferous taxon which often occurs in the same bed as *Archaeopteris*. Permineralized specimens range from small twigs and branches to trunks with a diameter of 1.5 m and height of nearly 20 m. In transection, many of them show a conspicuous pith with a ring of mesarch primary xylem strands at the periphery in contact with the secondary xylem. The latter is compact, and the wood is typically pycnoxylic (Fig. 4.2 A) as in modern conifers. Most of the trunk is made up of secondary wood and the tracheids are pitted in a characteristic manner (typical Cordaitean type). On the radial walls, the pits are arranged in groups separated by unpitted regions, and this pattern is repeated on adjacent tracheids. In a longitudinal radial section, continuous horizontal bands of pits are separated by unpitted bands (Fig. 4.2 B). This feature makes the genus immediately recognizable.

Fig. 4.1. *Archaeopteris* sp., reconstruction (After Beck 1962)

Phylogenetic Considerations of *Archaeopteris*

The arborescent habit, megaphyllous leaves, secondary growth, circular bordered pits, ray tracheids, collateral vascular bundles and probable heterosporous reproduction suggest *Archaeopteris* to be intermediate between the Psilophyte and the seed plant (Beck 1962). It had, however, advanced beyond the Lower and Middle Devonian psilophytic level and approached the Carboniferous gymnosperms. *Archaeopteris* is also similar to the pteridosperms in foliage and fructifications, and in its primary body to both pteridosperms and some coniferophytes. The secondary wood is remarkably similar to the coniferophytes. There is, thus, considerable evidence in support of its being a progymnosperm (one of the advanced genera). It had either evolved concurrently with (and in some ways parallel to) both primitive pteridosperms and coniferophytes, or was directly ancestral to some pteridosperms or coniferophytes. Further information is, however, necessary for its final determination (Beck 1962).

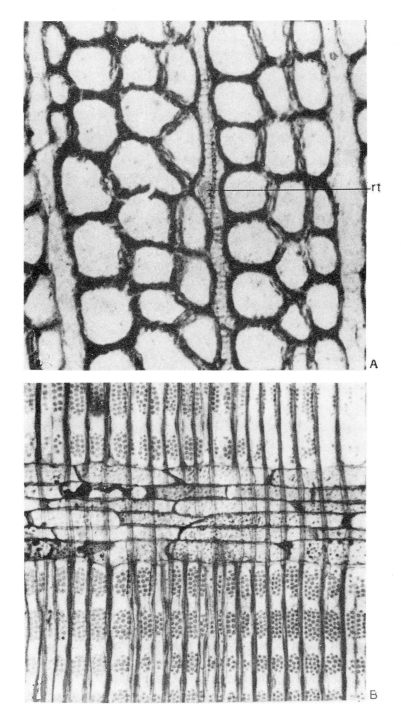

Fig. 4.2 A, B.   *Callixylon newberryi.* **A** Transection secondary wood shows ray tracheids *(rt)* . **B** Radial section, secondary wood with grouped pitting. (After Beck 1970)

5. Gymnospermopsida

According to Beck (1976), the Progymnospermopsida represents a network of Devonian age from which Cycadophytes and Coniferophytes evolved (Chap. 4). The class Gymnospermopsida (excluding Gnetopsida) forms a natural group of vascular plants comprising at least two divergent phyletic lines (Cycadophytes and Coniferophytes), which evolved certain distinctive characteristics (Stewart 1981).

In Cycadophytes (Chaps. 6–8) including Pteridospermales, Cycadeoidales and Cycadales, the plants have (a) large pinnate frond-like leaves, (b) stem with a well-developed pith and cortex, (c) manoxylic secondary xylem with very long tracheids of large diameter, several rows of bordered pits mostly on the radial walls, and high multiseriate wood rays, (d) the ovules and pollen organs are borne on modified leaves, and (e) a dicotyledonous embryo (wherever known) in a seed.

In Coniferophytes (Chaps. 13–16), including Cordaitales, Voltziales, Coniferales and Taxales, the plants generally have (a) spirally arranged simple leaves, (b) compact woody stems with pycnoxylic secondary xylem of tracheids and narrow vascular rays, a gymnosperm-type eustele where leaf traces diverge from sympodia and (c) heterospory in most species.

Besides these two groups, there are other Mesozoic seed plants whose relationship are unknown/uncertain (Chaps. 9–12).

6. Pteridospermales

The discovery of this group of gymnosperms from the Palaeozoic is one of the notable achievements of the palaeobotanists, and of enormous importance to phylogeny. It includes fern-like assemblage of seed plants which first appeared in the Upper Devonian. It extended to the Mesozoic through the Carboniferous and Permian (Fig. 1.1) and became extinct millions of years ago. Originally, they were presumed to be ferns and the name age of ferns given to the Carboniferous period was largely due to this misconception. Williamson (1887) first recognized, in the carboniferous flora, plants combining in their anatomical structures the characters of both ferns and cycads. Keeping this in view, H. Potónié (1899) named this group Cycadofilices to indicate its composite character. It was later discovered that most of these fern-like leaves were associated with true seed structures. Because of their similarities to the pteridophytes on the one hand, and spermatophytes on the other, the name Pteridospermae was proposed by Oliver and Scott (1904). It also included Cycadofilices (plants containing seeds).

A criterion for the separation of true ferns from the seed plants has been recognized (see Thomas 1955). The fossil leaves of Cycadales, Cycadeoidales (Bennettitales), Ginkgoales and Coniferales have cuticles which resist the action of strong acidic oxidizing agents. They could be separated and, after suitable treatment, examined for the shape of epidermal cells and structure of stomata (see Chap. 7). The fragments of fossil fern leaves dissolve completely when treated in a similar manner. Some of the Palaeozoic and most of the supposed Mesozoic pteridosperm fronds have cuticles similar to gymnosperms. The same treatment reveals the cutinized membranes in fossil seeds, pollen-bearing structures, and pollen grains. By comparing the form, size, appearance of epidermal cells, and the stomatal apparatus of reproductive organs with those of associated leaves, it could be determined from which leaves the fertile organs were derived. Cuticular studies thus helped in the reconstruction of original plants.

The fossil remains of most of these plants are, however, of fragmentary nature and only a few are completely known. The description is, therefore, based on isolated parts which are assigned to a large number of form genera.

Some of the features that generally characterize pteridosperms are: (a) Plants rather large with slender stems. There is a solid or a medullated protostele with mesarch (rarely exarch) primary xylem, occasionally polystelic. The bifacial vascular cambium produced secondary phloem and a limited amount of manoxylic secondary wood composed of tracheids with multiseriate pitting, especially on the radial walls. There is a prominent and characteristic outer cortex of longitudinally aligned fiber strands. (b) The leaves are mostly large, fern-like, and often multipinnate. (c) The ovules and microsporangia are borne on unmodified or only slightly modified foliage. The seeds are borne on the fronds in a variety of ways. In some, they are partially enclosed in a distinct *cupule*. The microsporangiate organs are terminally clustered sporangia which usually form large complex synangial fructifications (Andrews 1961).

The Pteridospermales are more diverse in their reproductive biology than any other group of vascular plants. They have been studied intensively, as they are central to our understanding of gymnosperm evolution.

The Palaeozoic Pteridospermales are generally classified into four families: Calamopityaceae, Lyginopteridaceae, Medullosaceae, and Callistophytaceae. It is presumed that the seed ferns originated from protostelic Aneurophytales (Middle Devonian) through the Calamopityaceae (Upper Devonian and Lower Carboniferous) where the other families diverged. The Callistophytaceae and Lyginopteridaceae are abundant in the upper Carboniferous along with Medullosaceae which continued into Permian (Fig. 1.1).

Lyginopteridaceae

Among the Palaeozoic pteridosperms, the Lyginopteridaceae represent some of the oldest (stratigraphically), simplest and most primitive forms (Pigg et al. 1986). The members occur in the Carboniferous (see Fig. 1.1). Several stem genera, ovules and different types of pollen organs have been discovered from permineralized remains alone (Taylor and Millay 1981). However, many questions still remain unanswered, concerning the morphology, affinities of disarticulated organs and reconstruction of whole plants.

Lyginopteris oldhamia (Calymmatotheca hoeninghausi)

This is the most familiar and best known of the pteridosperms. Different plant parts have been discovered at various times and given separate names. It illustrates a classic case of fossil plant reconstruction from 1828–1929. *L. oldhamia* is Upper Carboniferous plant discovered in England, Europe and North America, and is extremely common in coal balls from the coal mines of Lancashire and Yorkshire.

Habit. The frequently branched stem *(L. oldhamia)* is long, slender, ca. 3–4 cm in diameter, and soft-textured. It bore ca. 0.5m long spirally arranged leaves. The radial symmetry of the stem indicates that the plant grew erect, although it seems impossible that such a small stem could have supported a large crown of leaves without additional support. It is assumed that the plant reclined somewhat against other plants, or against steep cliffs, and grew in thickets or jungle-like associations, or that they were small scrambling lianas, or possibly shrubs.

Root.   The roots *(Kaloxylon)* are adventitious and grow from the leaf-bearing portion of the stem, as well as from the older parts from where the leaves had already fallen. The larger roots have a diameter of about 7 mm and produce numerous lateral roots. It has exarch protostele with considerable secondary growth. The diarch lateral roots arise exactly opposite to the protoxylem.

Stem.   Well preserved petrifactions of the stem show a large central pith of parenchymatous cells and clusters of thick-walled cells (sclerotic nests) with dark contents. The size of the pith, and the amount of the two types of tissues, with respect to each other, vary. Several mesarch (nearly exarch) primary xylem strands are present around the pith. There is a continuous ring of secondary wood (Fig. 6.1A) of large tracheids with numerous angular bordered pits irregularly arranged on the radial walls. The rays are 1–12 cells wide and four or five cells high. Occasionally, the amount of internal secondary wood varied. In well-preserved stems, the secondary phloem can be distinguished immediately outside the secondary wood, followed by a narrow band of radially aligned cells referred to as a periderm. The outer cortex consists of radially elongated fibre bands which form a regular anastomosing network *(Dictyoxylon* cortex) in their longitudinal course through the stem (Fig. 6.1 B). The windows of this network are filled with parenchymatous cells. The fibrous outer cortex makes *Lyginopteris* one of the easiest fossils to identify and must have given rigidity to the stem. In transection, the disposition of fibres gives the general impression of Roman numerals on a clock face (Fig. 6.1 A).

Leaf.   The leaves were borne in a 2/5 phyllotaxy. The frond *(Sphenopteris hoeninghausii)* forked 10 cm from its point of junction with the stem, and each of the two divisions was three-times pinnate (Andrews 1961). The petiole was slightly swollen at the base, and forked about half way up the frond into two arms. The pinnae were laterally arranged on the rachis below the fork as well as above it. Secondary pinnae bear two alternate rows of small-lobed pinnules (Fig. 6.2 D). The epidermis is cutinized. The lamina showed a mesophyll of palisade and spongy parenchyma. Stomata occur on the lower side. Figure 6.2 C shows a compression-impression fossil of *Sphenopteris*-type foliage terminating in clusters of cupules (with characteristics of *Stamnostoma*; see Stewart 1983).

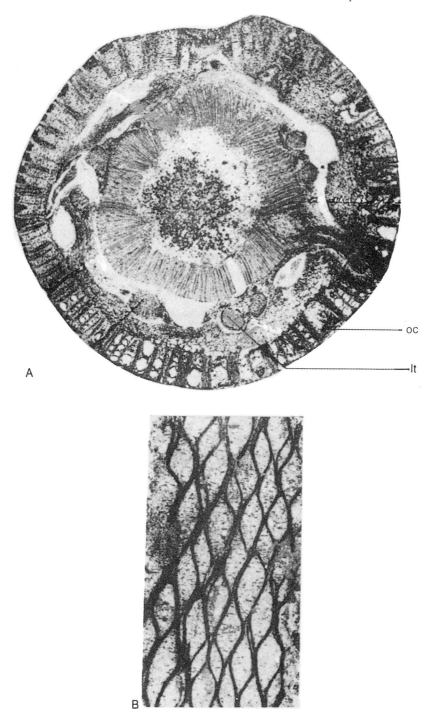

Fig. 6.1 A, B. *Lyginopteris oldhamia* stem. **A** Transection. *It* leaf trace. *oc* outer fibrous cortex. **B.** Outer cortex in tangential longisection. (After Andrews 1961)

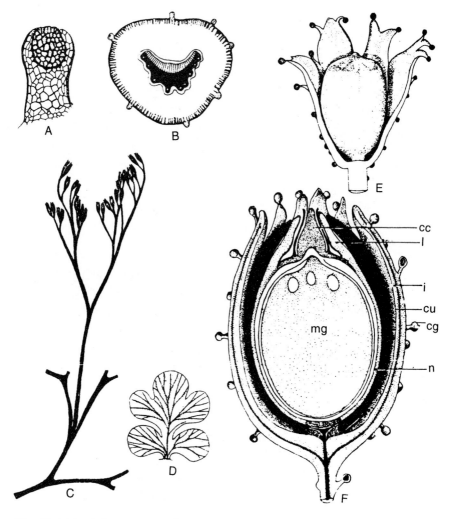

Fig. 6.2 A-F. **A** *Lyginopteris oldhamia,* capitate gland. **B** *Lyginorachis* sp., transection petiole. **C** Compression-impression fossil of *Sphenopteris* type foliage terminating in *Stamnostoma* cupules. **D** *Sphenopteris* pinnules. **E** *Lagenostoma ovoides* in cupule of *Calymmatotheca.* **F** *Lagenostoma ovoides,* longisection ovule in cupule (*cu*), capitate glands (*cg*) occur on cupule. *cc* central column. *i* integument. *l* lagenostome. *mg* megagametophyte shows archegonia. *n* nucellus. (**A** After Scott 1909, **B, C** after Long 1963, **D** after Arnold 1947, **E** after Oliver and Scott 1904, **F** after Long 1944)

The base of the petiole shows two strands. A leaf trace originates by tangential division of a primary wood strand, passes out through the secondary xylem (by way of a large ray.) and is associated with a small strip of secondary tissue on its outer side. The trace divides in the inner cortex of the stem; the two resultant traces traverse the cortex and ultimately unite again at the base of the petiole to form a V-, Y- or W-shaped bundle with the concavity towards the upper side (Fig. 6.2 B). The xylem is completely surrounded by phloem, and there are several protoxylem groups on the lower side. The hypodermis consists of thick-walled cells, and the inner cortex contains plates of stone cells. The isolated preserved fragments of petioles show "butterfly" traces and are usually placed in the form genus *Lyginorachis*.

Except the roots, all other parts of the plant have capitate glands which provide the clue to the connection between the seed and the vegetative organs. The glands are stalked or sessile. The spherical head contained a mass of cells which probably had a secretory function (Fig. 6.2 A). The glands are considered as emergences, formed by the epidermis and outer part of cortex.

Sporangia. Pollen-bearing organs have not been observed in organic connection with the stems or seeds of Lyginopteridaceae. However, most palaeobotanists now presume that *Crossotheca* is the likely microsporangiate fructification of *Lyginopteris*. The fertile organs attached to foliage are classified as *Sphenopteris*. In *Crossotheca*, the fertile branch tip is slightly expanded into a circular or oval limb, described as hair brush or epaulettes. On the margins of the lower surface these bear a fringe of suspended bilocular sporangia, 3 mm by 1.5 mm (Fig. 6.3). The sporangia are apparently devoid of annuli and contain spores with triradiate markings. *Crossotheca*-type microsporangia produce trilete prepollen of the dispersed type observed frequently in *Lagenostoma* pollen chambers (Stewart 1983).

Several authors have suggested that biotic pollination may have occurred in some Carboniferous pteridosperms. To attract pollinators (insects and animals), there could be some early type of nectary as glandular hairs, e.g. capitate hairs on the cupule lobes of *Lagenostoma* (Taylor and Millay 1979).

Seed. The seed *(Lagenostoma lomaxi)* is usually without a cupule. However, during early development, it was surrounded by a lobed cupule covered by capitate glands (Fig. 6.2 E). The cupules terminate the ultimate naked branches of the frond, which also bears sterile pinnules. The cupules are tulip-shaped or like the husk of a hazel nut. The envelope formed by the cupule has eight to ten lobes in the distal half. Reconstructions show that the lobes apparently open outward, which allowed the ovule to shed. The cupule enclosed a single erect ovule attached to it by its base.

The seeds are barrel-shaped, radially symmetrical and 5.5 mm long, 4.4 mm in diameter. There is a single vascularized integument which adheres to the nucellus, except in the apical region.

The nucellus shows a conical prolongation at its apex and is surrounded by a moat-shaped pollen chamber (Fig. 6.2 F). In well-preserved specimens, numerous pollen grains have been observed in this space. The integument is quite stout, there are nine vascular bundles, and there are no free lobes

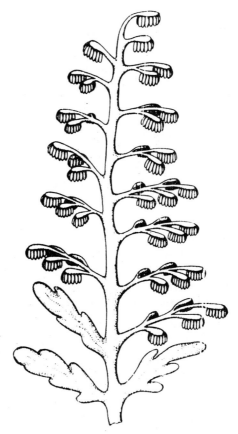

Fig. 6.3 *Crossotheca* sp., male fructification, fertile pinnule shows pendent sporangia. (After Andrews 1961)

at the apex. There is a simple hole through which the lagenostome (a pollen-capturing device) protruded slightly (Fig. 6.2 F). Long (1944) described a specimen of a closely related species, *Lagenostoma ovoides* Williamson, where the female gametophyte was preserved and showed several archegonia near the micropylar end. The central apical portion of the megagametophyte elongated into a "tent-pole" apparatus. It may have ruptured the megaspore membrane and floor of the pollen chamber, and exposed the archegonia to the microgametophytes. After pollination, it may have pushed the central column into the distal end of the micropyle and plugged it.

Most of the seeds discovered in English petrifactions are isolated. It is presumed that they were released from the cupule by a basal abscission mechanism (Andrews 1961).

Phylogenetic Considerations

For several years, the pteridosperms were presumed to be an evolutionary link between the ferns and cycads. There is, however, evidence that the origin of Filicopsida is independent of the Cycadophytes. Thus, any similarity between the two groups must be interpreted as parallel evolution (Stewart 1983).

The Palaeozoic pteridosperms display several evolutionary adaptations and variations which, in later geological time, became established in the vascular plants that evolved from them (Stewart 1983). Some major trends in evolution are: (a) The planated bilaterally symmetrical frond evolved from three-dimensional radially arranged telome trusses. (b) The evolution of (gymnospermous) eustele from protostelic ancestors; and manoxylic wood. (c) Cyclic synangia evolved from clusters of microsporangia which terminated fertile telome trusses. (d) A variety of pollen and prepollen types evolved from trilete spores. (e) The protective layers, around the megasporangium, evolved from adjacent telomes to form integumentary layers and cupules. (f) Special mechanisms such as the lagenostome and pollination drops evolved to direct pollen to the pollen chambers. Pollen tubes appeared for the first time.

7. Cycadeoidales

Foliage generally attributed to Cycadophytes (cycads and cycadeoids) has been observed from the Permian. In the Mesozoic there are several remains of foliage considered to be definitely of cycads and cycadeoids (they cannot be separated by morphological features alone). The leaves had characteristic leathery texture due to heavy cutinization of the epidermis, and abundant supporting tissue within the lamina and around the veins. This is because the foliage was exposed to arid and semi-arid climates for 200 or more million years and, therefore, could survive the adverse conditions of the Permian much better than the Pteridosperms (Arnold 1953). These tissues are able to resist decay, the compressed foliage often retains part of the cutinized epidermis which can be separated from the rock and studied (microscopically).

Extensive studies of the epidermal structure of cycadophytes have been made to understand the relations between living and fossil forms. Nathorst (1902, 1909), a pioneer in the investigation of cutinized material, developed a technique where the pattern of epidermal cells could be shown by colloidion films. Thomas and Bancroft (1913) studied and compared the epidermal characters of recent cycads with several Mesozoic forms. Florin (1931) observed two types of stomata in gymnosperms. These differ in the mode of origin of the guard and subsidiary cells: (a) Haplocheilic—the stomatal mother cell divides once, and the daughter cells form guard cells. The adjoining cells are the subsidiary cells and they often encircle it. (b) Syndetocheilic—the mother cell divides twice to form two lateral subsidiary cells, the third division produces the guard cells. The haplocheilic stomata are characteristic of Pteridosperms, Cycadales, Ginkgoales, Cordaitales, Coniferales and Ephedrales; the syndetocheilic of Cycadeoidales, Welwitschiales, Gnetales and some angiosperms. P. Maheshwari and V. Vasil (1961 b) report haplocheilic stomata in *Gnetum* (see Chap. 20). Harris (1932) showed that the epidermal cells of cycads vary in shape, and the depth at which the stomata lie below the surface also varies. The cuticle covers the entire outer surface of the epidermis, except the actual stomatal opening (Fig. 7.1 A-E). In Cycadeoidales, the cuticle which covers the guard cell extends backward under the subsidiary cell (Fig. 7.1 F,G), and is thickest where the cuticle of the subsidiary cell joins it.

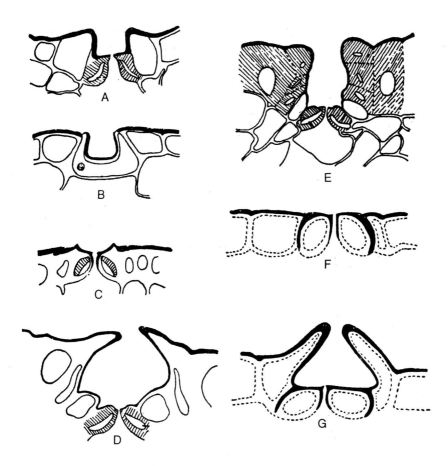

Fig. 7.1 A-G. Stomatal apparatus in cycads (**A-E**) and cycadeoides (**F, G**). **A, B** *Zamia muricata,* **C** *Stangeria paradoxa,* **D** *Cycas revoluta,* **E** *Dioon edule,* **F** *Pterophyllum rosenkrantzii,* **G** *Ptilophyllum pecten.* (**A-D, F, G** After Harris 1932, **E** after Florin 1931)

The cycadophytic stomatal apparatus often presents a confusing picture, in surface view, due to the outlines of the subsidiary and encircling cells which rise above the guard cells. In living cycads, the structure can be analyzed by sections and microchemical tests (Harris (1932). In fossils, these techniques cannot be applied, as only the surface cuticule of the guard cell is all that remains.

The cycadophytic leaf remains are placed in two Orders: (a) Nilssoniales (cycads)—the general structure is similar to living cycads, the stomata are haplocheilic, the guard cells have a thin deposit of cuticle, the epidermal cells are not oriented in rows and have straight walls, e.g. *Nilssonia*. (b) Bennettitales (cycadeoids) —the stomata are syndetocheilic, the guard cells are heavily cutinized on their outer and dorsal walls, and the stomata tend to be oriented at right angles to the veins. The epidermal cells are arranged in distinct rows and have walls with wavy outline. Some common frond genera are *Pterophyllum* (abundant), *Zamites* and *Otozamites*.

The Cycadeoidales, which coexisted with Cycadales during the Mesozoic, became extinct by the end of the Cretaceous. Some of the well-known taxa in this Order are: *Cycadeoidea* (Cycadeoidaceae), *Williamsonia*, *Williamsoniella* and *Wielandiella* (Williamsoniaceae).

Cycadeoidaceae

Cycadeoidea.   *Cycadeoidea* with more than 30 spp. flourished from the middle Triassic throughout the Jurassic to Early or Middle Cretaceous. They had a wide geographical distribution: USA, Mexico, Italy, Belgium, France, Germany, Austria, Poland and India, The first specimens were collected from the Isle of Wight and Isle of Portland. In America, most of the specimens are from the Black Hills of South Dakota. These are petrified trunks and the specimens are heavy, bulky and hard, and resembled beehives so much that the collectors were misled to interpret them as beehives, wasp's nests, corals, etc. The trunks are stout, globular, armoured with spirally arranged leaf bases embedded in a ramentum of flat tongue-shaped scales. Some of them were unbranched, while others branched near the ground level into a cluster of short trunks which resembled a bunch of pineapples (Fig. 7.2).

Stem.   A transection of the stele of *Cycadeoidea* shows a large pith in the centre surrounded by a ring of endarch primary xylem (Fig. 7.3 B), and a broad zone of manoxylic secondary xylem and phloem in nearly equal proportion. The wood comprises scalariform tracheids with circular bordered pits on radial walls. The ring is dissected by wide rays. In addition, several uni-or biseriate rays are also present. Each leaf trace is C-shaped at the point of origin. As it passes through the cortex, it splits into a number of mesarch bundles which form closed cylinder at the base of the leaf. There is no indication of girdling and the leaf traces directly supply a leaf base.

Delevoryas (1959, 1960) studied the formation of the cone vascular trace and its subtending leaf trace. In *C. wyomingensis*, the cone trace is derived from the fusion of four leaf traces deep in the cortex. In its outward course the large cone trace gives rise to four leaf traces, and one of these supplies the leaf subtending the cone.

Leaves.   Specimens with attached mature leaves have not been discovered, but it is assumed that they were spirally arranged and borne in a crown at the apex of the trunk (Fig. 7.3 A). The leaves were simple and pinnate, as in a bud containing young unexpanded leaves. They were assigned to *Ptilophyllum* or a *Dictyozamites* type.

Fig. 7.2   *Cycadeoidea marshiana* trunk with profuse branches. (After Wieland 1906)

Reproduction

The reproductive organs are bisporangiate cones (Crepet 1972, 1974) borne among persistent leaf bases (Fig. 7.4 A). More than 500 cones, one in the axil of every leaf, have been observed on a single trunk of *Cycadeoidea dartonii* and appear as rosette-like bodies. Figure 7.3 C is a tangential section of the trunk of *C. jenneyana* which shows leaves and reproductive branches in cross section. In *C. dacotensis* the cones are 6 cm long and 5–10 cm in diameter, borne on a peduncle of comparable length. In the upper portion of the peduncle there are 100–150 elongated hairy bracts which enclose the young reproductive organs.

The ovulate receptacle is subtended by a whorl of microsporophylls. A reexamination of the massive pollen-bearing region led Delevoryas (1963) to suggest that its structure may be fundamentally different from the frequently illustrated flower-like reconstruction by Wieland (1906). Presumably, it comprises a whorl of about 20 sporophylls, each with a pinnate structure and each pinnule bearing two rows of synangia. They appear around the

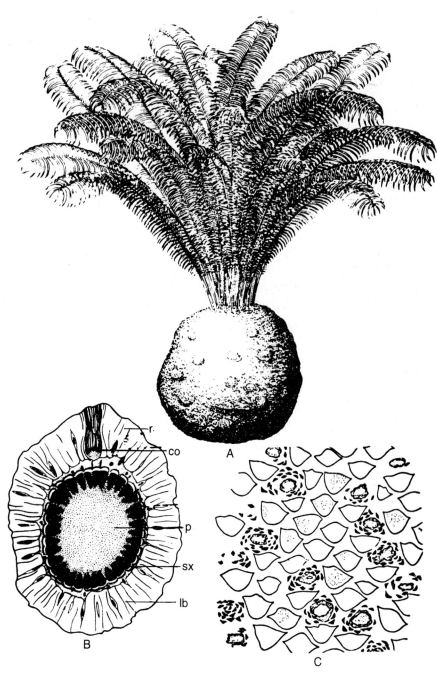

Fig. 7.3 A-C. A, B *Cycadeoidea* sp. **A** Reconstruction of plant. **B** Transection of trunk with a large pith (*p*), poorly developed secondary xylem (*sx*), and thick ramentum (*r*) with embedded cone (*c*) and leaf bases (*lb*). **C** *C. jenneyana*, a tangential section of trunk shows leaves and reproductive branches in cross sections. (**A** After Delevoryas 1971, **B** after Crepet 1974, **C** after Bierhorst 1971)

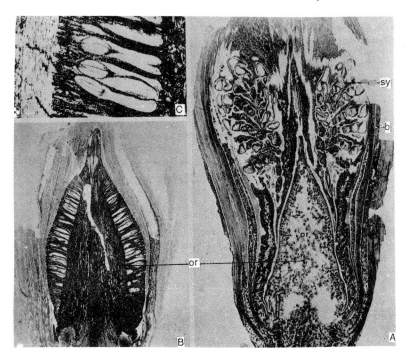

Fig. 7.4 A-C. *Cycadeoidea* sp. **A** Longisection cone shows ovulate receptacle (*or*), microsporophylls bear synangia (*sy*) and investing bracts (*b*). **B** Longisection ovulate receptacle bear stalked ovules separated by interseminal scales. **C** Stalked ovules and interseminal scales. (After Crepet 1974)

apical meristem of the cone (Fig. 7.4A; 7.5A). Crepet (1974) examined well-preserved permineralized specimens, and observed that each pinnate microsporophyll became revolute as it matured, with the synangium-bearing pinnae folded inward parallel to the plane of the pinna-rachis (Fig. 7.4 A). The recurved microsporophyll rachises were fused with one another above the ovulate receptacle, and at their tip. The latter was occasionally fused to its base. A synangium is a kidney-shaped structure containing 8–20 tubular sporangia (within the periphery of the synangium). Monocolpate grains have been observed within the sporangium. Thus, wind pollination is eliminated in *Cycadeoidea*, and selfing appears to be the chief mechanism; pollination by boring insects is also a possibility (Crepet 1972, 1974). The imposition of self-pollination was an important factor in the extinction of the group, which did not continue beyond Cretaceous (Crepet 1974).

The ovuliferous receptacle is centrally located, conical or dome-shaped and at the tip of the cone axis (Fig. 7.4 A, B). It comprised upright stalked ovules enveloped by club-shaped vascularized interseminal scales with expanded tips (Fig. 7.4 B, C). The orthotropous ovules are small (up to 3 mm long), have a tubular unvascularized integument which extended and surrounded a long micropyle. The ovules are enveloped except at the tip of the micropyle (Fig. 7.5 B). Crepet and Delevoryas (1972) observed ovules with remnants of a linear megaspore tetrad. However, stages in megagametophyte development have not been observed, which is quite unusual, as well-preserved cycadeoid seeds with dicot embryos have been discovered.

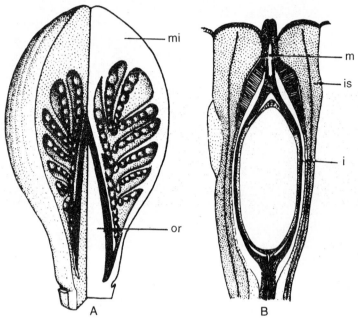

Fig. 7.5 A, B.   A *Cycadeoidea* fructification in median longisection. **B** *C. morieri*, ovule and adjacent scales in longisection. *m* micropyle, *i* integument, *is* interseminal scale. (**A** After Seward 1969, **B** after Crepet 1974)

Phylogenetic Considerations

The origin and relationship of cycadeoids have aroused considerable interest. Several vegetative characters are cycad-like, while the reproductive bodies are unique—the ovules are borne on long stalks interspersed with interseminal scales.

In their general morphology, the fronds of cycadeoids are remarkably similar to pinnate leaves of the cycads (except their stomatal structure). Their stem anatomy also resembles the cycads except in the absence of leaf girdles, and presence of abundant and compact secondary xylem with scalariform tracheids.

The cycadeoids have bisporangiate cones, while in cycads they are monosporangiate. The morphological nature of the stalked ovules and interseminal scales in cycadeoids is difficult to interpret. The whorls of microsporophylls with pinnate synangium-bearing branches are also different from spirally arranged microsporophylls with abaxial sporangia in cycads. However, both groups have monocolpate pollen.

Consequently, when vegetative characters alone are considered, a close relationship between cycads and cycadeoids seems likely but, when reproductive structures are also considered, the relationship appears remote. Therefore, this may be interpreted as two distinct groups with a possible common ancestor among the Palaeozoic pteridosperms (Chamberlain 1920). The presence of syndetocheilic stomata in pteridosperms is in favour of this interpretation.

8. Cycadales

The plants have woody, columnar, unbranched stems with a crown of large, pinnately compound leaves, which make it look like a palm tree. Mucilage canals are present in the pith and cortex, and the wood is of manoxylic type. The leaf trace is diploxylic. The plants are dioecious, the male and female cones are mostly terminal. The megasporophyll has two to eight large orthotropous ovules. The microsporophylls have microsporangia on the abaxial surface. The motile male gamete (spermatozoid) has a spiral band of flagella.

Fossil Cycads

The beginning of Cycadales is not known. The fossil remains provisionally assigned to Cycadales are meagre, in the Upper Palaeozoic. From the Triassic have been reported undoubted cycads which spread rapidly over the earth. Tertiary cycads mostly belonged to the modern genera, or were closely related to them (Arnold 1953). The most complete and well-preserved remains are from the Upper Triassic and Jurassic.

Gould (1971) described a well-preserved permineralized specimen of *Lyssoxylon*, a stem genus, from the Upper Triassic of Arizona. The stem has a large central pith, a cylinder of compact secondary wood, and an outer cortical region surrounded by persistent spirally arranged leaf bases. The stele has well-defined "growth rings", rays of two types, and tracheids with multiseriate alternate circular bordered pits of the araucaroid type, on the radial walls (cf. extant *Dioon spinulosum*). These characteristics indicate that the affinities of *Lyssoxylon* are with the Cycadales.

Delevoryas and Hope (1971), from the Upper Triassic beds of North Carolina, described a remarkably complete compression-impression fossil of a cycad stem with leaves and pollen cone attached. The stomata are haplocheilic, and other characteristic features are also cycadean. The attached fronds are comparable to the isolated cycad frond type *Pseudoctenis*. A new genus *Leptocycas* (Fig. 8.1 A) has been established. The reconstruction depicts a slender stem, ca.1.5 m tall, bearing a crown of leaves, some exceed 30 cm in length. Persistent leaf bases on the stem are absent. The general morphology of *Leptocycas* suggests that primitive cycads had slender stems; the bulky, fleshy stem of a typical extant cycad is derived.

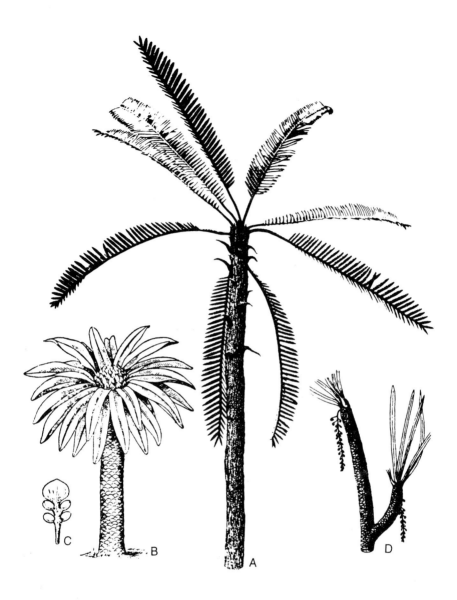

Fig. 8.1 A–D.   **A, B, D** Reconstruction of extinct Cycad plant. **A** *Leptocycas gracilis.*
B *Bijuvia simplex.* **C** *Palaeoycas*, an ovule-bearing organ. **D** Cycad bearing *Beania*
ovulate cones and *Nilssonia* type foliage. (**A** After Delevoryas and Hope 1971,
B,C after Florin 1933a, **D** after Harris 1961)

By making use of imaginary trunks, Florin (1933a) and Harris (1961)
also prepared reconstructions of extinct cycads. After an exhaustive study
of epidermal and stomatal structure of isolated compression-impression
fossils of leaves and reproductive structures, Florin (1933a) reconstructed
a plant of an Upper Triassic cycad and named it *Bijuvia simplex* (Fig. 8.1 B).

It showed *Doratophyllum* leaves around a cluster of spirally arranged *Palaeoycas* megasporophylls (Fig. 8.1 C; their stomatal arrangement and structure are comparable) which crowned the apex of an imaginary unbranched stout trunk ca. 3 m high.

Harris (1961) placed the reproductive organs and foliage on an imaginary branched stem which had an armour of leaf bases (Fig. 8.1D), and reconstructed a Jurassic cycad. It had crown of *Nilssonia tenuinervis* leaves (Fig. 8.1D), ovulate cones of *Baenia* sp. (described later), and pollen cones of *Androstrobus* sp. (described later). All the genera are from the Jurassic Cayton Bay beds of Yorkshire.

Nilssonia. This is one of the most common and widely distributed Mesozoic cycadean leaf types—slender, 60 cm long, 10 cm wide. The lamina is entire or dissected into pinnule-like segments and is supported on a strong mid-rib from which simple veins pass obliquely or at right angles to the margin (Fig. 8.2A). The veins are rarely forked. There is considerable

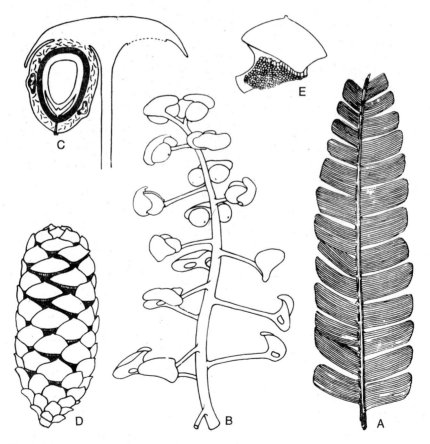

Fig. 8.2 A–E. Fossil cycads. **A** *Nilssonia compta*, leaf. **B, C** *Beania gracilis.* **B** Seed "cone". **C** Megasporophyll with one seed in longisection. **D, E** *Androstrobus manis.* **D** Microsporangiate cone. **E** Microsorophyll. (**A** After Andrews 1961, **B–E** after Harris 1941, 1964)

variation within a species. The stomatal characters display cycadean affinity. The genus is nearly worldwide in distribution from Triassic to Cretaceous.

Baenia. Two species are known. In *B. gracilis*, peltate sporophylls (1.5–2.5 cm) are arranged spirally in a loose cone (10 cm long) around a central axis, each bearing two ovules (Fig. 8.2 B,C) with their micropyles towards the cone axis. The mature seeds (detached seeds are also called *Baenia*) are 7 × 7 mm to 16 × 16 mm. On maceration, four cutinized layers have been observed. Their distribution indicates that the integument is fused to the nucellus throughout the lower two-thirds of the seed, and is free at the apex (Fig. 8.2 C). There is an outer fleshy, middle stony, and an inner fleshy layer. There is evidence of vascular bundles running longitudinally up to the level where the integument became free from the nucellus. The remains of pollen grains (identical with those of *Androstrobus manis*) have been observed inside the micropyle.

Androstrobus manis. The cones are compact, ca. 5 cm long and 2 cm broad (Fig. 8.2 D). Each cone has a large number of spirally arranged peltate microsporophylls, each is terminated by a rhomboidal scale and has numerous finger-like pollen sacs on the abaxial side (Fig. 8.2 D,E). They are arranged in sori with dehiscence slits along their inward-facing walls. The pollen isolated from the microsporangia are small, ovoid and monocolpate, as in extant cycads.

Living Cycads

The cycads first appeared in the Upper Triassic, reached the peak of diversification and development during Mid-Mesozoic, and have been on the decline since then. At present there are 11 extant taxa with more than 100 spp. They have a disconnected distribution (see Chap. 1) and form an inconspicuous member of the vegetation. Most cycads grow in exposed habitats, and are considered xerophytes. A few plants (*Zamia pumila, Ceratozamia*) grow in shady and moist woods. One species of *Zamia* is an epiphyte (Arnold 1953). The plants are slow-growing. A plant of *Macrozamia* takes ca. 100 years to reach a height of 1 m (Schuster 1932); *Encephalartos* takes 200–300 years to grow 1.5–1.75 m tall.

Root.   The root system of cycads consists of a tuberous, contractile taproot with narrow racemosely branched lateral roots, and swollen dichotomously branched coralloid roots. The latter have been variously referred to as aerial roots, apogeotropic roots, nodules, tubercles, or pneumatophores. Most of these terms are unsuitable, as the features they describe are either absent or inconsistent (Staff and Ahern 1993). The term coralloid is accepted as appropriate for mature invaded (by Cyanobacteria) roots. A new term, precoralloid, has been introduced to describe root which are uninvaded but have the potential to be so and develop into coralloid roots (Staff and Ahern 1993).

Coralloid roots represent the only known, naturally occurring symbiosis between plant roots and nitrogen-fixing cyanobacteria (*Nostoc, Anabaena* and rarely *Calothrix*—see Ahern and Staff 1994). Early claims of bacterial initiation have been dispelled by successful sterile culture experiments (see Ahern and Staff 1994). Reinke (1872) discovered coralloid roots in *Cycas revoluta*. Since then it has been recorded in all taxa and all species examined, including the new genus *Chigua* (Stevenson 1990b). These roots can reach up to 10 cm in diameter and 500 gm weight and contribute significantly to the nitrogen metabolism of the plant as well as nitrogen economy of the ecosystem (Halliday and Pate 1976).

Ahern and Staff (1994) investigated the origin and development of coralloid roots in *Macrozamia communis* and described, for the first time, the sequence of precoralloid initiation, maturation, cyanobacterial invasion and coralloid development at the morphological level. The different stages can be identified by specific characters. The earliest stage is initiation of young apogeotropic, papillose roots (a distinct papillose sheath covers the root)—the precoralloids. The active apices continue to produce papillose tissue, which is gradually replaced (as it matures) by a thin dermal layer with scattered lenticels. Cyanobacterial invasion has been observed at different stages of precoralloid maturation and stimulates the irreversible development of precoralloids into coralloids. Although continuous, these two coralloid regions may be recognized by their external morphology. New coralloid growth involves cessation of formation of papillose sheath, change in gravitropic response, proliferation of distinctive apical lenticels depicting transitional stage and differentiation of a cyanobacterial zone.

Stem. Cycads are pachycaulous plants with stems either aerial and columnar (arborescent) or subterranean (geophilous), tuberous, and fleshy. In *Bowenia, Stangeria, Zamia,* and some species of *Macrozamia* and *Encephalartos,* the stem is nearly spherical. In *Zamia pumila* and *Stangeria eriopsis,* the vertical shoots are subterranean, the shoot apex remains underground even though new shoots are added as a result of leaf development. The external surface of these stems have a wrinkled appearance in the distal regions. The height of the leaf scars decrease basipetally because both the fleshy taproot and the stem contract, pulling the shoot underground. This contraction results from the collapse of horizontal rows of cells in both pith and cortex. As this collapse of cells occurs (Stevenson 1981), the vascular cylinder (which appears straight in longitudinal section) becomes sinuously distorted.

Most cycads appear palm-like due to their columnar trunk and apical clusters of large pinnate leaves. The tallest cycad is *Macrozamia hopei,* which is ca. 20 m high, while the smallest *Zamia pygmaea* has an underground stem 2 or 3 cm in diameter and is ca. 25 cm high. The stem apex of cycads is often several mm across, and is the most massive among all vascular plants.

The phloem in cycads has not been studied in detail (see Paliwal 1992). The available data are based mainly on *Cycas circinalis* (Strasburger 1891) and *Microcycas calocoma* (Chrysler 1926).

The axial system of the secondary system consists of sieve elements, parenchyma cells and fibers. The sieve elements are long with inclined end walls, their faces merge with the radial lateral walls. The sieve areas have variable shape. They occur on end walls and the radial face of lateral walls. Parenchyma cells are of two types: those containing starch (occasionally druses of calcium oxalate), and the albuminous cells. In the non-functional phloem the starch-containing cells remain intact, while the albuminous cells collapse. Pits develop on the walls between adjacent parenchyma cells. According to Chrysler (1926), the pits are identical in shape and distribution to the sieve areas in the sieve elements.

In the cycads, the primary phloem of the leaf appears to be devoid of fibres. In the older petioles and stems, the protophloem is crushed. Metaphloem tissue contains wide sieve elements and narrow parenchyma cells. The tracheids are separated from the sieve elements by parenchyma. Within the transfusion tissue (located on the flanks of the bundle) the parenchyma cells, next to the phloem, are rich in cytoplasmic contents and have rather large nuclei. According to Strasburger (1891), they serve as intermediate cells comparable to the albuminous cells.

The ultrastructure of sieve elements of *Zamia pseudoparasitica* reveals that the ontogeny is identical to that of the Pinaceae (Parthasarthy 1975). However, unlike the necrotic nuclei reported by various authors, in *Zamia* nuclei have been observed at various stages of degradation. P-protein is absent (as in other gymnosperms). The other cell organelles also exhibit a similar behaviour during differentiation. Aggregates of ER normally occur on both sides of sieve areas in the mature sieve cells and appear to be interconnected by elements of ER traversing the sieve pores.

Leaf. The leaf length varies from 5 or 6 cm (*Zamia pygmaea*) to about 3 m (*Cycas circinalis*), and is tough, fibrous and pinnate, bipinnate in *Bowenia*. The leaflet in *Chigua* has a prominent mid rib, the lateral veins depart at an acute angle and branch dichotomously. Circinate vernation has been observed in *Bowenia* (young rachis and pinnae), *Cycas* (only pinnae), and *Stangeria* and *Zamia* (only rachis). All cycads produce periodic crowns of foliage and scale leaves. In plants with columnar trunks, the leaf bases persist and form a characteristic armour. In geophilous stems, the leaf abscisses from the very base so that the stem surface is covered by a smooth corky layer. The stomata in all the cycads are haplocheilic.

Reproduction

The cycads are dioecious, the male and female sporophylls are organized into strobili, except in *Cycas*, where the megasporophylls remain loose and do not form a cone. Meagasporophylls in *Chigua* are peltate, hexagonal. Raised area or a mound is present at each angle of the hexagon. The length of the cone varies from 2 cm (*Zamia pygmaea*) to 80–100 cm (*Macrozamia peroffskyana*). A ripe female cone of *Dioon spinulosum* is nearly 61 cm long, and weighs up to 28 kg (see Bierhorst 1971).

Thermogenesis is a widespread phenomenon in the group (Tang 1987a), occurring in the strobili in 42 species of cycads. Heat production generally follows a circadian rhythm; the temperature and its duration can be correlated with the size of the strobili. The male strobili have a high concentration of starch that may supply the energy for thermogenesis.

The strobili of all the species examined release odours that are sweet, resinous or musty (Tang 1987a). Heat production may help volatilize odours that attract insects for pollination. The cone odours of some of the taxa (*Cycas, Dioon*) are very strong and can be detected outdoors several metres away. The odours may be released throughout the day (*Cycas, Stangeria*), or may be noticeable only during heat production (*Bowenia, Ceratozamia, Zamia*). The odours may be sweet and fruity (*Bowenia, Ceratozamia, Stangeria, Zamia*), musty (*Cycas, Dioon, Microcyas*), or resinous, often fruity (*Encephalartos, Lepidozamia, Macrozamia*).

Tang (1987b) conducted wind and insect exclusion pollination experiment in a wild population of *Zamia pumila* in Florida. Cones from which insects (but not wind) were excluded did not produce viable seeds. Cones from which wind (but not insects) were excluded produced abundant viable seeds. Two beetle species (*Pharaxonotha zamiae, Rhopalotria slossoni*) have been identified which may be affecting pollination. Abundant adults and larvae of both beetles are present on the male cones. On the female cones adults and larvae are present occasionally. *Z. pumila* produces sugar and amino acid-rich micropyle droplets which serve as pollinator reward.

Pant and Singh (1990) observed (in cycads growing in Allahabad) that the male cones of *Cycas circinalis* alone are invaded by pollenivorous small beetles, and other insects including ants while the compact crowns of megasporophylls of *C. rumphii* and *C. revoluta* are ignored. The beetles inside the male cone continue to be active and produce subsequent generations. These insects do not play any part in the pollination of the plant. This is further confirmed by the abortive nature of seeds in the neighbouring female plants of *Cycas*.

The microsporophylls are flat with a large number of sporangia arranged in sori on the under surface. In *Zamia* the microsporophylls are peltate (cf. *Equisetum*).

The structure and development of male and female gametophytes are more or less uniform through the group; *Microcyas* is the only exception.

In the development of the male gametophyte, the stalk cell divides several times and forms a row of body cells which results in the formation of as many as 16 (Norstog 1990) or/and 22 (Caldwell 1907, Downie 1928) spermatozoids from a single male gametophyte.

Spermatozoid

The spermatozoids exhibit a developmental gradient in complexity and mobility; first-formed spermatozoids are better organized and more actively motile as compared to the last-formed spermatozoids (Norstog 1990).

All cycads have motile ciliated spermatozoids, first discovered by Ikeno (1896) in *Cycas*, and are the largest ciliate male gametes known in plants. They measure 180–210 μm in length in *C. revoluta*, 230 μm in length and 300 μm in diameter in *Dioon*, 332 μm in length and 180 μm in diameter in *Zamia floridana*, 180–300 μm in size in *Z. integrifolia* and 400 μm in diameter in *Z. chigua*.

Norstog (1967a, 1968, 1974, 1975, 1977) studied the ultrastructure of the male gamete of *Zamia*. The division of the body cell produces two spermatids, each with one blepharoplast (diameter ca. 10 μm) and surrounded by cytoplasmic asters. The bulk of the organelle consists of closely packed procentriolar bodies, each with a hub-and-spoke arrangement, and nine-fold symmetry (Mizukami and Gall 1966). During maturation, the blepharoplast forms an attenuated spiral band, and the procentrioles contribute directly to the formation of the flagella of the mature spermatozoid. The spermatozoids are oriented "back to back". The greater part of the spermatozoid in *Zamia* (and other cycads) is the nucleus surrounded by a shallow zone (ca. 10 μm) of cytoplasm. A flagellated spiral band occupies the anterior half of the cell. About 10 000–12 000 flagella, each ca. 60 μm long, are attached to this spiral band.

The spiral band (excluding flagella) is composed of three regions: (a) an outermost electron-dense layer (DL), (b) an intermediate lightly staining granular region, and (c) a four-layered complex of electron-dense tubules, fins and compartments termed Vierergruppe (VG). The VG of *Zamia* is similar to spermatozoids of *Marchantia* (Carothers and Kreitner 1967), *Pteridium* (Tourte and Hurel-Py 1967, Duckett and Bell 1971), *Equisetum* (Duckett and Bell 1969, Duckett 1973) and to the three-layered structures, Driergruppe, present in spermatozoids of *Polytrichum* (Heitz 1960, Paolillo 1965, Paolillo et al. 1968) and *Bryum* (Bonnot 1967).

An extensive system of microtubules is associated with the VG in *Zamia*. It is similar to a microtubule system termed spline (Carothers and Kreitner 1968) which occurs in *Marchantia* spermatozoids. However, microtubules are much more numerous in *Zamia* (ca. 60 000) than in *Marchantia* (only 17).

The flagella are subtended by a relatively long (ca. 4 μm) basal body. Much of this length is occupied by an internal system of fibres with a stellate pattern similar to the flagella of algae (Manton 1964), bryophytes

(Paolillo 1965, Carothers and Krietner 1967) and pteridophytes (Duckett and Bell 1969, 1971, Duckett 1973). The basal bodies terminate in the DL of the band.

The cytoplasm is characterized by a network of a granular ER. Mitochondria dispersed in the cytoplasm and are relatively small with fewer cristae [comparatively, the spermatozoid of *Pteridium* has a conspicuous localized system of mitrochondria with well-developed cristae associated with the flagellated band (Bell et al. 1971)].

Ovule

The general structure of the ovule in all cycads is essentially similar. The egg is perhaps the largest in the plant kingdom (P. Maheshwari and Sanwal 1963). A typical feature of the archegonium is its massive nucleus. In *Dioon* the egg nucleus is ca. 600 μm long (Chamberlain 1906), and in *Zamia umbrosa* 710 μm × 490 μm, 1330-μm-long archegonia (Bryan and Evans 1956).

Pollen Tube

The process of fertilization, embryogenesis and seed germination is more or less similar. The pollen tube in cycads is not involved in the conduction of sperm to the egg (unlike the pollen tubes of conifers, gnetophytes and angiosperms). It functions primarily to obtain nutrients (cf. a haustorial fungus) for the growth and development of the male gametophyte. The cycad pollen tube penetrates the nucellar cells by enzymatically destroying them. Choi and Friedman (1991) studied pollen tube growth and its relation to the nucellus in *Zamia furfuracea* using light and transmission electron microscopy (TEM). Following germination of pollen grain at the distal end, the pollen tube grows intercellularly through the subepidermal layers of the micropylar apex of the nucellus. Additional localized outgrowths of the pollen tubes, penetrating the walls of the individual nucellar cells, have been observed. Intracellular haustorial growth ultimately leads to a complete destruction of each penetrated cell, and also appears to induce degeneration of proximal unpenetrated nucellar cells. This pattern of intracellular penetration of the sporophyte by the male gametophyte in *Z. furfuracea* suggests that the heterotrophic and tissue-specific relationships that male gametophytes of seed plants have with their host sporophytes are much more diverse than had previously been known (Choi and Friedman 1991). According to Johri (1992), the investigation on *Z. furfuracea* leave no doubt about the haustorial role of the pollen tube (so far there is no such study on any angiosperm). The pollen tube growing in the nucellar tissue is interpreted as a vegetative structure since it is not concerned in the reproductive processes (Choi and Friedman 1991).

Professor Knut J. Norstog (Miami, Florida, USA) has devoted considerable attention to the reproductive biology of the cycads. According to him (in a personal discussion with Professor B. M. Johri; see Johri 1992), in *Microcycas* and *Zamia* the proximal (lower) pole of the pollen grain *does* produce a pollen tube ca. 1.5–2 mm wide and 3–5 mm long. This proximal extension of the microgametophyte expands very rapidly in about 3–4 days, breaking through the inner nucellar epidermis into the space of the archegonial chamber. This growth occurs at the expense of nucellar fluids. As the pollen tube elogates, the nucellus becomes dry and papery. During this period, the spermatids swim about in the pollen tube for a day.

Chamberlain (1906) showed pollen tubes quite clearly during fertilization in *Dioon edule* (see *Cycas* described later). According to Johri (1992), a type of siphonogamy has been initiated in the cycads and they should no longer be regarded strictly as non-siphonogamous.

The spermatozoids of cycads swim in a fluid medium (enclosed within the developing ovule) whose solute concentration is approximately 0.6 M (Norstog 1975). The spermatozoids remain motile for a relatively short time interval (15 min or less) when collected in pollen tube liquid or in 20% sucrose solution. *Zamia* spermatozoids in sucrose solution exhibit an initial burst of rapid swimming for ca. 5 min, followed by flexing movements of the anterior portion of the cells accompanied by slow flagellar beating. Finally, only vibration of the flagella is observed, which ceases after approximately 30 min. This short-lived activity of the spermatozoid is possibly due to the low volume of cytoplasm compared to the nuclear volume and to the absence of a direct mitochondrial association with the flagellated band of this massive cell. It may also be due to the unnatural conditions (such as sucrose solution) under which spermatozoids were observed (Norstog 1975).

The spermatozoid swims with a rotating motion, with its flagellated band directed forward. There is a wave-like beating of the flagella and a rapid flexing of the anterior of the cell and occasional contractions and extensions of this region. Such movement, more appropriately termed euglenoid (Norstog 1965a) than amoeboid (Webber 1901), may contribute to the passage of the spermatozoid through the neck canal of the archegonium. In *Zamia integrifolia*, at the time of fertilization, the four neck cells become turgid and separate and a passage (ca. 50 μM in diameter) is formed. The neck cells shrivel soon after fertilization. Actual passage of spermatozoids through the archegonial neck canal has not been observed. Spermatozoids of *Zamia* and other cycads (Lawson 1926) assume a special form after passage through the neck. The flagellated band remains in contact and several spermatozoids are seen within the egg cytoplasm of a single archegonium (Norstog 1975).

Embryo

The embryogeny (in one or two species in a few taxa) has been investigated adequately (see Dogra 1992). Information is available on embryogeny for

Bowenia (Lawson 1926), *Ceratozamia* (Chamberlain 1912), *Cycas* (Ikeno 1898, Treub 1884, Chamberlain 1919, 1935, Swamy 1948, De Silva and Tambiah 1952, Rao 1963), *Dioon* (Chamberlain 1910), *Encephalartos* (Saxton 1910, Sedgwick 1924, De Sloover 1963), *Macrozamia* (Brough and Taylor 1940, Chamberlain 1913, Baird 1939), *Stangeria* (Chamberlain 1916) and *Zamia* (Coulter and Chamberlain 1903, Bryan 1952).

Early divisions in the zygote occur in situ, are synchronous and in quick succession. Later divisions are not synchronous, some of the nuclei may even fail to divide, so that the final number of free nuclei is less than expected. The nuclei are evenly distributed in the proembryo. Ten free mitoses occur (highest in gymnosperms), before walls appear in *Dioon edule, Stangeria paradoxa, Encephalartos friderici-guilielmi, Cycas circinalis* and *Cycas* species (embryogeny of *Cycas* described later in this Chapter). More than eight free-nuclear divisions in *Zamia floridana*, and six free-nuclear divisions (lowest in cycads) occur in *Bowenia serrulata*. A decrease in the number of free-nuclear divisions (from ten to six mitoses) in the cycads has been observed.

An "advanced" trend (over proembryo with evenly distributed nuclei) has been observed when the free nuclei migrate from scattered position to the base of the proembryo (cf. conifers; see Dogra 1992). Further division mostly occur at the base. Thus the proembryo has two physiologically different regions—the upper inactive and the lower active.

Wall formation takes place in the proembryo after the last free-nuclear division. Depending on the position of the free nuclei at wall formation, the segmentation may be complete or partial.

In *Encephalartos friderici-guilielmi, E. villosus* and *Macrozamia spiralis*, the segmentation of nearly the entire egg cell results in the formation of a primary proembryo which has a dense, compact, active basal region. In *Macrozamia reidlei* wall formation occurs throughout the proembryo except in a small central region which contains free nuclei and cytoplasm. This region breaks down in later stages so that there is a hollow cavity in the centre. In *Cycas* the cellular proembryo formation is restricted to the periphery and the base. In *Stangeria, Zamia* and *Bowenia*, the proembryo cells are formed only at the base (as in conifers; see Dogra 1992).

The primary proembryo develops directly into late embryo without elongation of the suspensor. The internal division and the U.S.E.(U—upper tier, S—suspensor tier, E—embryonal tier) pattern of conifers (see Chap. 15) are absent.

The cellular primary proembryo gradually differentiates (morphologically) into a large inactive primary upper region, and a very active primary embryonal region. The primary upper region becomes gradually reduced to an inactive region enclosing a central vacuole. Later, these cells degenerate. The primary embryonal region consists of compact, densely staining, actively dividing uniform cells concentrated at the (basal) tip. This is the meristematic cell region of the embryo. The numerous proximal cells elongate to give rise to tubular cells which (in all cycads) form a very long, elongated cord. This carries the meristematic primary embryonal tip of the late embryo into the prothallial tissue. In *Zamia* and *Bowenia*, there is a layer of elongated cells around the compact terminal embryonal cells (cap cells). Cleavage polyembryony is absent in cycads.

The embryo is mostly dicotyledonous, in *Encephalartos* there are three cotyledons, and *Ceratozamia* has only one, which always develops on the side of the seed facing the ground. If such a germinating seed is rotated in a clinostat throughout the developmental period, it develops two cotyledons (Dorety 1908). The size of the mature seed varies — *Cycas* spp. and *Macrozamia denisonii* have the largest seed (ca. 6 or 7 cm), and *Zamia pygmaea* the smallest (ca. 5–7 mm). The seeds of all cycads germinate without a resting period.

Chromosome Number

According to Khoshoo (1962), among the cycads the karyotype of *Cycas* and *Microcycas* is most advanced. Vovides and Olivares (1996) observed that *Zamia loddigesii* forms a morphologically variable complex on the Yucaten Peninsula, Mexico. Several dipold chromosome numbers have been recorded in the species: 2n = 17, 24, 25, 26 and 27. The karyotypes differ, in individuals of the same population, in the number of metacentric and telocentric chromosomes present. Centromere fission as well as pericentric inversions and unequal translocations are probable mechanism for the karyotype variation. A correlation between higher chromosome number, increasing dryness of the habitats along with asymmetrical karyotypes suggest that karyotype evolution in *Z. loddigesii* is recent (Vovides and Olivares 1996).

The various taxa display well-marked affinity, yet it is difficult to arrive at a satisfactory classification of the genera (within the group). The classification, therefore, varies and the taxa are included in one, two or three tribes/sub-families/families. According to Johnson (1959), there are three families:

a) Cycadaceae   – *Cycas*
b) Stangeriaceae – *Stangeria*
c) Zamiaceae   – *Lepidozamia, Macrozamia, Encephalartos, Dioon, Microcycas, Ceratozamia, Zamia, Bowenia*

Cycadaceae

Cycas

Cycas is the most widely distributed taxon (see Chap. 1), and has 20 species (see Willis 1966).

Morphology

Cycas has a columnar stem with a cluster of pinnately compound large leaves and gives the plant the appearance of a palm tree (Fig. 8.3A). Normally, it is unbranched, but beyond a certain age, branching trunks are common; branching can also be induced by injury. A bonsai (see Chap. 23) of *Cycas revoluta* can be induced to branch more profusely than happens in nature. Sometimes, dwarfs hundreds-of-years-old may have nearly 20 crowns (Pant 1973). The branches develop from adventitious buds called bulbils, arising from the lower fleshy portion of old leaf bases. In the beginning there are a number of scale leaves around a small stem; later, crowns of leaves are produced. When they have grown larger, these bulbils develop branches on the stem. Occasionally, they may have even adventitious roots at their base. On separation they grow into new plants (vegetative propagation).

A   B

Fig. 8.3 A, B. A *Cycas revoluta*, female plant. **B** *C. circinalis,* columnar trunk shows alternate bands of large (foliage leaves) and small (megasporophyall) rhomboidal bases. (**A** After Strasburger 1930, **B** after Pant 1973)

For the most part the stem is covered with a thick persistent armour of regularly alternating bands of large and small rhomboidal leaf bases (Fig. 8.3B); the larger one is the base of the foliage leaf and the smaller of a scale leaf as well as of the megasporophyll (in a female plant). The approximate age of a plant can be calculated from the whorl of leaves and megasporophylls produced every year. However, the leaf bases at the base of the trunk may be shed off by abscission, leaving behind the corky surface of the stem which now appears narrower than the apical portion.

There are two types of leaves. (a) A cluster of scale leaves which alternate with green leaves and are often roughly felted. (b) Large unipinnate compound foliage leaves. The leaflets are tough, leathery, have a mid-rib, and show circinate vernation (Fig. 8.4 A,B). The rate of leaf development is slow but uniform.

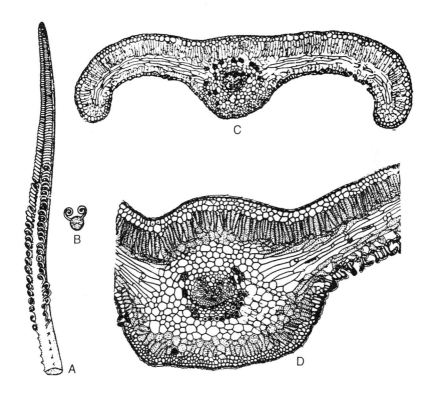

Fig. 8.4 A–D. *Cycas revoluta.* **A** Adaxial view of foliage shows acropetal unrolling of circinnate pinnae. **B** Transection rachis shows young pinnae in circinnate vernation. **C, D** Transection pinna. **C** Shows revolute margins. (**A, B** After Foster and Gifford 1959, **C, D** after Pant 1973)

In the beginning, there is a normal taproot system. Later, it is replaced by strong branched adventitious roots which develop a number of apogeotropic (negatively geotropic) roots. These roots grow horizontally and vertically, and have lenticels on their surface. There is an incursion of a blue-green alga (Cyanobacteria) which multiplies rapidly and forms a conspicuous zone in the cortex. The alga causes a distortion of the rootlets, giving rise to a mass of tubercles which look like corals, hence these roots are called coralloid (Fig. 8.5 C, D). They are greenish or brownish and are abundant in the seedling stage, especially in the nurseries. Two species of *Nostoc* and one of *Anabaena* (family Nostocaceae), *Oscillatoria* (Oscillatoriaceae) and diatoms have been reported (Fritsch 1945). The presence of numerous lenticels on the surface of these roots, (also on normal roots), abundant air spaces in the cortex, and the emergence of these roots above the ground suggest that they also serve as aerating organs.

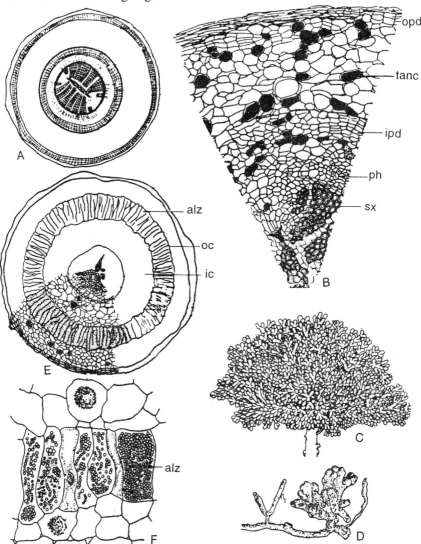

Fig. 8.5 A-F.   Root. **A, B, D-F** *Cycas revoluta*, **C**, *C. rumphii*. **A** Transection of diarch normal root (diagrammatic). **B** Inset from A. *ipd* inner periderm, *opd* outer periderm, *ph* phloem, *sx* secondary xylem, *tanc* tannin cell. **C, D** Coralloid roots. **E** Transection of triarch coralloid root, the algal zone (*alz*) is in the middle of the cortex. *ic* inner cortex. *oc* outer cortex. **F** Portion of coralloid root with algal zone. (**A–C** After Pant 1973, **D–F** after Wettstein 1935)

Anatomy

Root. A normal young root, at the point of attachment to the base of the stem, has up to eight protoxylem groups. The roots become progressively thinner, and the number of protoxylem groups becomes reduced to a diarch condition. The branch roots are mostly diarch. A typical periderm is produced and older roots show accessory cambia. The first cambium appears away from the xylem so that in an old root the primary and secondary xylem become separated (Fig. 8.5 A, B). The protoxylem shows characteristic spiral thckenings (Greguss 1955). The secondary xylem is manoxylic with abundant multiseriate rays. Pitting of the tracheids is similar to that in secondary xylem elements of the stem. Some tracheids in the wood of the root have peculiar delicate spirals (Greguss 1955). The vascular tissue is surrounded by a pericycle of starch cells. The endodermis shows typical casparian strips. Tannin-filled dark brown cells are interspersed all over. Druses, or rhomboidal crystals, occur in the root, either in isolation or in longitudinal series.

The anatomy of a coralloid root is comparable to that of a normal root (Fig. 8.5 E), except that (a) the development of vascular tissue is poor, (b) the secondary growth is absent, and (c) the cortex has a conspicuous greenish zone of radially elongate thin-walled cells with large intercellular spaces. These cells contain a blue-green alga (Fig. 8.5 E, F). Initially, these cells are more richly cytoplasmic than the adjacent cells.

Grilli (1963) studied (with EM) the blue-green alga/e in the root nodules of *Cycas revoluta*. The "infected" cells showed remarkable polymorphism, in the organization of the chromatoplasm, characterized by either a lamellar system or a reticulum, or both. The polymorphism in structure is probably related to the light intensity, age, and the depth of the root nodules in the soil. The ultrastructure of the heterocysts, and their abundance and development in relation to various environmental factors have been studied.

Stem. A young stem of *Cycas* has an irregular outline on account of persistent leaf bases (Fig. 8.6 A, B). A central broad pith, surrounded by a ring of numerous vascular bundles, is connected to the cortex by medullary rays. The stem always remains predominantly parenchymatous with scanty development of xylem (manoxylic). Numerous canals (clear circular areas) occur throughout the parenchymatous tissue (Fig. 8.6 C) and petiole, and are interconnected through leaf gaps. These canals contain mucilage, which is partly miscible with water and dries up into a hard, clear, somewhat crystalline mass. The parenchymatous cells have abundant starch.

The poor development of xylem is balanced by the thick and rigid armour of leaf bases and periderm. The primary vascular cylinder is an endarch siphonostele in adult plants, but mesarch in the seedling. The bundles are conjoint, collateral and open. The primary cambium is short-lived, but *Cycas*

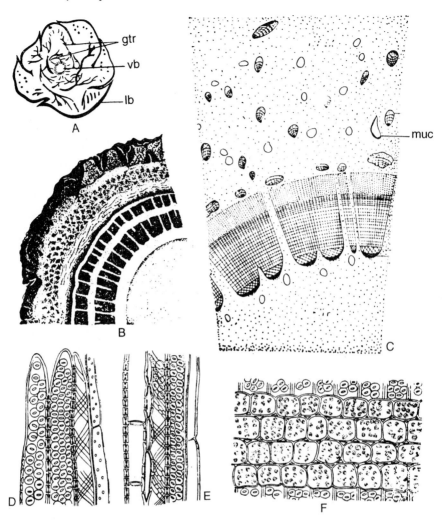

Fig. 8.6 A–F.   Stem. **A, B, D–F** *Cycas revoluta*, **C** *Cycas* sp. **A** Transection young stem (diagrammatic), note leaf bases, *gtr* girdle traces. *lb* leaf base. *vb* vascular bundle. **B** Transection of stem, one sector (diagrammatic) shows successive vascular rings and persistent leaf bases. **C** Transection of stem, sector (diagrammatic) shows vascular bundles, cortical strands and mucilage canal (*muc*). **D–F** Xylem in radial (**D, F**) and tangential (**E**) section. (**A** After de Bary 1884, **B, C** after Pant 1973, **D–F** after Greguss 1955)

is polyxylic (Fig. 8.6 B). As soon as the primary cambium has stopped functioning, a second cambium differentiates in the cortex and develops a stronger second ring of wood. Such rings are formed successively one after the other, but the outer rings become gradually weaker. Annual rings are absent. According to Terrazas (1991), observations of the vascular tissue of *Cycas* shoots provide evidence that the first vascular cambium (originating from procambium), as well as subsequent successive cambia, are simultaneously active (several previous workers, namely, Coulter and Chamberlain 1917, and Pant 1973, have stated that the original cambium is short-lived). The second cambium is established during the seedling stage and differentiates from the cortical cells. Cambial activity also occurs within phloem parenchyma cells of the first vascular cylinder. In all cambial layers conspicuous cell divisions occur and undifferentiated xylem and phloem cells are present. Tracheids in the first and successive vascular cylinders are generally of the same length. However, there is a trend towards increasing length within the successive cylinders possibly because successive cambia are long-lived (Terrazas 1991).

The metaxylem usually has scalariform, and occasionally multiseriate pitted elements. In older plants, the growth is slow and the protoxylem may also show scalariform thickening. The secondary xylem has bi- or tri- or multi-seriate bordered pits on radial walls, and rarely on tangential walls. These pits are in alternate rows, comparable to those in Araucariaceae (Fig. 8.6 D–F). The pits are oval or round (Fig. 8.6 D,E). Ultrastructural studies have shown that the torus is absent in the pit membranes of *Cycas* (Eicke and Metzner-Kuster 1961). The presence of tertiary spirals (Fig. 8.6 D, E) in tracheids (as in *Taxus*) has been reported. They are probably similar to spirals reported in the wood of the root by Greguss (1955).

Leaf.   A peculiar feature in *Cycas* (and allied genera) is the girdling of the leaf traces (Fig. 8.6 A). A leaf trace, after arising from the stelar cylinder, usually does not pass directly into the nearest leaf but turns round a semi-circle or "girdles" the stem horizontally, and enters the leaf almost opposite to its point of origin. It is joined at intervals by other traces, so that a number of them enter each rachis. The rachis is cylindrical, and the pinnae are inserted on it (Fig. 8.7 A). It has a thick-walled epidermis covered by a thick cuticle. and stomata are irregularly distributed. There is a broad outer zone of fibrous cells intermixed with short chlorenchyma, and an inner zone of large parenchymatous ground tissue with mucilage canals. Several collateral open vascular bundles are embedded in it. Usually, only two large and a variable number of small strands enter the base of the leaf. Immediately on entering, the two main bundles branch and form numerous bundles which become arranged like the inverted Greek letter omega—Ω (Fig 8.7A). Towards the tip of the rachis the number of bundles becomes reduced and they become arranged in a C-shaped arc. The centripetal xylem appears like a wedge, with the protoxylem at the apex facing the phloem (Fig. 8.7 B). An arc of cambium lies on its outer side (towards the phloem). A few centrifugal xylem elements are present on its inner side, usually in two or three separate groups or in a continuous arc (Fig. 8.7 B). The centrifugal xylem is separated from the protoxylem and adjacent centripetal xylem by a few parenchyma cells. On its outer side, towards the primary phloem, a few layers of radially seriated secondary phloem cells are present. Each vascular bundle is surrounded by

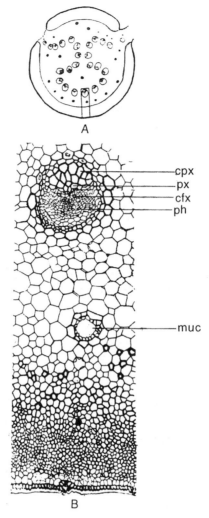

A

cpx
px
cfx
ph

muc

B

Fig. 8.7 A, B. *Cycas revoluta.* **A** Transection rachis shows inverted omega arrangement of bundles. **B** Enlargement of part of **A**. *cfx* centrifugal xylem. *cpx* centripetal xylem. *muc* mucilage canal. *ph* phloem. *px* protoxylem. (After Pant 1973)

a fibrous sheath. This characteristic diploxylic foliar bundle is seen in the rachis as well as in the pinnae (Fig. 8.4 C, D).

Leaf traces are endarch at the point of origin in the stem, become mesarch in the cortex, and farther out in the pinnae many traces are completely exarch. During their course, the centripetal xylem becomes progressively large, while the centrifugal xylem becomes reduced.

In their longitudinal course, the bundles of the rachis frequently dichotomize and anastomose. The bundles lying at the two open ends of the Ω-shaped arc, facing the grooves in the rachis, are the first to enter the pinnae. The next two bundles of the arc either directly enter the next set of pinnae or may branch, and the outer strand enters a pinna. This process continues till all the pinnae in the leaf have become vascularized.

The leaf has xerophytic characteristics. The upper and lower epidermes are heavily cutinized. There is a hypodermis or one or two layers of sclerenchymatous cells, followed by a palisade tissue, and finally the lower epidermis. The stomata are sunken and haplocheilic. Between the palisade and the spongy tissue, there are several layers of transversely elongated thin-walled colourless cells, the transfusion tissue, helpful in conduction. The vascular bundle of the median vein has a mesarch xylem, the phloem lies below the xylem, and there is a sclerenchymatous sheath around the bundle (Fig. 8.4 C, D).

The stomata, 75 μm × 34 μm, of cycads are generally larger than those of other gymnosperms. Pant and Mehra (1963) have studied the development in *Cycas circinalis*, *C. revoluta* (Fig. 8.8 A–F) and *C. rumphii*. In all these species, the stomata are haplocheilic. In a young stoma, the cells which surround the guard cell or the guard mother cells divide tangentially, form the subsidiary and encircling cells, and a radial division increases their number. In *C. revoluta*, the stomata are deeply sunken and amphicyclic, i.e. surrounded by two rings of cells (Fig. 8.8E). The first ring consists of four to eight subsidiary cells which partly overlap the stoma. The second ring of cells has a thicker layer of cuticle, and forms elongated finger-like processes overarching the deeply sunken stoma, resulting in a conspicuous epistomatal dome. In *C. circinalis*, *C. rumphii* and other species, the stomata have only an irregular ring of subsidiary cells around a stoma which is not deeply sunken. In *C. revoluta*, the guard cells show radiating striations on their surface.

A part of the guard cell wall is lignified (in lateral and polar lamellae), the rest is chiefly cellulosic. The rigid lignin lamellae help in the opening and closing of the stomatal aperture. When the guard cells become turgid, the cellulosic region stretches readily, while the lignified lamellae direct the stretching, resulting in the widening of the stomatal aperture. The cuticle has no function in the opening mechanism, and the subsidiary and the encircling cells are regarded as accessory to the stomatal apparatus.

Reproduction

All *Cycas* spp. are dioecious. The reproductive cycle starts after several years of vegetative growth.

Male Cone

The short-stalked male strobilus is large and compact (Fig. 8.9A). The length of a ripe cone is ca.40 cm (*C. revoluta*) to 60–80 cm, or even more (*C. circinalis*). Normally, the cone is terminal, and the apex of the main stem exhausts itself during its production. Then the growth of the trunk is continued by a lateral bud which arises from the base of the peduncle, pushes the cone to one side, and takes its place. This new shoot apex

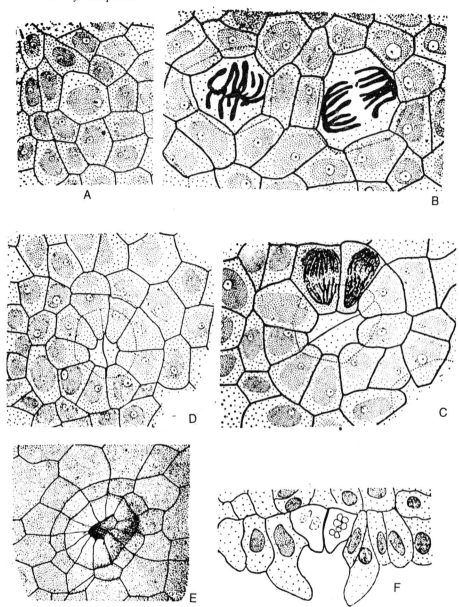

Fig. 8.8 A–F. *Cycas revoluta*, development of stomata. **A** Guard cell mother cell, lower epidermis of young leaflet. **B** Guard cell mother cells in division. **C** Young guard cells (dotted outline) slightly sunk below the adjoining cells. **D** Two guard cells (dotted outline) surrounded by subsidiary and encircling cells. **E** Mature stoma in surface view. **F** Vertical section of young stoma. (After Pant and Mehra 1963)

appears terminal, forms fresh crowns of scales, leaves, and later a cone. This process is repeated with the production of each new cone.

A male cone is an oval or conical structure with numerous spirally-

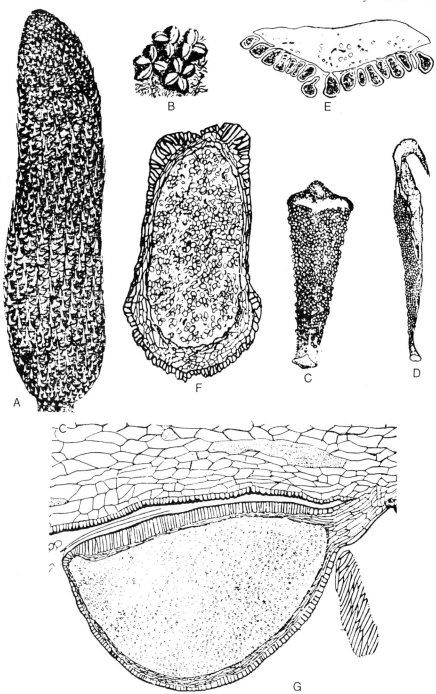

Fig. 8.9 A–G. **A** *Cycas revoluta,* male cone. **B–D** *C. circinalis.* **B** Sporangia in sori, longitudinal slits radiate from the center of each sorus. **C, D** Dorsal (**C**) and lateral (**D**) views of microsporophyll. **E–G** *Cycas* sp. **E** Transection of microsporophyll. **F, G** Tran- and longisection of microsporangium. (After Pant 1973)

arranged microsporophylls which are almost perpendicularly attached to the cone axis. Each microsporophyll is parenchymatous, has a hairy epidermis, stomata are present mainly on the lower surface, and numerous vascular bundles and mucilage ducts traverse through it (Fig. 8.9 E).

A mature sporophyll is hard and woody. It has a wedge-shaped, distally expanded, fertile portion which bears a large number of sporangia on its entire lower surface (Fig. 8.9 C, D). The sterile end tapers and has a pointed upcurved apex (Fig. 8.9 D). The sporangia are arranged in definite groups or sori (cf. ferns like those of the family Marattiaceae). Each sorus has three to six sporangia placed around an indusial papilla from which they are originally formed. The mature sporangia can be easily distinguished by the lines of dehiscence which radiate round the indusial papilla (Fig. 8.9 B). A mature sporangium is an oval sac with a short, and stout stalk. The wall consists of a thick-walled epidermis called exothecium (Fig. 8.9 F, G). A cuticle is present on its outer surface. The middle zone consists of three or four layers of thin-walled cells and an inner lining of tapetum, which mostly consists of small cells with large nuclei and thick granular cytoplasm. The sporangium is packed with pollen grains (Fig. 8.9 F, G).

Microsporangium

In *Zamia* and *Stangeria* the archesporium has a hypodermal origin (Pant 1973). One, sometimes several hypodermal cells undergo periclinal divisions, resulting in the primary parietal cells on the outer side and primary sporogenous cells on the inner side. These cells continue to divide and give rise to a several-layered wall, central mass of sporogenous tissue, and a tapetum between the two. In *Cycas* (Swamy 1948, De Silva and Tambiah 1952), the sporogenous cells repeatedly divide mitotically, enlarge, become rounded and form microspore mother cells (mimc). The tapetal cells with large nuclei and thick granular cytoplasm breakdown, and the nucleated cytoplasm is absorbed by the mimc. Some of the inner cells of the sporangial wall may also be absorbed. Each mature mimc has a large nucleus and a number of starch grains (Fig. 8.10 A).

Microsporogenesis.   The mimc undergo reduction divisions (Fig. 8.10 B–E). After meiosis I, a ring-like thickening appears on the callose wall of the mother cell in the equatorial plate. Usually, the wall laid down after meiosis I seems to be thicker than the one which developes after meiosis II. It is likely that the abundance of starch, especially on the equatorial plate, obstructs a clear view of the cell plate (Fig. 8.10 B–E).

The cytoplasm of each microspore contracts and secretes a membrane which forms its independent wall (Fig. 8.10 F, G). Thus, each member of a tetrad is contained within its own chamber. The original wall of the mother cell, which appears to be made up of pectic substances, collapses and no trace is left after a week of division. The pollen grain of cycads

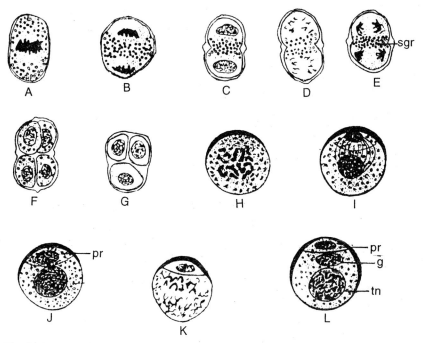

Fig. 8.10 A–L.   A–J *Cycas rumphii*, development of microspores and male gametophyte.
A, B Microspore mother cells, Meiosis I. **C–E** Meiosis II, starch grains (*sgr*) on equatorial
plate. **F, G** Microspore tetrads. **H–J** Delimitation of prothallial cell (*pr*) and larger cell.
K, L *Cycas* sp. **K** Larger cell in division. **L** Cross section of pollen grain at shedding
shows prothallial cell, generative cell (g) and tube nucleus (*tn*). (**A–J** After De Silva and
Tambiah 1952, **K, L** after Pant 1973)

shows an oval germinal furrow or sulcus on its distal face (Millay and
Taylor 1976). The furrow develops by a modification of the exine in the
region and can be differentiated into a peripheral and a central zone (Audran
1965, 1971). According to Gullvag (1966), in *C. revoluta* the exine is made
up of two layers. The outer one comprises orbicules and/or a sculptured
granular layer, while the inner layer is lamellated. The newly formed
microspores are uninucleate. Excluding the exodermis, the sporangial wall
now consists of three or four layers of thin-walled polygonal cells; most of
the tapetum has already been consumed.

Male Gametophyte.   The wall of the uninucleate pollen grain shows the
exine and intine. The latter is the last layer to be laid down and is of
variable thickness (Audran 1965), thinnest at the proximal or prothallial
cell end, and thickest at the distal end of the grain. Numerous Golgi vesicles
take part in its formation. The pollen contains abundant starch grains. The
nucleus divides assymetrically, forming a small lenticular prothallial cell,
and a large central cell (Fig. 8.10 H–J). The prothallial cell persists, unlike
in other gymnosperms. It remains attached to the lower region (proximal)

of the pollen grain wall. The antheridial initial (the central cell acts as an antheridial initial since there is only one prothallial cell), divides again to form a small antheridial cell and a large tube cell (Fig. 8.10 K, L). The former is attached to the intine at the site of the prothallial cell. The tube cell is usually vacuolate and has a large nucleus. The pollen grains are shed at the three-celled stage (prothallial cell, antheridial cell, and tube cell).

The microsporangia dehisce by longitudinal slits. The lines of dehiscence of various sporangia radiate out from the centre of the sorus towards the tip of the sporangia, and are usually marked by unthickened cells of the exothecium. After the sporangia open, the cone axis elongates and the tightly packed sporophylls separate from each other; this helps in the release of pollen. In a strobilus, the microsporophylls mature progressively from the apex to the base, and it takes several days for all the pollen in a strobilus to be shed. The numerous light pollen grains are easily blown by wind.

Megasporophyll

Instead of a female strobilus in *Cycas*, the megasporophylls form a loose crown at the apex; the apical growing point continues further growth. Each megasporophyll is a leaf-like structure, 15–30 cm long, and pinnately dissected in its upper region. In the lower portion, one to five pairs of ovules are borne on either side of the sporophyll. The sporophylls and ovules are covered with yellow hairs (ramenta) which fall off as the ovules ripen. The seeds develop a soft orange colour. In various species there is a gradual reduction in the expanded part of the sporophylls. In *C. siamensis, C. pectinata* and *C. revoluta*, the megasporophyll is large with elongated pinnae (Fig. 8.11 A, B, D). In *C. rumphii, C. beddomei and C. circinalis*, it is narrow and the pinnae are reduced to almost mere serrations (Fig. 8.11 C, E, F). In *C. normanbyana*, the sporophyll is highly reduced and bears only two ovules (Fig. 8.11 G).

Ovule.   The earlier stages of development of the ovule in cycads have not been fully studied. This may be due to the difficulty in locating very young ovules since they are deeply embedded in the crown of the leaves, and are not easily visible (from outside). For the collection of such ovules, one must dissect the entire crown. In these slow-growing plants, most of the ovules (on a plant) are at the same stage of development. Therefore, by "destroying" one plant, only one developmental stage can be obtained. The megasporophylls emerge from the crown when the ovules are already in the free-nuclear stage of the gametophyte. Owing to such difficulties, the details of megasporogenesis have been worked out for only a few cycads (P. Maheshwari and H. Singh 1967).

The large orthotropous unitegmic ovule may be up to 6 cm long and

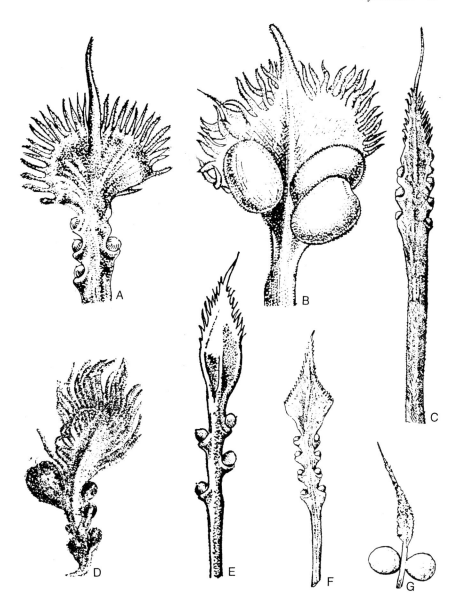

Fig. 8.11 A–G. Megasporophylls. **A** *Cycas siamensis.* **B** *C. pectinata.* **C** *C. rumphii.* **D** *C. revoluta.* **E** *C. beddomei.* **F** *C. circinalis.* **G.** *C. normanbyana.* (After Pant 1973)

4 cm in diameter. The integument is fused with the nucellus, and is free from it only at the apex. It can be differentiated into three layers: (a) the outer fleshy layer which becomes variously coloured at maturity, (b) middle stony layer, and (c) an inner fleshy layer which is resorbed and becomes papery even before maturity. The vascular supply at the base of the ovule divides into three strands, the middle one supplies the base of the sporangium,

and the two laterals enter the integument and divide again, The thicker strand supplies the outer fleshy layer, and the thinner ramifies into the fleshy inner layer (Fig. 8.12 F). The integument is open at the tip to form the micropyle. Due to disorganization of some of the nucellar cells at the apex, a concavity (the pollen chamber) is formed. The epidermal cells of the nucellus divide periclinally to form a nucellar cap (Pant 1973). A mature nucellus is beak-shaped. Its epidermis is heavily cutinized and has stomata, although it is enclosed by a thick integument.

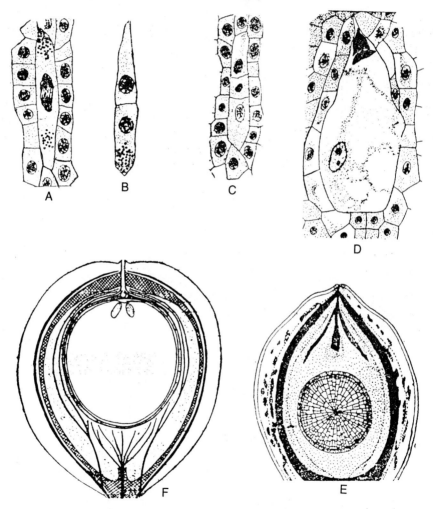

Fig. 8.12 A–F. Megasporogenesis and female gametophyte. **A–D** *C. rumphii.* **A** Megaspore mother cell. **B, C** Dyad (**B**) and triad (**C**). **D** Functional chalazal, and two micropylar degenerated megaspores. **E, F** *Cycas* sp., longisection of ovule (diagrammatic). **E** Cellular female gametophyte. F Mature ovule. (**A-D** After De Silva and Tambiah 1952, **E, F** after Pant 1973)

Megasporogenesis. Generally, the young nucellus has one to several hypodermal archesporial cells which divide periclinally to form the inner primary sporogenous and the outer primary parietal layer. The latter and its derivatives repeatedly divide periclinally to form a massive parietal tissue above the primary sporogenous cells. The latter may divide so that the sporogenous tissue increases. One of the sporogenous cells functions as the megaspore mother cell. One or more megaspore mother cells (mgmc) become distinct by their prominent nuclei and dense cytoplasm (Fig. 8.12A). The elongated cells show a thick cell wall (under the light microscope). The mgmc undergoes reduction divisions, but in *C. rumphii* the upper dyad fails to divide, so that a row of three cells is formed (Fig. 8.12 B–D, De Silva and Tambiah 1952).

Female Gametophyte. Repeated free-nuclear divisions occur in the functional chalazal megaspore, and the nuclei become distributed in the peripheral layer of cytoplasm (around a large central vacuole). The actual number of free nuclei before centripetal wall formation have not been counted in *Cycas* (Pant 1973). Gradually, the entire female gametophyte becomes cellular (Fig. 8.12 E). A further investigation will provide detailed information on the development of the female gametophyte.

A nutritive tissue, the spongy tissue or tapetum, of densely cytoplasmic cells differentiates around the sporogenous cells. The cell walls of the spongy tissue are acetolysis-resistant, and perforated. The walls of the innermost tapetal cells form a thin sheet of resistant material with the inner surface covered with sudanophilic droplets, or merely an aggregation of droplets. These droplets coalesce to form clusters on the tapetal cell wall. According to Pettitt (1966), the droplets are responsible for the deposition of material on the outer surface of the megaspore membrane. Eventually, the spongy tissue degenerates and lies compressed between the female gametophyte and the outer tissues of the ovule. Pettitt presumes that this layer has been interpreted as the megaspore membrane in cycads (and *Ginkgo*). When the female gametophyte has become cellular (after the degeneration of the spongy tissue), an "endospermic" jacket of large and multinucleate cells develops around the gametophyte.

Archegonium

At the micropylar tip of the gametophyte (prothallus), two to six archegonia differentiate. The archegonial initial (Fig. 8.13 A) divides to form a primary neck cell and a central cell. The former divides anticlinally, resulting in two neck cells (Fig. 8.13 B, C). They divide once again, just before fertilization, to form four neck cells. The second wall is much thinner than the first one. This is a regular feature in *Cycas* (Norstog 1972). By the time the nucleus of the central cell is ready to divide, the neck cells become large and turgid and project into the archegonial chamber. The central cell

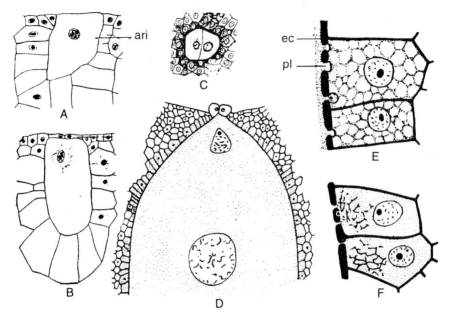

Fig. 8.13 A–F. Archegonium. **A–C** *Cycas rumphii.* **A** Archegonial initial *(ari).*
B Archegonium. **C** Two neck cells of archegonium in transection. **D–F** *C. circinalis.*
D Longisection (part of) archegonium. **E** Plugs *(pl)* of egg cytoplasm *(ec).* **F** Cells (devoid
of contents) show dark-staining bodies. (**A–C** After De Silva and Tambiah 1952, **D–F**
after Rao 1961)

continues to grow for several months before its nucleus divides. The ventral
canal nucleus moves up into the neck of the archegonium (Fig. 8.13 D),
where it soon disorganizes. Occasionally, however, it enlarges and simulates
the egg nucleus. The maturation of different archegonia in a gametophyte
is synchronous. The egg cytoplasm shows a number of proteid vacuoles
which increase after fertilization. In *C. rumphii* they are rich in proteins
(De Silva and Tambiah 1952).

The gametophytic region around the archegonia grows upwards and forms
a depression in the middle, the archegonial chamber.

The archegonia occur singly, and each has a well-developed single-
layered jacket. There is usually a thick wall with simple pits between the
jacket cells and the central cell of the archegonium. Stopes and Fujii (1906)
observed distinct plasmodesmatal connections between the walls of the
jacket cells and the egg. They refute any "communication" through simple
pores between the inner layer of the jacket and the egg cells, as claimed by
Ikeno (1898) in *Cycas.* According to Swamy (1948), the pores become
occluded by plug-like thickenings, and the jacket cells become depleted of
their contents (Fig. 8.13 E, F). The thick cell wall forms the so-called egg-
membrane during the post-fertilization stages. The egg membrane thickens
as the egg matures. It retains its form and remains connected with the

suspensor of the embryo when the latter is fairly well advanced. There are prominent pits on the thick wall. Extension of the egg cytoplasm come in contact with the cytoplasm of the jacket cells, through the plasmodesmata present in the pit areas.

Pollination

The male cones mature when the young megasporophylls have emerged on the female plant, and the ovules are at the free-nuclear gametophytic stage and ready for pollination. The single integument is drawn out into a short micropylar tube, which secretes a sugary exudate (pollination drop) close to the nucellus of the ovule. The wingless pollen grains usually float into the pollination drop, which retracts soon after. The pollen grains show a germinal furrow, which closes in dry weather and is wide open in high humidity. After landing on the pollination drop, the pollen grains imbibe water and other substances from the exudate, become heavy and sink down the micropyle to reach the nucellus, where they germinate.

The male cones of *Cycas* growing in warm climate emit a strong odour (described earlier); several insects visit them when the pollen ripens (see Chamberlain 1935, Pant 1973). Thus, these insects may also be carriers of pollen in cycads, although this has to be confirmed by critical field observation. Following pollination, the ovules enlarge and reach their full size. Unpollinated ovules remain small and eventually dry up. In the female plants of *C. rumphii* (growing in the Peradinya Botanic Garden, Sri Lanka), some of the ovules could develop to full size even when no male plant of the same species was growing in the vicinity. This is due to pollination by pollen of other cycads like *Encephalartos* and *Macrozamia*. The pollen grains germinate in the pollen chamber and the male gametophyte develops to an advanced stage although sperms are not formed (Le Goc 1917). Such ovules do not form embryos. Pant (1973) occasionally observed fully developed ovules in the megasporophylls of *Cycas revoluta* and *C. rumphii* growing in the Botanical Garden of the Allahabad University, where the nearest male plants of *C. circinalis* were about 5 km away. The stimulus for the development of these ovules was quite likely provided by the pollen grains of some other plant, possibly an angiosperm or conifers, like *Pinus, Thuja, Cupressus* (Pant 1973).

Niklas and Norstog (1984) studied the implication of aerodynamics on reproduction, and the pattern of pollen deposition on megasporophylls of *Cycas* (and megastrobili of *Dioon* and *Zamia*). A characteristic air-disturbance pattern around megasporophylls influences the quality and arrangement of wind-borne pollen grain deposition on the surface of megasporophylls and ovules. The largest number of pollen grains adhere on the windward profiles. The megasporophyll deflects the air flow passing over it leeward, distal surfaces, where pollen grains accumulate. The preferential concentration of pollen on the distal portions may help pollination. Water dislodges the adhering pollen grains, which flow along the glabrous ovule-bearing margins of megasporophylls and collect near or on the micropyle. The authors suggest two phases in cycad pollination: (a) the transport of wind-borne pollen grains to megasporophyll/megastrobili (*Cycas, Dioon, Zamia*); (b) subsequent transport of adhering pollen to ovules by water and/or passive sifting (*Cycas*) or insect activity (*Dioon* and *Zamia*).

After the pollen grains land in the pollen chamber (situated in the upper beak-shaped portion of the nucellus), a group of cells lying below begin to degenerate and extend the lower limit of the chamber. This post-pollination extension is the lower pollen chamber or intermediary chamber.

The intine of the pollen grain ruptures the exine at the distal end, grows into a tube in the region of the furrow, and the tube nucleus migrates into it (Fig. 8.14 A). The pollen tube grows laterally and intercellularly into the tissue of the nucellus, instead of towards the archegonia as in other gymnosperms. During its course through the nucellus, the tube often branches horizontally, derives nourishment as it digests its way, and accumulates

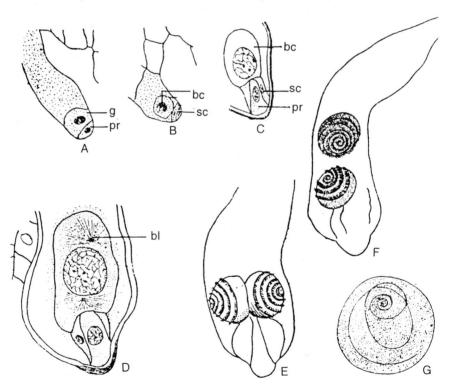

Fig. 8.14 A-G. Development of antherozoids. **A–D** *Cycas rumphii*, pollen tube with prothallial (*pr*) and generative (*g*) cell. **B** Generative cell divides to form stalk (*sc*) and body (*bc*) cells. **C** Prothallial cell extends into the stalk cell. **D** The body cell shows two well-developed blepharoplasts (*bl*). **E, F** *C. revoluta*. Ciliated antherozoids, before (**E**) and after (**F**) free movement. **G** *Cycas* sp. Antherozoid with ciliated spiral band. (**A–D** After De Silva and Tambiah 1952, **E, F** after Miyake 1906, **G**, after Pant 1973)

considerable amount of starch for the development of the male gametophyte. The pollen tube does not transport the male gametes but establishes only a haustorial system.

In the beginning the grain-end of the pollen tube is enclosed in the exine but, eventually, it enlarges and cannot be contained therein. This proximal end of the pollen grain contains the prothallial, stalk and body cells (or male gametes). The broad, short (Fig. 8.14 C-F) proximal end grows towards the female gametophyte. The branched pollen tube continues invasion of the nucellus and the haustorial activity increases.

The antheridial cell divides periclinally; the cell adjacent to the prothallial cell is sterile and designated the stalk cell, the other cell is the body cell (Fig. 8.14 A-C). The stalk cell is surrounded by a distinct cell wall. It enlarges considerably, and accumulates starch and other materials. Simultaneously, the prothallial cell bulges and its upper end penetrates through the stalk cell, which forms a life-belt-like ring around the prothallial cell. The body cell (Fig. 8.14 C, D) begins to enlarge and elongate along the longitudinal axis of the pollen grain. The nucleus of the body cell also enlarges, and two small granular bodies (the blepharoplasts) appear, one on each side of the prothallial cell (Fig. 8.14 D). Cytoplasmic rays (also called astral rays) radiate from the blepharoplasts. The astral rays are the microtubules which extend from the matrix or the surface of the blepharoplasts. The latter comes to lie on the two poles of the body cell along the longitudinal axis of the pollen grain. The body cell enlarges, becomes spherical, and the blepharoplasts move through 90° and come to lie at right angles to the long axis of the pollen. The blepharoplasts increase in size and become very prominent.

The mitotic spindle of the body cell lies at right angles to the long axis of the pollen tube, so that the two sperms lie side by side (Fig. 8.14 E) enclosed in the wall of the body cell till they mature. The blepharoplast forms a part of clock-spring-band-bearing cilia, around the gradually or abruptly tapering distal part of the sperm (Fig. 8.14 F, G). This band in *C. revoluta* shows a right-hand coiling (as seen from the apex of the sperm). The spiral band has six turns (gyres), and in a radial section (*C. circinalis*) appear stair-like. The sperms start rotating movement while still enclosed in the body cell. The sperms of *C. revoluta* are 180–210 μm in length, and visible to the naked eye. They are released by the rupture/dissolution of the outer walls of the sperm mother cells into the basal end of the pollen grain. By their cilia they move freely in the fluid of the turgid pollen grain-end and remain active for several hours.

The nucellar tissue between the pollen and archegonial chamber collapses, and the proximal end of the pollen grain pushes down through this opening and "hangs" freely in the dry archegonial chamber (Fig. 8.15 A). A bunch of pollen grains grow downward and disorganize it completely between the intermediary chamber and the female gametophyte. The proximal end of

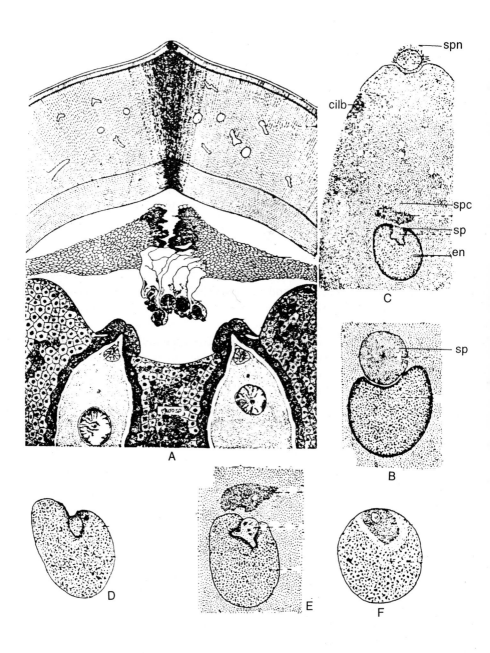

Fig. 8.15 A–F. Fertilization. **A** *Cycas* sp., longisection upper part of ovule (semidiagrammatic). **B–F** *C. revoluta*, successive stages in the fusion of egg and sperm nuclei. *cilb* ciliary band. *en* egg nucleus. *sp* spermatozoid. *spc* spermatozoid cytoplasm. *spn* supernumerary spermatozoid. (**A** After Pant 1973, **B–F** after Ikeno 1898)

the pollen grain lies in a longitudinally continuous cavity in the middle of the nucellus (Fig. 8.15 A).

Fertilization

The proximal end of numerous pollen grains hang into the archegonial chamber (Fig. 8.15 A). Finally it bursts and forms a short pollen tube through which it discharges the male gametes into the archegonial chamber. The gametes swim for about 15 min with a forward and circular motion; the band of flagella form the anterior end. The large spermatozoids enter the egg through the opening between the neck cells within a few minutes of their release. The male gametes often become distorted during the passage, since the opening is narrow. On entering the egg cytoplasm, the flagellate band of the male gamete is cast off, and is left at the tip of the egg cytoplasm (Fig. 8.15 B), where it gradually dissolves and is eventually represented by dark-staining granules. The naked sperm nucleus shrinks as it moves downward to unite with the egg nucleus. The male nucleus penetrates deep into the larger egg nucleus before its membrane disappears (Fig. 8.15 B–F). In *C. circinalis* the two sets of chromatin can be easily distinguished from each other by the different morphology of their threads. The spiral band usually persists in the egg cytoplasm long after fertilization. Occasionally, a sperm may get close to the egg nucleus with its ciliary coat, which is cast off just before uniting with the egg nucleus. After a sperm entered an archegonium (Fig. 8.15 C–F), it is sealed off by a dark-staining bubstance which is probably derived from a degenerated sperm. Often one or two, sometimes up to five, additional sperms penetrate the cytoplasm of an egg cell; only the first sperm fertilizes the egg nucleus. While still far from the fertilized nucleus, the supernumerary sperms degenerate without casting off their spiral bands or sheaths. Rarely, they may even penetrate close to the egg nucleus and cast their coats there.

Embryogeny

The zygote nucleus divides in situ followed by several free-nuclear divisions (Fig. 8.16 A-E). The nuclei are distributed throughout the young proembryo. In later stages, the free nuclei mostly concentrate at the base of the proembryo (Fig. 8.16 F) and only a few nuclei are present in the upper thin cytoplasm. Subsequently, only the nuclei at the base divide while the upper nuclei show signs of degeneration. At the time of wall formation, there are 512 free nuclei in *C. circinalis* (Rao 1963). Cells are formed only in the lower portion of the archegonial cavity (Fig. 8.16 F-J), the nuclei present in the upper area remain free. During later stages, the free nuclei, along with the circumjacent cytoplasm, degenerate and form a plug. Following wall formation, the cells at the base divide and function as embryonal cells. The upper cells (just above the embryonal cells) differentiate into a suspensor and the uppermost layer of cells as the buffer cells, the latter form a few layers of tissue around a central vacuole (Pant 1973).

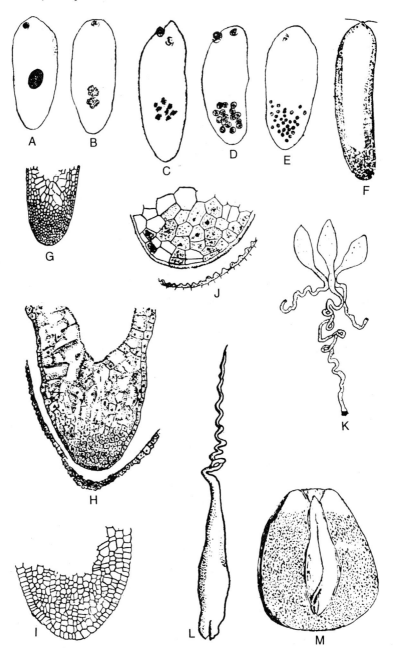

Fig. 8.16 A-M. Embryogeny. **A-E, J-M** *Cycas* sp. **A-E** Zygote, 2–, 8–, 16– and 32–nucleate proembryo. **F, G, I,** *C. circinalis,* **H** *C. rumphii.* **F** Young proembryo shows segmentation. **G-J** Apical end of young proembryos. **K** Three embryos at different stages of development. **L** Dicotyledonous embryo. **M** Longisection seed. (**A-E** After Swamy 1948, **F, G, I** after Treub 1884, **H, K** after De Silva and Tambiah 1952, **J** after Pant 1973, **L** after Schuster 1932, **M** after Richard, from Engler and Prantl 1926, see Pant 1973)

The thick inner tangential walls of the jacket cells persist till an advanced stage of proembryo. The suspensor penetrates the chalazal end to the thick wall which is very prominent and is called the egg membrane.

Differentiation of Embryo.   After the proembryo becomes cellular, the cells of the upper region elongate to form a very long suspensor. It grows rapidly, and thrusts the basal embryonic region through the egg membrane and archegonial jacket deep into the gametophytic cells, which have abundant food reserve. This is due to enzymatic digestion of the gametophytic cells when they come in contact with the embryonic cells. Due to rapid elongation, the suspensor becomes much coiled and twisted (Fig. 8.16K). Usually, more than one archegonium in an ovule is fertilized, and the development of multiple zygotes leads to simple or archegonial polyembryony (Fig. 8.16 K). At first, all the zygotes in an ovule appear to develop with equal vigour, the embryos project into the gametophytic tissue, and their suspensors become closely intertwined with each other. Due to competitive growth, ultimately only one embryo develops while the rest abort at various stages of development (Fig. 8.16 K, L). The suspensor of the aborted embryos remain attached to the tough egg membrane, and persist for a long time. The suspensor of the successful embryos is, therefore, a composite structure of the coiled suspensors of the entire group of embryos in an ovule. When stretched, the suspensor may be up to 10 cm long.

Seed

The studies on late embryogenesis in cycads are incomplete. A number of cells are formed in the embryonic region before the differentiation of various organs. The maturation of the embryo in seed takes over a year after fertilization. The seed is shed at any stage during this period, and the development of the embryo is completed on the ground. To begin with, the embryo only increases in size without differentiation into its organs. Then the coleorhiza develops at the micropylar end of the embryo. The latter shows internal differentiation into two polar meristems, epidermis, cortex, procambium and pith; the hypocotyl is rather short in cycads. The number of cotyledons varies from one to three, closely adpressed to one another and appear as a single structure. The coleorhiza is partly derived from the suspensor and becomes quite "hard" in a mature embryo. It is morphologically equivalent to the root cap. At maturity several outer layers of the root cap become especially thick-walled to form a distinctive caplike structure. There are abundant mucilage cells in the tissues of the embryo.

The attractive, red or orange fleshy seeds contain abundant food reserve (Fig. 8.16 M). Quite likely, they are dispersed by birds and rodents. The cycad seeds do not have a resting period and the viability is short. Germination of the seed is epigeal.

Chromosome Number

Cycas circinalis, C. media var *basaltica, C. revoluta* var *revol,C. revoluta* var *taiwaniana* and *C. siamensis* have similar karyotypes of 2n = 22. There are 2 median-centromeric, 4 long submedian-centromeric, 4 short submedian-centromeric and 12 terminal-centromeric chromosomes. Kokubugata and Kondo (1996) compared the similar karyotypes with each other by using (fluorescent staining method with double counter-staining reagents binding for opposite base-pairs) chromomycin A_3 (CMA) specifically for guanine-cytosine and 4′, 6-diamidino-2-phenylindole (DAPI) specifically for adeninethymine. These techniques are useful to investigate the chromosome variability and to mark the species-distinctive regions on chromosome for cytotaxonomy. A CMA band has been observed at the terminal region of each of the 4 largest submedian-centromeric chromosomes. The 12 terminal-centromeric chromosomes display CMA bands at the terminal and pericentral region (2 of the chromosomes carry a CMA band in the interstitial region of the long arm in different positions in each of the 4 species). All the chromosomes exhibit DAPI at the centromeric region.

Similarities of the karyotypes, the CMA and the DAPI fluorescent patterns, observed in the species of *Cycas,* suggest that speciation of the genus may have occurred without any major karyotypic change. *Cycas* thus appears to be a monophyletic group in morphological and anatomical (Stevenson 1990a), and cytological characters (Kokubugata and Kondo 1996).

Temporal Consideration

As compared to the temperate gymnosperms, the reproductive cycle of cycads occurs at different times of the year. The cones in the South Indian *C. circinalis* are initiated probably in April, and show micro- and megaspore mother cells during June-July (Rao 1963). The ovules have a free-nuclear gametophyte at the time of pollination (8 months after initiation) in December. Fertilization occurs in May-June (5–6 months after pollination), followed by a very slow development of the embryo, in which the cotyledons appear during November-December. The seeds with an immature embryo are shed during May and June (1 year after fertilization). The embryo matures and attains full size, and the seeds germinate during September-October (4 months after shedding). Thus, this taxon takes 2 years and 5 months to complete its life cycle, although there is no winter rest. In *C. rumphii,* the ovules are pollinated in May (De Silva and Tambiah 1952), fertilization occurs 13 months later (i.e. the following June), and the seeds are shed in January but the embryo matures only by March. Consequently, the time lapse between pollination and fertilization is much longer.

The general reproductive cycle of cycads is:

1. There is a long time gap from the time of ovule initiation to pollination (the ovules are in an advanced stage of development at the time of pollination, as compared to conifers).

2. There is a short time lapse between pollination and fertilization. Most of this time seems to be taken up in the maturation of archegonia and filling up of their cytoplasm.

3. The embryo takes a long time to mature, partly on the plant and partly after shedding.

Phylogenitic Considerations

Morphological studies suggest that the immediate ancestors of the living cycads are extinct. No extant taxa can be considered to be the forerunner, and they cannot be arranged into a series considering all the characters because every taxon shows primitive and advanced characters. *Cycas* is regarded as a primitive form, because its megasporophylls are loosely arranged and do not form a strobilus, and its circinate young leaves are fern-like. The compact male cones, pitted secondary wood, and single-veined pinnules display advancement. According to Gaussen (1950): (a) *Cycas* shows 12 primitive and 70 advanced features. (b) *Zamia* is the most advanced in the group. However, the species with tuberous stem have scalariform tracheids (filicinean feature). Gaussen (1950) lists 67 primitive and only 33 advanced characters in *Zamia*. (c) *Encephalartos* has fern-like leaves (it was classified with the ferns until the cycadean nature was revealed after observation of the cones), the stomata are least specialized, the epidermal cells of the leaf have undulate lateral walls (cf. ferns) and presence of scalariform wood, and centripetal xylem in the cone axes. (d) The male and female gametohytes of *Microcycas* are presumed to be the most primitive in seed plants because of the largest number of male gametes (16 to 22) and archegonia (64 to 200). The vegetative anatomy, and the cones are considered to be advanced.

A genus which appears advanced in some respects may thus be primitive in others. There is a similarity (in the group) above the generic level. Each of the ten (eleven) extant taxa of the Cycadales presents a terminal branchlet of a phylogenetic tree where all connecting links have disappeared (see Arnold 1953). Therefore, the interrelationships in modern cycads are difficult to visualize. According to Chamberlain (1920), the cycads did not give rise to any other plant group, and will probably become extinct in the next geological period.

9. Caytoniales

This is a Mesozoic Pteridosperm which is often described as a distinct Order. In some classificatory systems, three families are included in this Order: Caytoniaceae, Corystospermaceae and Peltaspermaceae (see Stewart 1983).

Caytoniaceae

Several spp. of leaves (*Sagenopteris*), pollen organs (*Caytonanthus*) and ovule-bearing organs (*Caytonia*) are known in association, but not in organic connection, from Greenland, Sardinia, western Canada, the eastern USSR, England and Siberia. The leaves, known for over a century, are scattered geographically, and stratigraphically from Upper Triassic to Lower Cretaceous rocks (Fig. 1.1). The reproductive organs were first described by Thomas (1925) from Mid-Jurassic rocks at Cayton Bay in Yorkshire. There is a similarity between the epidermal cell structure (specially stomata) of *Sagenopteris* with that of the fruit axes, and the presence of caytonanthus type pollen in the nucellar beak cavity of *Caytonia* show that the three organs are congeneric.

Leaf. *Sagenopteris phillipsi* leaves have a slender petiole which bears four lanceolate, palmately compound, terminal leaflets (Fig. 9.1 A). Each leaflet has a prominent mid-rib and reticulate venation (Fig. 9.1B). Haplocheilic stomata (Fig. 9.1 C) are present on the lower surface. The entire leaf was shed by an abscission layer (Harris 1951).

Pollen-Bearing Organs. The structure of *Caytonanthus* is unusual. It has a dorsiventral rachis which bears opposite or sub-opposite, irregularly branched pinnae. The ultimate branches bear terminal, pendant, tubular synangia (Fig. 9.2 A). Each synangium has three or four sporangia which, on dehiscence, separate from each other except at the tip (Fig. 9.2 B). The synangium is radially symmetrical (Harris 1935) and differs from bilateral symmetry in extant flowering plants. The pollen grains are winged (Fig. 9.2 C).

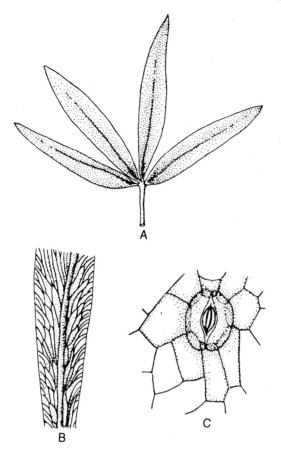

Fig. 9.1 A-C. *Sagenopteris phillipsi.* **A** Palmate leaf. **B** Leaflet shows reticulate venation. **C** Stoma. (**A, B** After Thomas 1925, **C** after Harris 1964)

Raunsgaard and Fries (1986) studied by light microscopy, TEM and SEM, the pollen grains extracted from *Caytonanthus kockii* (lowermost Jurassic of Scoresby Sound, East Greenland) and *C. arberi* (Middle Jurassic of Yorkshire, England). The pollen grains are small, ovate, bisaccate, and have a nearly smooth surface except for the granular sulcus area.

In transmitted light the internal structure of sacci appears reticulate with elongated meshes (as in several conifers). Ultrastructural studies with TEM reveal: (a) A spongy infilling of the sacci formed by elongated, cylindrical, solid elements that radiate from the central corpus, and branch irregularly. A similar infilling of the sacci (this appears to be an unique feature) has also been observed in other Mesozoic pteridosperm. (b) The outermost part of the pollen wall is thin, homogenous and has a smooth surface, while the innermost layer (endoexine) is thick, and has a laminate to thready structure. The thick laminate endexine and the sacci distinguish the pollen of *Caytonanthus* from the angiosperms and group them with gymnosperms.

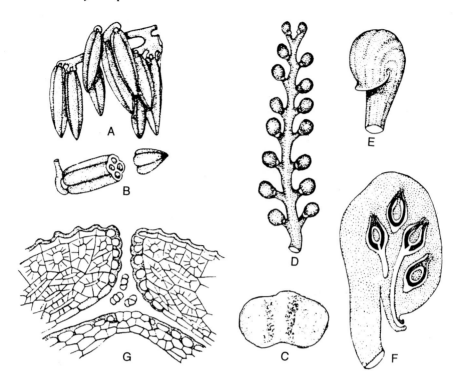

Fig. 9.2 A-G.   A, B *Caytonanthus kochi*. **A** Microsporophyll. **B** Cross section synangium shows four microsporangia. **C** *Caytonanthus*, bisaccate pollen grain. **D, E** *Caytonia nathorsti*. **D** Reconstruction megasporophyll. **E** Young cupule shows lip and opening. **F** *C. thomasi*, longisection cupule shows position of ovules. **G** *Caytonanthus* type pollen grains in micropyle of *Caytonia* ovule. (**A, B** After Harris 1937, **C, E** after Harris 1964, **D** after Thomas 1925, **F** after Harris 1933, **G** after Harris 1951)

However, the ultrastructure of the sacci of the pollen differs from coniferopsid pollen.

Ovule-Bearing Organs. *Caytonia* consists of an axis ca. 4 cm long, which bears two rows of lateral ovule containing cupules (Fig. 9.2 D). A cupule is ca. 4.5 mm in diameter, globose, recurved and has a lip-like projection adjacent to the pedicel above an opening (Fig. 9.2 E); it closes at maturity. The cupule encloses a U-shaped row of ovules which are in a single cavity in *C. sewardi* or in separate ones, as in *C. thomasi* (Fig. 9.2 F). The number of ovules varies; in *C. sewardi* there are 8 in a single row, in *C. nathorstii* 15, and in *C. thomasii* about 30 in a double row. Each ovule is elliptical, orthotropous and has a short micropyle which points toward the opening below. Some of the ovules and the wall on which they are borne are quite fleshy. Pollen grains have been seen within the micropyle of the ovules (Fig. 9.2 G). The opening must have been very small. It is likely that the pollen reached the micropyle by means of a "pollen drop mechanism", as in many extant gymnosperms. The pollen grains trapped in the drop floated up the channels to the seeds, and the two lateral bladders on the pollen grains may have acted as flotation devices. It is also possible that at pollination time it was a much more open structure. Later, the opening became small by differential surface growth.

There is a single integument, which is free from the nucellus. The cuticle covering the nucellus is exceptionally thick. There is no evidence of vascular system in the integument.

Phylogenetic Considerations

The relationship of Caytoniaceae is not at all clear, and it appears to be quite isolated. Few fossils have created such a stir among morphologists as Caytoniaceae when it was first discovered by Thomas (1925). He emphasized similarities with both pteridosperms and angiosperms. He assumed that it provides a clue to the origin of flowering plants. He considered the flange on the cupule as a stigma, the cupule itself as a kind of carpel and the *Caytonanthus* synangium as the angiosperm stamen (despite its radial symmetry, and lack of a filament and connective). However, the pollination in *Caytonia* is still at the gymnosperm level (Harris 1964). Thus, a decision on its phylogenetic relationships should be taken only after we know more about the nature of its cupules (Sporne 1965).

10. Glossopteridales

During the Lower Carboniferous, the flora of the North and South Hemispheres was more or less similar. By the Upper Carboniferous and Lower Permian, there was a completely different flora in South America, much of the southern part of Africa, Australia, the Indian Peninsula, Antarctica, New Zealand and other smaller land masses. These regions together formed the supposed continent of Gondwanaland, separated by the Tethys Sea from the other continent of the Northern Hemisphere.

The Gondwana era started with a cold environment and was followed by a warm moist climate which supported a new type of vegetation. This flora is often called the glossopteris flora because of the abundance of the leaves known as *Glossopteris*. They appear in the Upper Carboniferous, extend into the Triassic, and decline thereafter (Fig. 1.1). Leaves with characteristics of *Glossopteris* have also been reported from the Jurassic of Mexico (Delevoryas 1969). The geology of the Gondwana system is best known in India, where several studies of the flora have been conducted.

Glossopteridaceae

The name *Glossopteris* was proposed by Brongniart (1828), for the leaves which were lanceolate/tongue-shaped, 3 to 40 cm in length. For a long time, it was presumed to be a fern, as the leaves resembled a modern taxon such as *Polypodium musaeifolium* (Sporne 1965).

Evidence accumulated by Gould and Delevoryas (1977) and Pant (1977) clearly shows that *Glossopteris* was a large tree. Gould and Delevoryas made a detailed reconstruction of the plant (Fig. 10.1A). The trunk is nearly 6 m tall, with gymnospermous wood of araucarioxylon type. It is supported by a root system of the vertebraria type, conspicuous growth rings are present in roots, trunks and branches. The leaves are in spirals or whorls, probably on short shoots (Pant 1977). The plants are deciduous (Surange and Chandra 1976), as shown by the presence of a large number of isolated leaves and very few reproductive organs.

Root. The permineralized plant remains from the Upper Permian contain silicified remains of *Vertebraria*, the detached roots of a *Glossopteris* plant

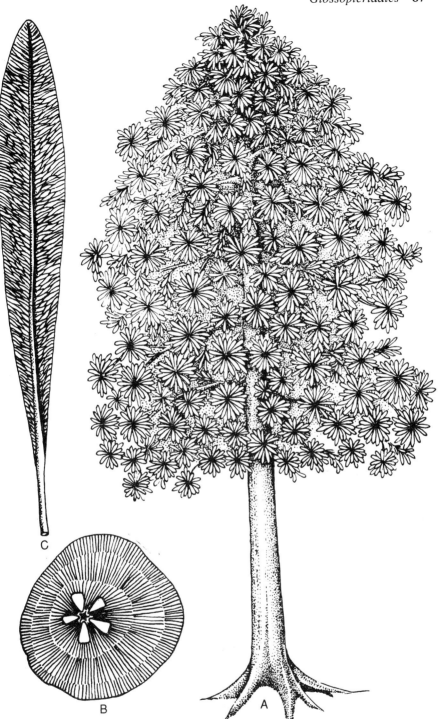

Fig. 10.1 A-C. **A** *Glossopteris* tree, reconstruction. **B** *Vertebraria,* transection.
C *Glossopteris* sp., leaf (reconstruction) shows mid-rib and reticulate venation.
(**A, C** After Gould and Delevoryas 1977, **B** after Gould 1975)

(Gould 1975). They can be easily identified by their characteristic wedge-like sectors (Fig. 10.1A) which radiate from the centre of the axis. This is the secondary xylem, arranged around a central polyarch protostele, with the protoxylem strands alternating with the radiating arms (Fig. 10.1B). The cavities between the radiating arms suggest that *Vertebraria* is the root system of a plant which flourished in a semi-aquatic environment. The roots branched commonly (indicated by the presence of branch root traces and smaller roots in the rock specimens). The secondary xylem is identical to the pycnoxylic wood of *Araucarioxylon arberi* trunks. Both have opposite/alternate circular bordered pits in five or six vertical rows. They may be in groups of two to seven, crowded on the radial walls so that the pits are hexagonal. Uniseriate rays are 1–20 cells high. The various observations undoubtedly show that silicified roots of *Vertebraria* and stems of *Araucarioxylon arberi* are congeneric (Stewart 1983).

Leaf.   *Glossopteris* and *Gangamopteris* are the most notable elements of the flora. Studies on epidermal structure confirm the existence of numerous species. According to Surange (1966), a fair degree of speciation can be expected during the 50 million years of their existence on earth. *Glossopteris* had a prominent mid-rib and reticulate venation (Fig. 10.1C). In a few species, a distinct petiole was present. *Gangamopteris* was similar, but lacked the midrib. *Glossopteris* leaves are characteristically hypostomatic (stomata on lower surface), the stomata are haplocheilic and irregularly placed between the veins. Internally, the leaf is differentiated into an upper and lower epidermis, and a mesophyll of palisade and spongy parenchyma. The mid-rib is made up of closely spaced branching and anastomosing vascular bundles. The lateral bundles which form the veins are associated with sclerenchyma strands in the lamina.

There are different types of staminate and ovuliferous structures which belong to this group of plants. They were borne separately. It is not known whether they were borne on different parts of the same plant or on separate plants. More than 30 genera of fructification are attributed to *Glossopteris* and *Gangamopteris*.

Male Fructification.   Pollen organs have been recovered from Permian deposits of South Africa, India and Australia (Surange and H.K. Maheshwari 1970, Lacey van Dijk and Gordon-Gray 1974, 1975, Gould and Delevoryas 1977). Surange and H.K. Maheshwari reconstructed the microsporophyll *Eretmonia* as a unit. It has a stalk which bears an expanded, nearly triangular, distal lamina. Two branches arise, just beyond the mid-point, on the adaxial surface of the stalk each bearing whorls of arberiella-type sporangia (Fig. 10.2A,B). Venation pattern and epidermal features show their undoubted glossopterid affinities. Abundant *Arberiella* sporangia and striatites type of bisaccate pollen grains with transverse striations on the corpus (Fig. 10.2C)

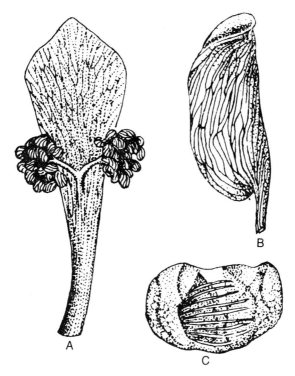

Fig. 10.2 A-C. **A** *Eretomonia* sp., fertile leaf with two clusters of microsporangia. **B** *Arberiella,* a glossopterid microsporangium. **C** *Striatites* type of pollen of *Glossopteris.* (**A** After Surange and H.K. Maheshwari 1970, **B, C** after Pant 1977)

are dispersed among the permineralized specimens of *Glossopteris* (Gould and Delevoryas 1977).

Glossotheca is another pollen organ which shows clusters of *Arberiella* sporangia (Surange and H.K. Maheshwari 1970). The spatulate unit has a single branch, which arises from near the mid-vein, and bears paired laterals terminating in clusters of sporangia.

Ovulate Structures. The ovulate organ consists of a stalked fertile head/capitulum. In a few specimens it is attached to modified leaf-like bracts. Gould and Delevoryas (1977) report it to be attached (in one specimen) to true leaves, either in the axil of a bract or to its upper surface. The fertile bracts may sometimes bear several capitula, as in *Lidgettonia* (Fig. 10.3A) or *Denkania* (Fig. 10.3 B).

In *Lidgettonia,* the fertile leaves are shorter than the sterile leaves, spathulate, and have a round apex. The veins spread from the petiole of the leaf into the lamina, where they fork and anastomose. The petiole of the leaf bears two rows, each containing three or four disc-like capitula on the abaxial side (Fig. 10.3A). Each capitulum is peltate and stalked. There are six or seven ovules in a row on the underside of the capitulum, near the

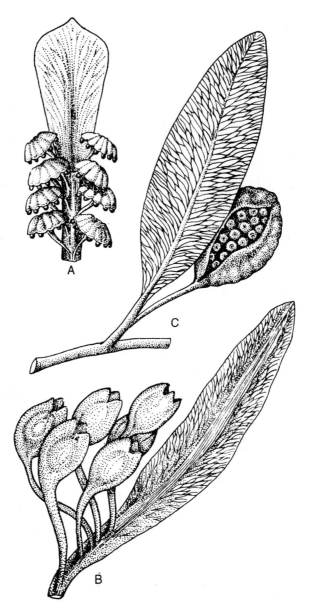

Fig. 10.3 A-C.   **A** *Lidgettonia mucronata,* fertile leaf shows disc-shaped capitula in
two rows. **B** *Denkania indica,* adaxial surface of glossopterid leaf with a row of pedicels.
Each has a terminal cupule which contains a single ovule. **C** *Dictyopteridium* sp.,
ovuliferous capitulum with subtending leaf (seen from underside). (**A** After Surange and
Chandra 1972, **B** after Surange and Chandra 1971, **C** modified from Surange and Chandra
1975, and Gould and Delevoryas 1977, see Stewart 1983)

undulating margin in *L. mucronata.* In *Denkania indica,* the pedicels arise
in a row from the adaxial surface of the subtending glossopterid leaf

(Fig. 10.3 B). Each pedicel terminates in a cupule containing a single ovule whose impression can be seen inside.

In *Arberia,* the more divided capitula appear to be axillary, while in the entire forms such as *Dictyopteridium* they are surface-attached (Fig. 10.3 C; 10.4 B). Thus, these plants have leaf-borne reproductive organs, which is unique among gymnosperms. It is likely that they originated in the axillary position and were later incorporated to the leaf surface (Pant 1982).

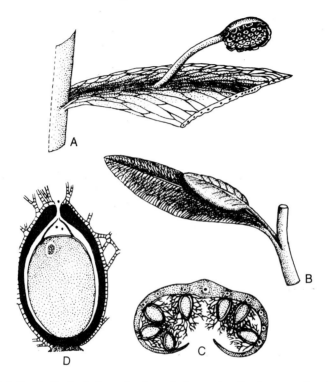

Fig. 10.4 A-D. **A** *Ottokaria bengalensis,* capitulum and stalk adnate to subtending leaf. **B-D** *Glossopteris (Dictyopteridium).* **B** Ovuliferous capitulum with subtending leaf. **C** Transection ovule-bearing capitulum. **D** Single ovule from capitulum in longisection. (**A** After Schopf 1976, **B-D** after Gould and Delevoryas 1977)

The capitulum could be entire or divided, and always bears the ovules on the lower surface which may be enfolded (Fig. 10.3 C). There is a wide range of variation in the number and arrangement of their ovules. In *Scutum* there are up to 75 ovules which are spaced randomly, while in *Senotheca* the capitulum is reduced and the ovules are spaced closely in two rows. *Denkania* has only one ovule on each capitulum which appears cupule-like (Surange and Chandra 1972). In *Dictyopteridium,* the edges curve over and enclose the ovules in a mass of hairs (Fig. 10.4 C), while in *Ottokaria* the space between the ovules is filled by a filamentous structure. The ovules are small and numerous, sessile, ovoid to pyriform (Fig. 10.4 C), and point towards the centre of the enclosing capitulum. The integument is thicker around the micropyle. The nucellus is free from the integument except at the chalazal end (Fig. 10.4 D). The pollen chamber occasionally contains bisaccate pollen grains with transverse striations on the corpus (a characteristic of *Glossopteris* pollen). A few ovules have been observed with megagametophytes containing a single archegonium. The embryo has not been observed in any ovule. In addition to the filaments which fill the spaces between the ovules, there are conical webs of filaments which extend from each micropyle. This suggests that pollen-drop liquid, which completely filled the structure, directed the pollen towards the micropyles.

It is not certain whether the glossopterids are a single group of plants; the nature of the ovules and their mode of attachment are quite varied. Surange and Chandra (1976) presumed there are two groups of glossopterids: a) With strobiloid receptacle: in *Dictyopteridium* the strobiloid ovule-bearing receptacle is borne in the axil of a stalked fertile bract that covers it like a protective spathe (Fig. 10.3 C). The receptacle and its fertile bract are subtended by and attached to the petiole of a *Glossopteris* leaf. In *Scutum,* according to Surange and H.K. Maheshwari (1970), the structure at the end of the pedicel is a cone of spirally arranged ovules. Later, a *Scutum* sp. was also observed to have a bilateral receptacle bearing the ovules, which were covered on one side by a protective scale leaf with glossopteris type of venation (Surange and Chandra 1972). Both receptacle and its subtending covering scale are borne on the pedicel. Earlier, Plumstead (1952, 1956) reported that the bivalved cupule of *Scutum* is attached by a pedicel to the midrib of the leaf. The fructifications are bisexual, i.e. half of the cupule nearest the adaxial surface bear carpels, while the other half is presumed to bear microsporangia. The affinity of the genus with the angiosperms (due to the presence of carpels by the reproductive structure of *Glossopteris)* had been claimed, but was seriously questioned and ultimately rejected (see Thomas and Spicer 1986). In *Ottokaria*, a Permian genus, the ovulate fructification is attached to the foliage of *Gangamopteris* (Fig. 10.4 A), and the ovules are borne in cones.
b) With foliar receptacle: i.e. ovules borne on modified leaves, e.g. *Lidgettonia* (Fig. 10.3 A) and *Denkania* (Fig. 10.3 B).

According to Schopf (1976), *Ottokaria* has a flattened capitulum with a marginal frill bearing the ovules on the undersurface and not in a cone. Gould and Delevoryas (1977) also present evidence, from permineralized specimens from the Permian of Australia, to suggest that the ovule-bearing structure of their *Glossopteris* specimens (including *Scutum* and *Ottokaria*) is foliar.

Phylogenetic Considerations

It is difficult to recognize the ancestors of the glossopterids because of the ovules and their different modes of attachment. There is also no clear relationship with any geologically younger group. According to Schopf (1976), they probably evolved from the Cordaitales; or from earlier Northern Hemisphere pteridosperms (Gould and Delevoryas 1977). Surange and Chandra (1975) place some of them in Pteridospermales and others in Glossopteridales. Pant (1986) suggests that the glossopterids can be either included in a broadly defined group of Pteridospermopsida, or kept altogether isolated, since their ancestors or descendants are presently unknown. It is, therefore, best to consider the glossopterids as a highly successful group of gymnosperms which dominated large areas of vegetation in their time and environment. They subsequently became extinct, due either to climatic change or the migration of more vigorous plants into the habitat (Thomas and Spicer 1986).

11. Pentoxylales

The Pentoxylales is a small group of plants of relatively recent discovery. The stem genus *Pentoxylon* and ovulate cone *Carnoconites* have been described by Srivastava (1946), leaf genus *Nipaniophyllum* by Sahni (1948), and micro-sporangiate organs *Sahnia* by Vishnu-Mittre (1952). Sahni (1948) reconstructed *Pentoxylon* (Fig. 11.1 A) and proposed the name Pentoxyleae for the group.

Pentoxylon

Habit. The habit of the plant is unknown. It was probably a shrubby plant which grew beside aquatic surroundings (Bose et al. 1984) and had erect branched leafy shoots. The latter, after a few seasons of growth, flopped to the ground or on to other stems and made a thicket.

Stem. The stem (Fig. 11.1 A) is dimorphic (cf. Ginkgoales, Coniferales) and the long shoot is an axis ca. 1 cm in diameter and with 5–7.5-mm-thick short shoots. Both long (Stewart 1983) and short shoots are covered with an armour of closely aggregated, spirally arranged scale and foliage leaf bases (Figs. 11.1 A; 11.2 A). *P. sahnii* varied from 2 to 3 mm across, and frequently had five closely aggregated steles (which gives the name to the genus and the group) arranged in a ring around a central ground tissue (Fig. 11.2 B). The number of steles is mostly five, occasionally six. According to Vishnu-Mittre (1957), the number of steles varied along the length of stem. The variation in the number of steles (4–10) is due to branching and the branches anastomose (Stewart 1983). Each stele has a tangentially elongated mesarch primary xylem mass, and a cambium which persists throughout the year in the young stem. In the older stem, the secondary wood is eccentric due to excessive development towards the centre of the stem (Fig. 11.2 B). The vascular steles in the long shoot present a polystelic condition (cf. Medullosaceae). There are five additional, much smaller steles, probably of the short shoots (Sahni 1948). They are almost entirely of secondary wood.

The wood is pycnoxylic made of compact tracheids which have uni- or

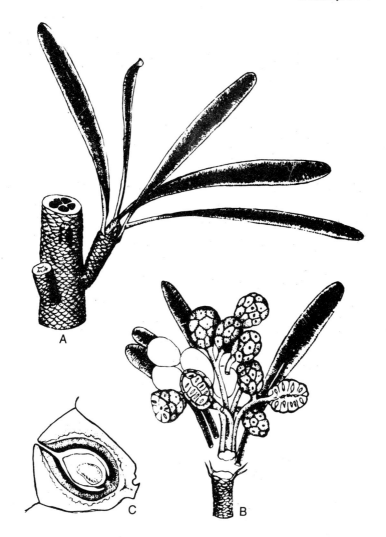

Fig. 11.1 A-C. *Pentoxylon sahnii.* **A** Reconstruction shows long and short shoot with spirally arranged leaf bases, and leaves on short shoot. **B, C** *Carnoconites.* **B** Reconstruction, short shoot bearing cones. **C** Longisection ovule. (After Sahni 1948)

bi-seriate circular bordered pits crowded on the radial walls. The wood rays are uniseriate and one to five cells high. The growth rings are well defined.

Leaf. *Nipaniophyllum raoi* is strap-shaped, 7 cm long and ca. 1 cm wide, with a prominent mid-rib (Fig. 11.1 A). The lateral veins are mostly unbranched and without anastomoses. The slightly sunken syndetocheilic stomata (cf. Cycadeoidales) are confined to the lower surface. The petiole is winged. There are about six leaf traces, arranged in an arc, which enter the base of each leaf. Traces to a leaf arise in pairs, one from each protoxylem

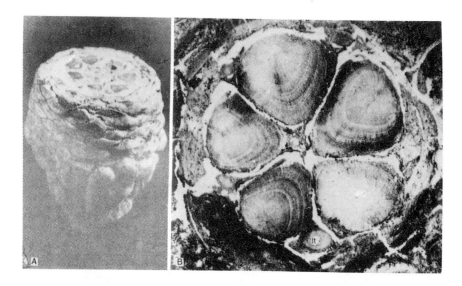

Fig. 11.2 A, B. *Pentoxylon sahnii.* **A** Long shoot shows spirally arranged leaf bases. **B** Transection long shoot shows five steles. *It* leaf trace. (**A** after Stewart 1983, **B** after Sahni 1948)

strand of adjacent steles (Stewart 1976). The paired traces divide in their upward and outward course through the cortex to a leaf base, and form six to nine bundles arranged in a tangential row (cf. Cycads; Stewart 1976).

Microsporangiate Organs. *Sahnia nipaniensis* were borne terminally on short shoots which resemble *Pentoxylon*. It has ca. 24 microsporophylls arranged in a single whorl round a conical receptacle, and are fused at the base into a disc. Each microsporophyll is filiform at its free end and bears short, spirally arranged branches which terminate in unilocular sporangia borne singly or in groups. The pollen grains are monocolpate (Cycadophyte type).

Ovulate Cones. *Carnoconites compactum* and *C. laxum* have been described. The peduncle usually dichotomizes or may remain unbranched. Each branch terminates in a mulberry-like single cone (Fig. 11.1 B). The latter varies in length from ca. 1.8 cm (*C. compactum*) to ca. 3 cm (*C. laxum*). Each cone has ca. 20 sessile erect ovules attached to a central receptacle. Each ovule has a single integument, which is free from the nucellus, and the micropyle is directed outward (Fig. 11.1 C). The integument has a thick sarcotesta and an inner sclerotesta (appearing platyspermic in a cross section). Each ovule is supplied by a trace at its chalazal end, which arises from a ring of vascular bundles in the cone axis.

Phylogenetic Considerations

This enigmatic group displays a unique combination of characters. The polystelic stems resemble some of the Palaeozoic Medullosaceae, yet the secondary wood is pycnoxylic (coniferous). The leaves show both cycadean and cycadeoidean features in the leaf trace anatomy and stomata. Both the pollen- and seed-bearing organs are stachyosporous (borne on stems). The microsporangiate organs morphologically resemble cycadeoids, and the pollen is cycadophytic. The ovulate cones are unlike any gymnosperm and their structure is peculiar to themselves. Most palaeobotanists agree that this group should be given a status similar to Cycadeoidales and Cycadales. Investigations during the last 30 (or more) years have not clarified the exact affinities of this group. According to Rao (1981), at the present level of our knowledge it cannot be referred to any known group of plants, exclusively. It can only be regarded as an isolated synthetic group which shows a mixture of features in common with Pteridospermales, Cycadales, Cycadeoidales and Coniferales. The correct phylogenetic placement must await further information (or arguments).

12. Ginkgoales

Ginkgoales is an ancient group which appeared in the Permian, and was well-represented and nearly worldwide in distribution in the Mesozoic. Their maximal diversity occurred during the Middle Jurassic, and remains have been collected from many countries: Alaska, Greenland, Scandinavia, Siberia, Mongolia, England, Europe, China, Japan, Australia, New Zealand, India, tip of South Africa and South and North America. These plants flourished for over 150 million years. By the beginning of the Oligocene (Tertiary), only 2 out of ca. 19 genera survived; *Ginkgo adiantoides* is one of these. This species became extinct by the Pliestocene (Tertiary), and Ginkgoales has been represented by the extant *Ginkgo biloba*—the "living fossil". It appeared in the Jurassic, and became extensively distributed in the Tertiary. In later periods, it was abundant in the more northern latitudes. At present, its natural occurrence is restricted to a small inaccessible region in southeastern China (see Chap. 1).

Fossil Taxa

Abundant compression-impression leaf remains have been collected, the reproductive organs are meagre. There are about seven or eight Mesozoic leaf genera included in Ginkgoales (Harris 1935, 1974). *Ginkgoites* and *Baiera* have the maximal number of species. These two taxa are indistinguishable in their cuticular and stomatal characteristics. However, *Baiera* lacks a petiole and its lamina is comparatively more wedge-shaped than that of *Ginkgoites*. The latter has a distinct petiole with two traces which diverge into the basal edge of a bilobed lamina. The fossil leaves, which cannot be distinguished from those of the extant *Ginkgo,* are included in this taxon, but the species are different. Those leaves which can be distinguished by their morphological and anatomical characteristices (size and shape of epidermal cells, distribution of stomata, structure of subsidiary cells, mesophyll and distribution of resin bodies) are placed in the genus *Ginkgoites*.

Generally, the primitive leaves were linear and deeply dissected. The leaves of *Arctobaiera* show a deeply dissected to entire margins (Fig. 12.1 A, B). *Sphenobaiera* (Lower Permian to Lower Cretaceous) has a characteristic dissected lamina (Fig. 12.1 C) with several dichotomised

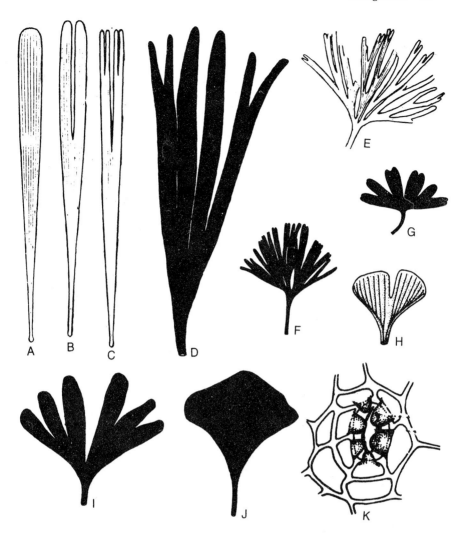

Fig. 12.1 A-K. Leaves and stomata. **A, B** *Arctobaeira.* **C** *Sphenobaiera.* **D, E** *Baiera spectabilis.* **F** *Ginkgoites minuta.* **G** *G. taeniata.* **H** *Ginkgoidium.* **I** *Ginkgo digitata.* **J** *G. lamariensis.* **K** *Ginkgoites lunzensis,* stomatal apparatus. (**A-C** after Florin 1951, **D, F-H** after Harris 1935, **E** after Schenk 1867, **I, J** after Brown 1943, **K** after Andrews 1961)

veins. *Baiera* (Middle Jurassic to Lower Cretaceous) had wedge-shaped leaf with an indistinct petiole (Fig. 12.1 D, E). The extent of lobing of lamina and variations in venation pattern in *Ginkgoites* spp. (Late Triassic) can be arranged in a series (Harris 1935). The leaf in *G. minuta* is highly dissected by equal dichotomies (Fig. 12.1 F), in *G. taeniata* the unequal dichotomies make each half of the leaf appear three-lobed (Fig. 12.1 G), and in *Ginkgoidium* the lamina is entire except for a median notch, and prominent marginal and unbranched secondary veins (Fig. 12.1 H). Thus, there is a general trend from deeply dissected Jurassic leaves to those of the Tertiary, which tend to be entire except for a median sinus (see Stewart 1983). This series may have an evolutionary significance, as shown by Mesozoic species of *Ginkgo*: *G. digitata* (Early Mesozoic) has a deeply dissected wedge-shaped petiolate leaf (Fig. 12.1 I), *G. lamariensis* (Cretaceous) has undissected wedge-shaped leaves (Fig. 12.1 J), and in *G. adiantoides* (Tertiary) the leaves are indistinguishable from those of extant *G. biloba*.

The stomata are haplocheilic. There are four to six subsidiary cells each with a blunt papilla which tends to overarch the guard cells as in *Ginkgoites lunzensis* (Fig. 12.1 K).

There are very few verified reports of reproductive structures. Numerous dispersed ovules of *Allicospermum* (Harris 1935) —similar to those of cycads and *G. biloba*—are associated with Jurassic and Cretaceous *Ginkgo*-like remains. *Karkenia* is an ovulate fructification from the Cretaceous of Argentina (Archangelsky 1965). It is a short stalk crowded with 100 or more ovules associated with *Ginkgoites tigrensis*. Paired ovules joined by a pad of tissue with *Ginkgo*-like stomata have been reported from Yorkshire Jurassic beds (Harris 1976), where abundant *Ginkgoites huttoni* occurs. In the same beds, small pollen-bearing catkins are present. Each consists of a stalk (ca. 5 mm long) to which are attached rather lax stalks with pairs of microsporangia at the tips. The pollen grains are monocolpate like those of *G. biloba*.

The *Ginkgo* line can be traced to the Lower Permian *Trichopitys* (often placed in its own order/family). Florin (1949) investigated *T. heteromorpha,* known since 1875 from the Lower Permian of southern France, and interpreted it to be the earliest member of the group (see Phylogenetic Considerations). It has spirally arranged dichotomously branched leaves without lamina (Fig. 12.2 A). There are small branched ovuliferous trusses (branch system, Fig. 12.2 B) in the axil or on the upper surface of the leaf base. Each ultimate branch bears a terminal, recurved ovule (Fig. 12.2 C), unlike *Ginkgo*.

The male organs of *Trichopitys* are not known. However, *Sphenobaiera furcata* (Triassic) bears clusters of microsporangia at the branch tips of a bifurcating axis, which are borne in turn on short shoots, along with leaves similar to *Trichopitys*.

Ginkgoaceae

Ginkgoaceae is a monotypic family. The deciduous leaves are fan-shaped with parallel veins. The tree is dioecious; male flowers are catkin-like while the female is long-stalked with (usually) two ovules. The male gametes are motile, and the fruit is drupaceous.

Ginkgo

Morphology

Ginkgo biloba is a tree more than 30 m high and exceeds 1.5 m in diameter.

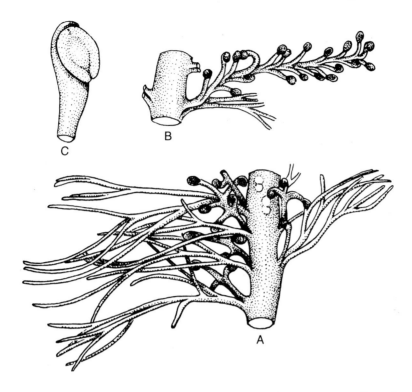

Fig. 12.2 A-C. *Trichopitys.* **A** Portion of shoot with sterile telome trusses (leaves) and axillary ovule-bearing shoot. **B** Ovule-bearing shoot. **C** Ovule. (After Florin 1949, 1951)

It resembles a conifer in its general habit. The crown becomes broad, irregular with age, and shows a variable pattern of branching. The main axis branches profusely. There are two kinds of shoots: (a) long shoots which elongate rapidly and bear scattered leaves, and (b) dwarf shoots which grow slowly and bear a terminal cluster of leaves; the older portion is covered with leaf scars of previous years (Fig. 12.3 A, D, F). Bud scales cover young spur (dwarf) shoots. Frequently, the dwarf shoots become more active and turn into a long shoot, while (more rarely) the terminal growth of a long shoot may be retarded and resemble a lateral spur shoot (Gunckel and Wetmore 1946). This suggests that there is not much fundamental difference between them. The apical meristems of the two types of shoots are also essentially similar; the difference between them depends on the duration of cell division and cell elongation in the stem tissues derived from the shoot apex (Foster 1938). There is experimental evidence to show that the dwarf shoots are formed due to the inhibitory effects of auxins produced by the tissues of the long shoots (Gunckel et al. 1949).

The shape and venation of the deciduous foliage leaves are unique. The

Fig. 12.3 A-F. *Ginkgo biloba.* **A** Dwarf shoot bearing male strobilus. **B, C** Microsporophylls. **D, F** Dwarf shoot bearing young strobili (**D**) and seeds (**F**). **E** Longisection female strobilus, (**A-E** After Ganguli and Kar 1982, **F** after Andrews 1961).

petiole is long, smooth, black and slender, traversed by two collateral vascular bundles. The lamina is broadly wedge-or fan-shaped, variously lobed, and the venation is conspicuously dichotomous (Fig. 12.4 D). The leaves resemble those of *Adiantum* (maidenhair fern) in form and venation, hence the popular name maidenhair tree. The old leaves are shed in autumn, when they change colour to a golden yellow; the new leaves appear in spring. There is considerable difference in the lobing of the leaves on the same tree. They may be nearly entire or two-lobed due to a conspicuous, often deep, apical notch. They are mostly bilobed on long shoot and entire on short shoot. In seedlings the leaves have several notches which give a palmatifid appearance (as in the extinct taxa).

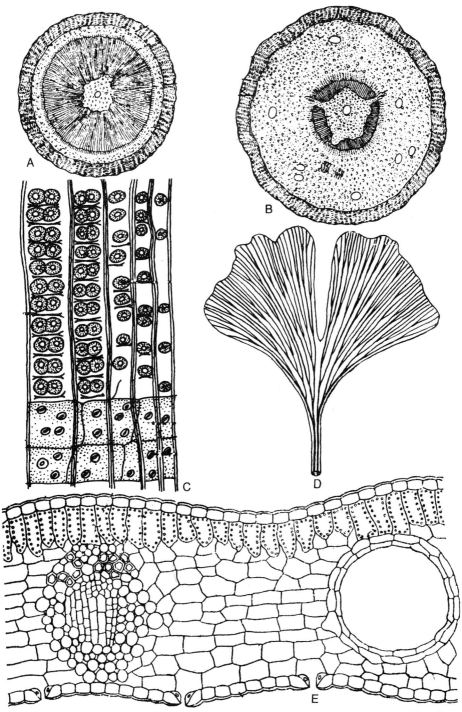

Fig. 12.4 A-E. *Ginkgo biloba.* Stem anatomy. **A, B** Transection long (**A**) and dwarf shoot (**B**). **C** radial longisection, secondary wood shows circular bordered pits, bars of sanio, and ray cells with cross-field pits. **D, E** Leaf. **D** Single leaf shows venation. **E** Vertical section (**A, B, E** After Ganguli and Kar 1982, **C, D** after Stewart 1983)

Anatomy

Root. The young roots are usually diarch. The endodermis has conspicuous thickenings on its radial walls and there is a broad pericycle. Older roots may be tetrach or hexarch. A radial longisection shows that the spiral elements of the protoxylem are followed successively by tracheids with (a) reticulate pitting, (b) transversely elongated simple pits and (c) bordered pits.

Secondary growth occurs, but annual rings are not pronounced. The tracheids have thinner walls. Some xylem parenchyma cells include crystals, and thick-walled fibres are abundant in phloem. The medullary rays are one to several cells high and often show crystals.

Stem. The shoot apex consists of superficial apical initials, a prominent subapical zone of central mother cells, and a zone of rib meristem around it on its proximal side (Foster 1938).

A young stem comprises an epidermis, a parenchymatous cortex and pith, and vascular cylinder. The stele is an endarch siphonostele. A long shoot has a thick zone of wood and comparatively narrow pith and cortex (Fig. 12.4 A). In the dwarf shoots, the ring of wood is narrow and there is a broad cortex and pith (Fig. 12.4 B). The primary xylem consists of a number of separate strands. The phloem forms a broad zone. The double leaf traces are very conspicuous.

A radial section of the wood shows tracheids with numerous round and opposite pits separated by bars of Sanio (Fig. 12.4 C). The uniseriate rays are one to three cells deep in the long shoots, and much deeper in dwarf shoots. Occasionally, a few xylem parenchyma cells contain calcium oxalate crystals. In secondary wood the tracheidal overlap is not extensive, and several tracheids end at the same level. This makes the wood very brittle, and in high wind storms the trees suffer heavily. When a tree is felled. the log frequently shatters as it strikes the ground (see Bierhorst 1971.

The secondary phloem is composed of sieve elements, parenchyma strands and fibres in the axial system and rays in the radial system (Srivastava 1963a). The albuminous cells associated with the sieve cells lack starch and are crushed in the old phloem. They are connected to the sieve cells by one-sided sieve areas. Lateral connections between the sieve elements and albuminous cells could not be traced, as callose was not present in sufficient quantity. These cells also have plastids but do not store normal, detectable starch. The fibers are elongated, tapering, tangentially flattened and non-septate. They have narrow lumen, the wall is thick, lamellated and appears to be cellulosic. They do not stain positive with phloroglucinol and HCl and are strongly birefringent under polarized light (see Paliwal 1992).

Outside the phloem there is a ring of sclereids, probably pericyclic, and an inner ring of thickened cells, whose walls appear to be gelatinized. Throughout the plant numerous mucilage canals occur in the pith and the cortex.

Leaf. The leaf has a double trace. A transection of the petiole shows two endarch vascular bundles. The primary xylem of the stem branches sympodially when the leaf traces are given off. The two traces to any leaf therefore arise independently from two different primary strands. They divide at the base of the blade and the resultant four strands fork repeatedly to form the dichotomous system of veins which occasionally anastomose in the lamina (see Stewart 1983). The venation of each of the two halves of the leaf is completely independent. Mucilage canals are present even between the veins of the leaf.

A vertical section of the lamina shows: (a) a thick cuticle, (b) stomata mostly on the lower surface of the leaf, (c) a distinct palisade only in the leaves on the long shoots (Fig. 12.4 E), (d) mucilage canals and (e) usually endarch vascular bundles with traces of centripetal xylem represented by one or two tracheids. The bundles are surrounded by a sheath of thick-walled cells.

On the lower epidermis, stomata occur irregularly scattered between the veins. They are haplocheilic, surrounded by four to six subsidiary cells, each with a blunt papilla which projects over the guard cells (see Stewart 1983). The characteristic accessory cells of the stomata are also recognizable in the extinct taxa.

Reproduction

Ginkgo is dioecious.[1] The male cones are pendant and catkin-like, borne on short shoots in the axil of normal leaves or scale leaves (Fig. 12.3 A). The ovulate cones are borne in groups at the apex of the dwarf shoot (Fig. 12.3 D, F). They are reduced, and each shoot bears two ovules on a long peduncle in the axil of a scale leaf (Fig. 12.3 D, E).

Male Cone

A male cone comprises 40–50 microsporophylls (Fig. 12.3 A). Each microsporophyll has a terminal knob, which contains a mucilage sac (Fig. 12.3 B, C), and there are two (occasionally three to seven) pendulous microsporangia which dehisce by longitudinal slits (Fig. 12.3 C).

Microsporangium. The strobili are initiated in summer, and appear as small papilae in the axils of bracts. Wolniak (1976) examined more than 300 cones, and observed: (a) There is no acropetal or basipetal progression in sporangial development; (b) There is no correlation between the size of the sporangium and its development. There is, however, variation in

---

[1]In the Botanical garden at Insbuck (Austria), on a female tree of *Ginkgo*, a branch from a male tree was grafted. The male cones developed and produced fertile pollen grains, pollination and fertilization occurred normally, and numerous (apricot-coloured) ripe seeds developed on the female tree (BMJ, pers. observ. 1957).

development, and in a sporangial pair the microsporocytes are ontogenetically similar. Earlier stages have not been observed but there is evidence of a single hypodermal cell which divides by anticlinal and periclinal divisions. The outer cells form a wall of four to seven layers, and the inner cells give rise to a large group of sporogenous cells. A tapetum surrounds the sporogenous tissue, and a peritapetal membrane has been reported (Pettitt 1966).

An endothecium differentiates, which is an exception, since in gymnosperms only an exothecium is reported. The endothecium develops from one to three layers of subepidermal cells, becomes thick-walled and develops fibrous thickenings (Fig. 12.5).

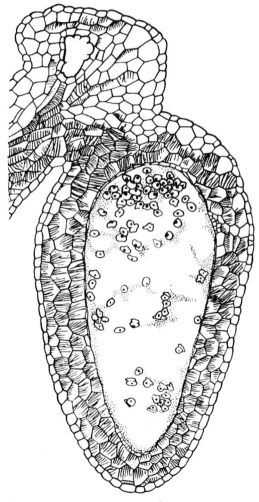

Fig. 12.5 *Ginkgo biloba.* Longisection microsporangium and a part of microsporophyll, several layers of wall show endothecial thickenings. (After Jeffrey and Torrey 1916)

Microsporogenesis. Meiosis in microspore mother cells coincides with the opening of the bud scales of the spur shoot. The distribution of starch in the microsporangium is specific, and first appears at the archesporial stage. Prior to meiosis, starch reappears in the sporogenous cells and then in tapetal cells. During meiosis I starch grains accumulate at the equatorial plate at metaphase, move to the opposite poles during anaphase and telophase, and become equally distributed after the completion of meiosis. After telophase I, besides starch, entire plastids and mitochondria shift to the equatorial region of the microspore until nuclear divisions cease. Microtubules and ER proliferations appear to stabilize the organelle distribution through meiosis II. Tetrads are formed by centripetal wall formation, and the organelles become distributed equally (Wolniak 1976). A callose wall has been observed during microspore mitosis (Gorska-Brylass 1968).

Male Gametophyte. The microspore nucleus cuts off two prothallial cells (Fig. 12.6 A, B); the first cell towards the wall is ephemeral while the second persists. The antheridial initial divides and forms a smaller antheridial cell, which remains attached to the intine, and a larger tube cell, which becomes vacuolate and has a conspicuous nucleus (Fig. 12.6 C, D). The antheridial cell divides periclinally to form the stalk cell (toward the pollen wall) with a distinct wall, and the body cell (Fig. 12.6 E). The stalk and body cells persist in situ. The persistent prothallial cell remains active and grows into the stalk cell which lies next to it. The stalk cell thus appears to form a jacket around the protruding prothallial cell (Fig. 12.6 D-F).

The microsporangium dehisces by a longitudinal slit along the inner face. The pollen is shed at the four-celled stage: two prothallials, one antheridial and a tube cell.

Ovule

The peduncle bifurcates and bears on each branch a single sessile ovule with a fleshy collar around its base (Fig. 12.3 E). The morphology of the collar has been variously interpreted (see Chamberlain 1935); it does not grow after pollination.

Usually there are only two ovules on each peduncle, occasionally three, four or more. Whatever the number of ovules, the peduncle always has twice the number of vascular bundles.

The morphology of the meristem which gives rise to the ovule needs a critical reinvestigation. The ovule is orthotropous with a beaked nucellus which has a heavily cutinized epidermis. The nucellus has a well-differentiated strand of elongated cells and extends almost to its entire length (Fig. 12.7 A). Its degeneration forms a narrow, deep pollen chamber (De Sloover-Colinet 1963). The inner cells degenerate first followed by the epidermis (Fig. 12.7 B, C). There is a single integument, which is free from the nucellus at the apex. Two unbranched vascular strands supply the base of the integument.

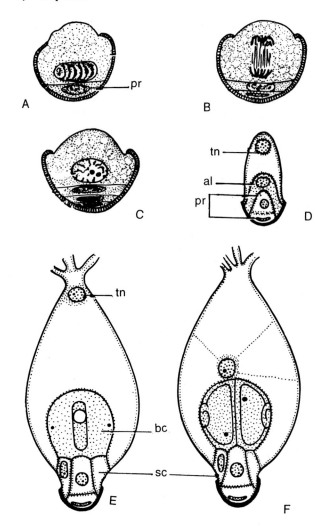

Fig. 12.6 A-F. *Ginkgo biloba.* Male gametophyte. **A-F** Development of male gametophyte. *al* antheridial cell, *bc* body cell, *pr* prothallial cell, *sc* stalk cell, *tn* tube nucleus. (**A-D** After Chamberlain 1935, **E. F** after Favre-Duchartre 1956)

Megasporogenesis.   One or more megaspore mother cell/s become distinct by their prominent nuclei and dense cytoplasm. Due to a considerable thickening of the middle lamella (Fig. 12.8 A), and the development of a double-layered wall, the wall of the mother cell becomes thick and two-layered (Stewart and Gifford 1967). The latter has densely staining outer layer which resembles the circumjacent nutritive tissue, and an inner layer which is similar to the middle lamella except for a tighter arrangement of the fibrillar structure. The young megaspore mother cell is spherical, and has a large nucleus in the centre. Its cytoplasm occasionally shows a small vacuole, relatively scanty endoplasmic reticulum (Stewart and Gifford 1967),

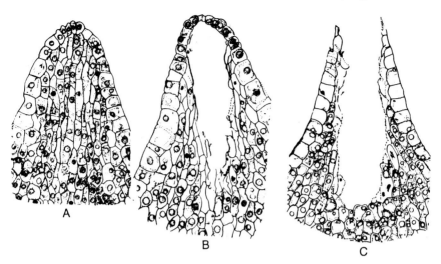

Fig. 12.7 A-C. *Ginkgo biloba.* A-C Longisection nucelli show successive stages of development of pollen chamber. (After De Sloover-Colinet 1963)

and randomly placed starch-bearing plastids, mitochondria and dictyosomes. The mother cell elongates as it matures. The cytoplasm in the micropylar half shows a large and complex system of ER (Fig. 12.8 B), while it is relatively meagre and there is no definite pattern in other parts of the cell. However, all over the cell, ER is only sparsely associated with ribosomes. In the mature mother cell the micropylar ER becomes reticulated with several "loops" and "circles" (Fig. 12.8 C). A vacuole appears below the nucleus of a mature mother cell, and several small vacuoles in the micropylar part. From the micropylar half, the plastids and mitochondria shift laterally to the chalazal end of the maturing mother cell (Fig. 12.8 D); the micropylar ER may have a role in the migration of these organelles. In a mature cell the plastids with a few mitochondria are restricted to the region below the nucleus, the mitochondria lie just below the plastids (Figs. 12.8 D; 12.9). Other organelles, like dictyosomes, lipid droplets, dense bodies bounded by unit membrane, and an occasional multivesicular body, do not show polarity in their distribution.

The nuclear envelope of the mother cell shows a large interruption in the lateral wall of the cell (Fig. 12.9), and remains identifiable until the onset of meiosis I. Starch accumulates only in the chalazal region of the megaspore mother cell. Following meiosis I, plastids and mitochondria become restricted to the chalazal dyad. Generally, a linear tetrad of megaspores is formed. In a triad—upper undivided dyad and two megaspores—it is the functional chalazal megaspore which has most of the plastids. The starch grains are consumed during the enlargement of the megaspore. It appears that starch begins to accumulate, much in advance, at the site of the functional megaspore.

Female Gametophyte.   The female gametophyte develops from the haploid

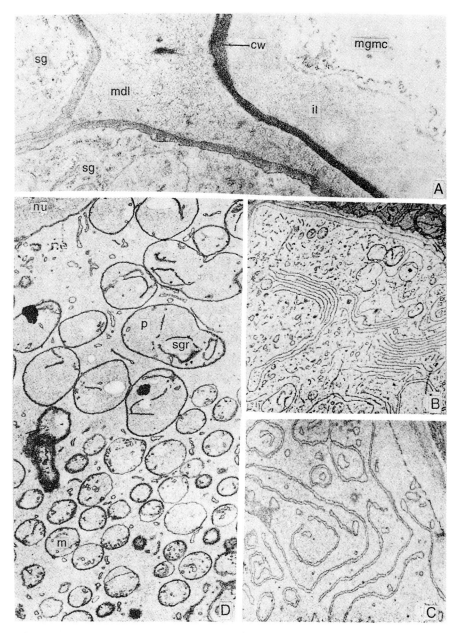

Fig. 12.8 A-D. *Ginkgo biloba.* Electron micrographs, megaspore mother cell. **A** Portion of megaspore mother cell (*mgmc*), outèr (*cw*) and inner (*il*) layer of mgmc wall, middle lamella (*mdl*), and spongy cells (*sg*). **B, C** Part of micropylar half of mother cell. **B** Extensive ER. **C** Reticulate ER. **D** Part of chalazal region of mother cell; note starch (*sgr*)- bearing plastids (*p*), near the nucleus (*nu*), and mitochondria (*m*) lower down. (After Stewart and Gifford 1967)

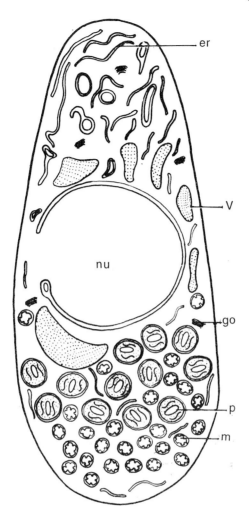

Fig. 12.9. *Ginkgo biloba.* Megaspore mother cell (diagrammatic). *er* endoplasmic reticulum. *go* golgi apparatus. *m* mitochondria. *nu* nucleus. *p* plastid. *v* vacuole. (After Stewart and Gifford 1967)

chalazal megaspore. The gametophytic tissue, however, is not uniformly haploid, as shown by cytological studies and cytophotometric measurements of the DNA content (Avanzi and Cionini 1971). At the beginning of the cellular stage, 5% nuclei of the gametophyte show 1C DNA content, 50% 2C, more than 40% 4C, and the remaining nuclei 8C or higher DNA content. This variation in the DNA content is attributed to endopolyploidy. The cells with 4C content are mostly located in the outer region of the gametophyte. The nuclei with higher DNA content (8C or more) degenerate in the young gametophyte. In older stages, most cells contain 2C DNA.

Free-nuclear divisions (Fig. 12.10 A-C) occur in the megaspore for about 4 weeks. According to Favre-Duchartre (1958), there are 13 successive

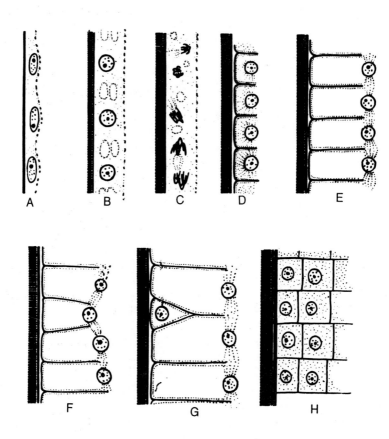

Fig. 12.10 A-H. *Ginkgo biloba.* Longisection of peripheral region of female gametophyte. **A, B** Cytoplasm with free nuclei. **C** Nuclear divisions **D-G** Formation of alveoli and initiation of walls. **H** Cellular gametophyte, note the thick wall. (After Favre-Duchartre 1956, 1958)

mitotic cycles, so that more than 8000 nuclei are formed. The divisions are initiated at the chalazal end, and proceed towards the micropyle. The prothallial cytoplasm, throughout cenocytic phase, adheres to the megaspore membrane. Walls are laid down at the end of the 14th mitotic wave. The gametophyte remains colourless throughout the free-nuclear stage. Typical alveoli are formed (Fig. 12.10 D-G), followed by cellularization (Fig. 12.10 H). The gametophyte becomes green (due to the presence of chlorophyll) and starch is synthesized. The female gametophyte of *Ginkgo biloba* is the only seed plant with a chlorophyllous gametophyte. The relative transluscence of the integumentary tissues of the ovule permits sufficient light to induce the synthesis of chlorophyll (Friedman and Goliber 1986). The plastids do not contain an organized thylakoid membrane system (Pettitt 1977). When cell formation begins, the female gametophyte has a light green colour, attributed to chlorophyll. EM of chloroplasts demonstrate stacking of thylakoid membrane in the grana. Plastids located deeper in the gametophyte have fewer thylakoid membranes and may also show prolamellar bodies.

The presence of chlorophyll has been confirmed by Burgerstein (1900). He showed that an alcoholic extract of the gametophyte fluoresces red; Carothers (1907) suggested it may be capable of photosynthesis. A measurement of photosynthetic active radiation (PAR) indicates that a gametophyte (growing within an ovule) can receive significant quantities of light, i.e 70 μ mol photons m^{-2} s^{-1} (Friedman and Goliber 1986). This unique ability to produce chlorophyll and perform photosynthesis results from its exposure to sufficient levels of light, and an inclination to react to this stimulus by the development of functional photosynthetic apparatus. The entire gametophytes were dissected free from the ovules. They were capable of gross photosynthesis (not net photosynthesis) under experimental conditions. On a dry weight basis, the maximum rate of carbon fixation, in near-saturating light intensities, was 3.64×10^{-3} μ mol CO2 g^{-1}s^{-1} (Friedman and Goliber 1986).

A mature prothallus contains water, nearly 60% of the whole mass (Favre-Duchartre 1956). Reserve food accumulates in both the fertile and sterile prothalli, and is therefore independent of fertilization. In addition to starch, a mature prothallus also contains lipids, and its density decreases sharply from the epidermis inwards (Fig. 12.11 A). The concentration of lipoptoteins increases during the maturation of the prothallus. The starch grains vary from 4–5 μm in diameter in the epidermal cells to 15–20 μm in the inner cells. They are simple, irregular in outline, and have a central hilum of indefinite shape (Fig. 12.11 B-F).

The apical part of the gametophyte regularly forms an unusual structure, the "tent pole" (Fig. 12.11 A). The central portion of the micropylar region grows by active cell divisions to form a column (the tent pole) which grows into the nucellus, or may even reach the pollen chamber.

At the micropylar end of the cellular gametophyte, one or two cells enlarge and function as archegonial initials. Each initial divides transversely and forms a large central cell and small neck cell of the archegonium (Fig. 12.12A). At maturity there are four swollen neck cells (Fig. 12.12 B), which project upward and may have secretory role (Lee 1955). In the beginning, the four neck cells are in one tier but later they may show a two-tiered arrangement (Favre-Duchartre 1956). In a female gametophyte the maturation of different archegonia is synchronous. The central cell nucleus divides to form a small ventral canal cell and the egg cell (Fig. 12.12 A, B). The mature egg cell is spherical, and measures ca. 500 μm in diameter with its nucleus ca. 100 μm across (Camefort 1965a). The chromatin is confined to small globule so that the nucleus appears almost colourless with Feulgen reaction (Fig. 12.12 B).

The single-layered archegonial jacket comprises uninucleate cells. Several simple pits occur on the wall between the jacket cells and the central cell

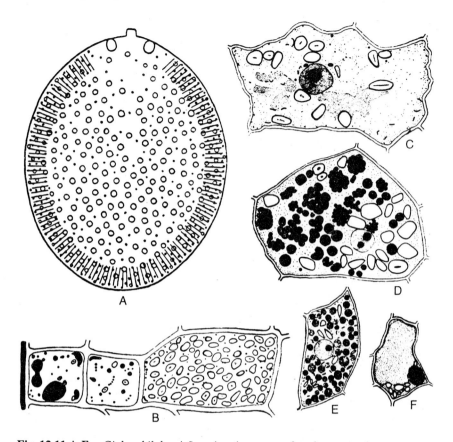

Fig. 12.11 A-F. *Ginkgo biloba.* **A** Longisection mature female gametophyte, note two archegonia and tent pole at the micropylar end, and distribution of starch (open circles), lipids (black dots) and lipo proteins (radial lines). **B** Three cells from the margin, outer layer is sudanophilic. **C-F** Cells from different layers of the gametophyte, at various stages of development, show accumulation of starch, lipids and lipoproteins. (After Favre-Duchartre 1958)

(Hirase 1895). Occasionally, lobes of the cytoplasm of the central cell may project into these cells. Plasmodesmata are also present in these pits (Stopes and Fujii 1906), and have been confirmed by ultrastructural studies (Maugini and Fiordi 1970). Probably, the nutrients are absorbed through the jacket. Avanzi and Cionini (1971) measured the DNA content of the jacket cells cytophotometrically. The large uninucleotate nuclei had DNA content corresponding to 2C, 2C–4C, 4C or 4C–8C, due to endoduplication. Cionini (1971) characterized the DNA in jacket cells by HCl hydrolysis curves and observed two types of DNA complexes: (a) Feulgen-stainable after 5 min. and (b) after 7 min. of hydrolysis. This is probably related to the functional activity of jacket cells or formation of nascent DNA during endoduplication.

The passage of materials from the jacket into the egg, and to the coenocytic (later cellular) proembryo has been studied using electron microscopy

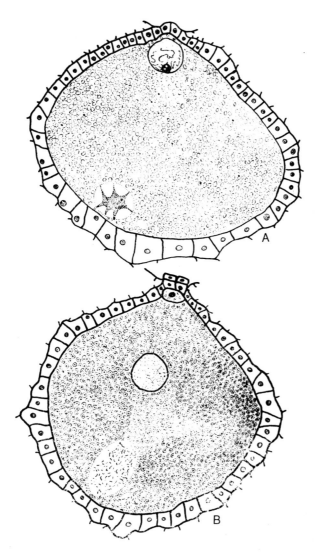

Fig. 12.12 A, B. *Ginkgo biloba.* development of archegonium. (After Favre-Duchartre 1956)

(Maugini and Fiordi 1970). The wall between the jacket and central cell of the archegonium is fairly thick, and has prominent pits (Fig. 12.13 A). The cytoplasm of the central cell forms short and blunt projections at the site of the pits (Fig. 12.13 B, C). The cytoplasm is separated only by the thin pit membrane, and they remain in contact through plasmodesmata. Soluble materials move across this contact. Granular material, which lies outside the plasma membrane of the egg cell, is deposited around the cytoplasmic projection of the egg (Fig. 12.13 C). According to Maugini and Fiordi (1970), the granular material represents the temporarily stored nutritive

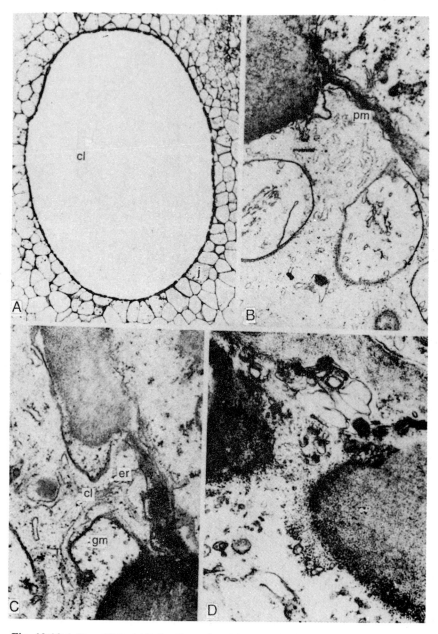

Fig. 12.13 A-D. *Ginkgo biloba.* Central and jacket cell. **A** Longisection shows thick and pitted wall between central (*cl*) and jacket layer (*j*). **B** Pit region shows plasma membrane (*pm*) in contact with the pit region of the cell wall. **C** Later stage, central cell cytoplasm (*cl*) extends into the pit, laterally disposed granular material (*gm*), endoplasmic reticulum (*er*) in the cytoplasmic appendage. **D** Degenerated jacket cell, the contents pass into the proembryo through open pit. (After Maughini and Fiordi 1970)

material on its way from the jacket cell to the egg cell. The plasma membrane of the coenocytic proembryo shows small invaginations and short and irregular microvilli. The increased surface provides a greater absorptive area for nutritive materials. Starch and protein granules, stored in the gametophyte, are translocated to archegonium (in a soluble form) through the pits. When the proembryo is free-nuclear, or immediately after cell formation, the pit mambrane breaks down at places. Through these passages, the mitochondria, plastids, dictyosomes, portions of endoplasmic reticulum and nuclei (whole or in parts) flow (from degenerated jacket cells) into the egg cytoplasm (Fig. 12.13 D). This is perhaps due to a sudden lowering of pressure inside the egg. The contents of the jacket cells partially accumulate between the cell wall and the plasma membrane of the proembryo.

The egg cytoplasm has the usual organelles like ER, plastids, Golgi bodies, ribosomes and a large number of mitochondria. Observations with an electron microscope (Camefort 1965a) have revealed the following cytoplasmic formations (so-called proteid vacuoles): (a) The morphology of small inclusions is somewhat different from others. They are completely enclosed in the double membrane of ER whose components stay together. An enveloping vacuole is thus absent. (b) Microbodies are present in abundance, have dense contents enclosed by a single membrane of ER. Their morphology is similar to certain lysosomes in animal cells. (c) Vesicular bodies occur seldom and comprise a mass of vescicles enclosed by a single membrane.

The amyloplasts in a mature egg are distributed at the periphery. They are enclosed in a layer of endoplasmic reticulum, in addition to their own membranes (Camefort 1965a). The amyloplasts continue to fragment until the egg is mature.

Pollination

The ovules are pollinated soon after megaspore formation. A pollination drop is secreted at the micropyle of the ovule. The wind-borne pollen, after landing on the pollination drop, imbibes nutrients from the fluid, becomes heavy, sinks down the micropyle, and reaches the nucellar tip. Unpollinated ovules drop from the tree about 4 weeks after anthesis.

Post-Pollination Development of Male Gametophyte. On the return of favourable weather after the winter, the pollen grains germinate in the pollen chamber (Favre-Duchartre 1956, De Sloover-Colinet 1963), The exine splits along the germinal furrow, the pollen tube emerges at the distal end, and the vegetative (tube) nucleus may pass into it. The tube grows laterally and horizontally into the massive nucellus and branches profusely. The pollen tube grows in between the nucellar cells, in a hypha-like manner (Fig. 12.14 A). The branched pollen tube in the nucellar tissue has been studied from sections and dissections (Friedman 1987), and undoubtedly has a haustorial function (see Chap. 2).

The proximal end of the pollen grain contains the prothallial, the stalk and body cells/male gametes. In the beginning it is enclosed in the exine, but later enlarges and bursts out of the exine. The short, broad end of the "tube" is usually unbranched (Fig. 12.14 B-F), and a bunch of them grow into the nucellus towards the female gametophyte. Consequently, the nucellus becomes completely disorganized between the intermediary chamber and the female gametophyte. At the grain end, the pollen tubes appear to

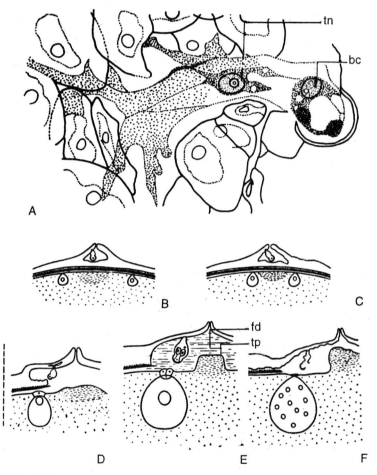

Fig. 12.14 A-F. *Ginkgo biloba*. **A** Haustorial ramifications of pollen tube between nucellar cells. *bc* body cell. *tn* tube nucleus. **B-F** Diagrammatic representation of upper portion of nucellus and female gametophyte, longisections. **B-D** Pollen tube and degenerated inner portion of nucellus. **E** Pollen tube close to the archegonium. *fd* fertilization fluid, *tp* tent pole. **F** Post-fertilized shrivelled nucellus. (**A** After De Sloover-Colinet 1963, **B-F** after Favre-Duchartre 1958)

lie in a longitudinally continuous cavity in the middle of the nucellus (Fig. 12.14 D-F). Thus, the pollen tube has: (a) a vegetative portion which grows laterally into the nucellus and is haustorial in function, and (b) a fertile portion which carries the motile gametes.

The occurrence of motile/ciliate sperms in *Ginkgo* was discovered by Hirase (1895).[2] This was a landmark in the history of (embryological) study of gymnoserms.

The ultrastructure of male gametes has been investigated by Gifford and Lin (1975). The body cell remains spherical during maturation (Fig. 12.15 A, B). Two blepharoplasts, from which cytoplasmic/astral rays radiate, appear on one side of the body cell (Fig. 12.15 A, B). These rays are the microtubules, which extend from the matrix or the surface of the blepharoplasts. The latter move through 90° and come to lie at a right angle to the long axis of the pollen tube (from the very begnning). The blepharoplasts

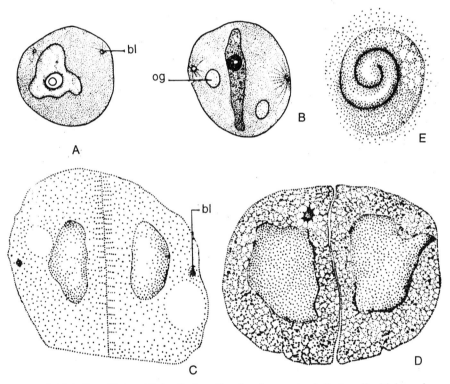

Fig. 12.15 A-E. *Ginkgo biloba.* Body cell and male gamete. **A** Body cell with irregular nucleus; two blepharoplasts (*bl*) at opposite poles. **B** Body cell with a lens-shaped nucleus and two osmiophilic globules (*og*), one on either side. **C** Division of body cell. **D** Two male gametes, one gamete shows a beak-like extension. **E** Gamete with three gyres of ciliate band (top view). (**A, B** After De Sloover-Colinet 1963, **C-E** after Favre-Duchartre 1956)

---

[2]This tree, in the Botanical Garden, Tokyo, is still in a healthy state.

increase in size and become prominent. The entire surface of the matrix is covered by a single layer of probasal bodies (Fig. 12.16 A, B). The total number of probasal bodies on one blepharoplast is 1000. Each probasal body shows a hub-and-spoke arrangement in the centre and nine-fold symmetry. A central tubule is present along the entire length of the probasal body. Microtubules/astral rays/cytoplasmic rays originate from the interior of the blepharoplast, pass between probasal bodies, and extend into the cytoplasm. The probasal bodies separate from each other and form the basal bodies of flagella. The matrix of the blepharoplast comprises dense and less dense regions, the latter appearing to be infiltrated by a network of tubules (Fig. 12.16 C, D).

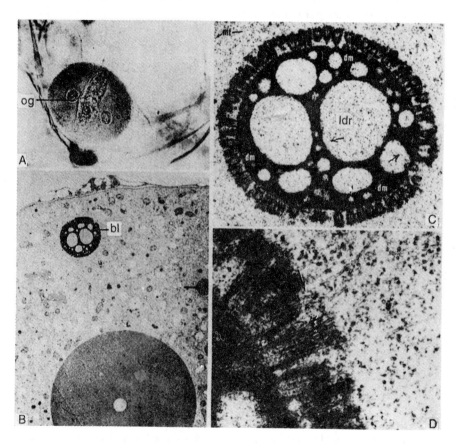

Fig. 12.16 A-D. *Ginkgo biloba.* electron micrographs. **A** Whole-mount (portion) of grain-end of pollen tube; the body cell shows lens-shaped nucleus and two prominent osmiophilic globules (*og*) on either side of nucleus. **B** Osmiophilic globule and blepharoplast (*bl*) in (a part of) body cell. **C** Numerous probasal bodies; microtubules in less dense regions (*ldr*—arrows). **D** Peripheral portion of blepharoplast; microtubules in groups extend from probasal body. (After Gifford and Lin 1975)

During its maturation, the nucleus of the body cell becomes lens-shaped (Figs. 12.15 B, 12.16 A). A vacuole, which is an osmiophilic globule (it has no membrane around it), appears on either side of the nucleus (Figs 12.15 B; 12.16 A) and is attached to the blepharoplasts (Lee 1955). A dumbell-shaped lipoprotein granule arises in the cytoplasm close to the nuclear membrane. In addition, the cytoplasm of the body cell shows other organelles, like protoplastids (numerous, electron-dense, showing some lamellar developments), mitochondria, small vacuoles, lipid bodies, abundant ribosomes, and relatively sparse ER and dictyosomes (Fig. 12.16 B). The body cell divides, the lipoprotein granule splits into two, and each half is incorporated into a spermatozoid. The mitotic spindle of the body cell lies at right angles to the long axis of the pollen tube, so that the two sperms, enclosed in the wall of the body cell, lie side by side (Fig. 12.15 C, D) until they are mature. The vacuole (osmiophillic globule), earlier attached to the blepharoplasts, shifts to the opposite pole of the cell and becomes attached to the sperm nucleus. Later, it acquires granular contents due to diffusion of chromatin from the nucleus (Lee 1955). The blepharoplast forms a part of the tapering distal part of the sperm. The spiral band has three turns/gyres (Fig. 12.15 E). There is simultaneous division of the nucleus of the body cell in the pollen tube and of the central cell in the archegonium.

Fertilization

In zooidogamous gynmosperms, autolysis of the nucellus, megaspore wall, and the gametophytic tissue around the archegonia form an archegonial chamber (Pettitt 1977). The turgid nucellar cells abruptly release their vacuolar contents, a fluid is produced which forms a pool and floods the archegonial chamber and the space above it (Fig. 12.14 E). The male gametes are released in this fluid (Lee 1955). The spermatozoids, with the band of flagella at their posterior end, swim about in the chamber with a forward and circular motion. The four neck cells open out as the egg cell pushes through the disintegrated ventral canal cell to form a beak. The egg nucleus may elongate and extend into the beak. As soon as the spermatozoid becomes attached to the projection of the egg, the elongated nucleus withdraws toward the centre and the beak of the egg retreats. According to Lee (1955), only the head of the sperm (made of a vacuole-like structure and cytoplasm) flows through; most of its body is left behind, outside the archegonium, and disorganizes immediately. However, Favre-Duchartre (1956) observed portions of the ciliate band of the sperm inside the egg. Extra sperms are prevented from entering into the egg cytoplasm by a thickening of the plasma membrane of the egg (Lee 1955).

During karyogamy, the paternal chromosomes (12) become short. They can be stained clearly by Feulgen reaction as soon as the male pronucleus penetrates the female pronucleus. They mix with the maternal chromosomes (12) during the first somatic prophase (Favre-Duchartre 1958).

Several pollen grains may germinate in a pollen chamber. However, as soon as one pollen tube has discharged its spermatozoids, the body cells in the other tubes frequently fail to divide and degenerate. When two spermatozoids are discharged, two eggs can be fertilized, but a second embryo rarely develops.

Both fertilized and unfertilized ovules are shed from the tree at irregular intervals. The ovules may be sterile or may enclose one or two embryos ranging from the coenocytic to the dicotyledonous stage. Unfertilized ovules shed from the tree can undergo fertilization and subsequent embryonic development.

Embryogeny

The zygote nucleus divides in situ followed by several free-nuclear divisions (Fig. 12.17 A). The nuclei become distributed throughout the young proembryo. Sometimes, evanescent walls (Fig. 12.17 B) appear during the free-nuclear period. During later stages, the nuclei become distributed almost evenly in the cytoplasm. Wall formation takes place when there are 256 free nuclei; the newly formed cells fill the entire proembryo (Fig. 12.17 C). Within the female prothallus, the cellular proembryo develops continuously; it is considerably influenced by the prevailing temperature. The cells at the base divide and function as embryonal cells, while the upper cells elongate to form a massive suspensor (Fig. 12.17D). There is, however, no well-defined suspensor; it is a micropylar region of elongated cells.

Differentiation of Embryo.   There is a cap-like structure formed by the upper cells (towards the neck of the archegonium) which is pierced through by the radicle at the time of germination (Favre-Duchartre 1958). The embryo develops (slowly) after the seeds have been shed.

The mature embryo is dicotyledonous, occasionally three cotyledons are present. The embryo may reach its maximum size in 3 months after fertilization, if the development is accelerated in an incubator. In nature, the embryos may remain healthy for 12 months if the ovules are preserved in a damp place. This is because the embryo is surrounded by the prothallial cells, which provide necessary humidity for its survival (Favre-Duchartre 1958). There is no after-ripening requirement, and the seeds germinate whenever a suitable substrate is available.

The most conspicuous change in a female gametophyte is the deposition of reserve food like fat, starch and protein. Unlike other gymnosperms, the accumulation of food reserves in *Ginkgo* occurs before fertilization (cf. cycads; Favre-Duchartre 1958).

Abundant starch as reserve food accumulates in the gametophyte even when the archegonia are still immature. Lipids occur as variously-shaped globules which decrease sharply from epidermal cells inwards. Lipoproteins appear nearly 2 months before fertilization, at first in the inner layers and

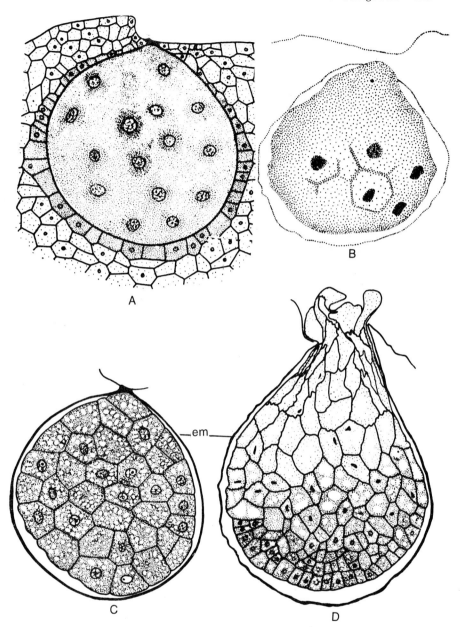

Fig. 12.17 A-D. *Ginkgo biloba.* Proembryogeny. **A** Longisection free-nuclear proembryo with the jacket layer and part of gametophyte. **B** Proembryo in non-median section, formation of evanescent walls. **C** Cellular proembryo. **D** Proembryo, the upper cells elongate to from the suspensor. *em* egg membrane. (**A, C, D** After Lyon 1904, **B** after Favre-Duchartre 1956)

later towards the periphery. In a mature gametophyte, there are four zones of storage cells: (a) The peripheral zone of three or four layers of isodiametric, vacuolate cells containing numerous round and densely aggregated lipid droplets in the cytoplasm. Where the density of the lipid droplets is low, the cytoplasm shows dictyosomes, a few mitochondria with short cristae, ER and chloroplasts. ER comprises rounded or irregular vescicles. Small starch grains occur embedded in the chloroplasts (Dexheimer 1973). (b) The middle zone contains lipoproteins and starch. The cells are large and vacuolate. The lipoprotein inclusions are irregular (diameter 5–50 μm). They are either enclosed in the central vacuole or are limited by a fine membrane in the periphery of the cell. Amyloplasts are abundant in these cells while other cytoplasmic organelles like ER, dictyosomes, mitochondria and ribosomes are rather sparse. (c) The inner zone contains starch. these cells are very large and show a thin parietal layer of cytoplasm. (d) The central zone cells do not have any reserve food material (Dexheimer 1973).

Seed. The seed coat is contributed both by the chalaza and integument. The integument differentiates into three zones: (a) outer (parenchymatous) sarcotesta, (b) middle (sclerenchymatous) sclerotesta and (c) innermost (thin-walled) endotesta. The sarcotesta is 5–6 mm thick in the equatorial region, and 2–3 mm in the micropylar and chalazal region; it is the only "live" portion of the integument. The epidermis has a ca. 15-μm-thick cuticle which is interrupted above the stomata. The cells contain chloroplasts, some of them also have druses of monohydrate calcium oxalate crystals. The bulk of the integument is formed of large turgid cells, the latter further enlarging towards the sclerotesta, while the number of chloroplasts and the size of starch grains decrease (Favre-Duchartre 1958). The stomata comprise two kidney-shaped cells filled with starch grains; the frequency is about eight stomata per mm².

The sclerotesta is hard and lignified. It forms the shell, which is slightly flattened laterally, and usually has two ribs (facing each other) which represent the suture. It is unlignified in the micropylar half, and forms as many valves as there are ribs.

The endotesta has a withered appearance. In the micropylar half it adheres to the sclerotesta and is free from the nucellus. It is parchment-like, has a golden-brown colour, and is thin and translucent. In the chalazal half, the endotesta unites with the nucellus, adheres to the prothallus, but is always separate from sclerotesta (Favre-Duchartre 1958).

The seed coat is vascularized. Several unbranched vascular bundles are arranged in a ring, the traces enter the chalaza, pass through the basal plate of thick-walled cells (see Schnarf 1937).

The mature seed is the size of a small apricot. The seed coat has an outer orange-coloured fleshy portion rich in butyric acid, and emits an odour like rancid butter. Inner to the fleshy layer is the stony layer, followed by the innermost papery layer. The dicotyledonous embryo is in the centre of the gametophyte, the so-called endosperm. The two cotyledons are normally equal and hypogeal, and have stomata mainly on the adaxial surface.

Germination. During germination, the plumule is pushed out of the testa by elongation and arching of the cotyledonary base. A portion of the cotyledon remains inside the seed, enlarges and persists through the first season, and functions as a haustorial organ. The first two or three leaves on the seedling are small and scale-like. The young stem stops its rapid elongation during the first winter after it has formed a close crown of leaves and large terminal bud.

Chromosome Number

In *Ginkgo biloba,* the haploid number of chromosomes is 12. There is evidence of karyotypic changes within the species. The basikaryo type is more or less constant; there are differences in the exact location of centromeres and in number and position of the satellites (see Khoshoo 1962).

Temporal Considerations

The duration of the life cycle in *G. biloba* is 14 months (Favre-Duchartre 1958).

The ovules are pollinated (in Paris) at the megaspore stage in the second half of April (first year). The pollen grains begin to germinate 3 weeks after landing in the pollen chamber.

In the female prothallus, the coenocytic phase continues until the middle of June, and archegonial initials can be observed by the end of the month. Fertilization begins early in September. Both fertilized and unfertilized ovules are shed from the tree at irregular intervals from October to April (second year).

The cellular proembryo develops within the prothallus. As soon as the embryo matures, it germinates in June (second year), in adequate water and at suitable temperature, without a period of dormancy.

Phylogenetic Considerations

The fossil record provides only a few clues as to when the Ginkgoales appeared first, and the ancestral group from which they have been derived (see Stewart 1983).

Trichopitys is one of the Palaeozoic genera which may have affinities with the Ginkgoales. However, there is no evidence of differentiation of *Trichopitys* axes into short shoots (a feature of *Ginkgo biloba*). Florin (1949) contemplates that as short shoots evolved, the ovulate branches present on the long branches of *Trichopitys* were transferred to the short shoots, and the number of ovules on a branch became reduced to one or two. Some support for this hypothesis can be derived from abnormal specimens of *G. biloba*, where several ovules are formed on an axillary branch system quite similar to that of the fertile truss of *Trichopitys*. If *Trichopitys* represents an early stage in the evolution of Ginkgophytes, then the relationship of the group appears to be with the conifer type. This is because the fertile shoot of *Trichopitys* is in the axil of a foliar unit (cf. Cordaitales) and is not a megasporophyll with ovules (Pteridospermales). On the other hand, evidence from the abnormal production of ovules on leaves of *G. biloba* does not support this view. Embryologically, too, they are similar to cycads (pollen grains, male gametophyte, motile sperms, development of female gametophyte). This implies that the origin of Ginkgoales from the conifer type is far from settled. According to Stewart (1983), it is best to presume that the Ginkgoales originated from a Palaeozoic ancestor, until there is more evidence from the fossil record.

While we know a good deal about various aspect of biology of *Ginkgo biloba*, priority attention should be paid to the gaps in our knowledge. A detailed and critical investigation should be made by applying newer techniques. A fascinating area of research would be the examination of living material through cine films. Microcinematography will bring out the sequenees of development (such as in the development of motile male gametes) which are otherwise difficult to follow. The scope in this method of study is endless.

13. Cordaitales

For a long time, the Cordaitales have been regarded as the dominant gymnosperm group of the Palaeozoic era. After the discovery of Pteridospermales, it became apparent that both the groups were present in the Palaeozoic. Although recorded from the Lower Pennsylvanian deposit (Good and Taylors 1970), cordaites became an important element of the Carboniferous in the middle Pennsylvanian dunes, and flourished into the Permian. The leaves of *Cordaites* are reported from the Permo-Carboniferous of Siberia, China, India, Australia and South America (see Stewart 1983). Leaves, stems, roots, cones and seeds of these plants are particularly abundant in the Mid-Pennsylvanian location of Iowa and Kansas in the U.S.A. The order includes a single family, Cordaitaceae.

Cordaitaceae

Cordaites is an organ genus for detached foliage of the Cordaitaceae. According to Arnold (1967), the name may also be employed, without typification, for the entire cordaitean plant.

Habit. *Cordaites* is a tree which attained a height of nearly 30 m and a diameter of more than 1 m at its expanded base. It had a crown of branches near the apex, which bore spirally arranged sessile leaves (Fig. 13.1A). Cridland (1964) reconstructed a cordaitean plant as a small tree ca. 5 m high (Fig. 13.1B). According to him, it had stilt roots (comparable to those of mangroves), and it inhabited swamps, estuaries and along seashores.

Root. The roots, known as *Amyelon,* are profusely branched and shallow, and form a pad of stilt roots which support the stem. A transection shows a central exarch actinostele with protoxylem strands at the tip of each arm of the stele. It is surrounded by a thick layer of secondary xylem. A well-developed periderm is present. As it grows it produces an outer layer of phellem of empty radially arranged cells (Fig. 13.2); lenticels are also present. A broad zone of aerenchymatous phelloderm is formed around the stele. As the root elongates, the protostele gives way to a siphonostele by medullation of the metaxylem to form a pith. The protoxylem points persist

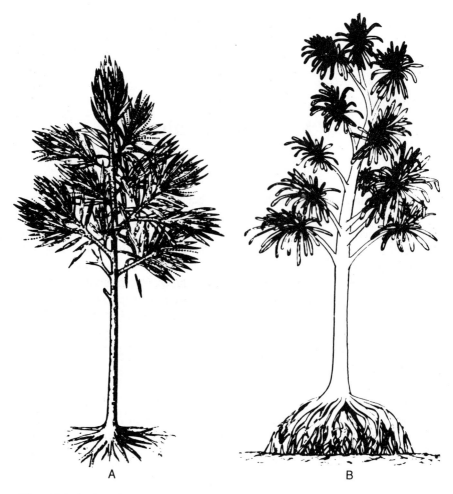

A   B

Fig. 13.1 A, B. Cordaitean plant, reconstruction from Carboniferous (**A**) and Pennsylvanian (**B**) era. (**A** After Scott 1909, **B** after Cridland 1964)

at the margin of pith and lie adjacent to the secondary xylem (Cridland 1964).

Stem. A transection of *Pennsylvanioxylon* (Figs. 13.3 A; 13.4 A) and *Mesoxylon* (Fig. 13.4 C) shows a large pith surrounded by pycnoxylic secondary wood. In the former, the distinction between primary and innermost secondary wood is apparent only in radial section. The latter also shows the innermost narrow protoxylem elements with spiral thickenings, followed by larger spiral and reticulate elements, and finally by tracheids with scalariform bars (Fig. 13.3 B). The compact secondary wood of *Pennsylvanioxylon* and *Mesoxylon* is similar in structure and comparable to *Araucarioxylon*. Bordered pits are present on radial walls in two or more rows. They are alternate and densely crowded, so that their borders appear

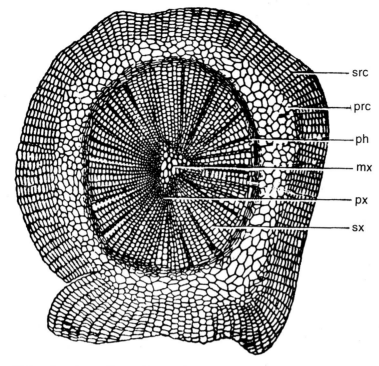

Fig. 13.2. *Ameylon radicans.* Transection root. *mx* metaxylem, *prc* primary cortex, *ph* phloem, *px* protoxylem, *src* secondary cortex, *sx* secondary xylem. (After Scott 1909)

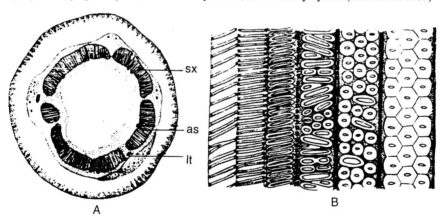

Fig. 13.3A, B. *Pennsylvanioxylon.* **A** Transection young branch. *as* axial sympodium. *lt* leaf trace. *sx* secondary xylem. **B** Wood, radial section. (After Takhtajan 1956)

hexagonal (Fig. 13.3 B). The pore of the pit is a transverse or inclined elliptical slit. The rays in the secondary wood are uniseriate and vary in height.

A well-preserved specimen of *Mesoxylon* shows a layer of secondary phloem (Fig. 13.4 C) which consists of radially arranged sieve cells, parenchyma and fibres. In a young stem, the cortex shows a hypodermal system of vertical strands of anastomosing fibres and secretory sacs. In older stems, an early formation of periderm in the cortex is evident. It apparently replaces the cortical tissues as the stem increases in diameter.

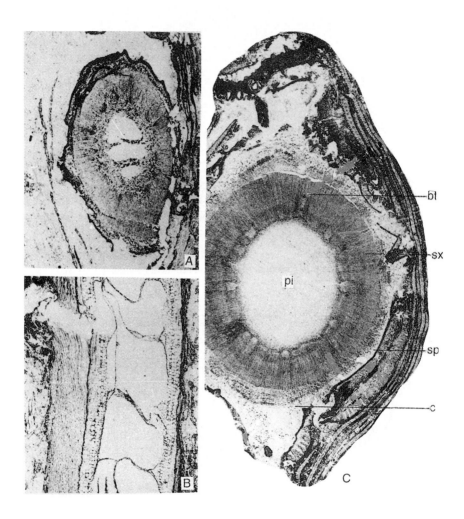

Fig. 13.4A-C.   A, B *Pennsylvanioxylon,* oblique (**A**) and longisection (**B**) stem shows septate pith. **C** *Mesoxylon,* stem transection. *bt* branch trace. *c* cortex, *pi* pith, *sp* secondary phloem. *sx* secondary xylem. (Courtesy Dr. G.W. Rothwell)

The large parenchymatous pith, in a well-developed stem, shows conspicuous transverse lens-shaped cracks which alternate with the diaphragms of parenchyma cells (Fig. 13.4 A, B). Such septate pith casts are named *Artisia*.

Leaf. The leaves, *Cordaites*, are spirally arranged, generally 15–20 cm long; unusually, ca. 1 m long and 15 cm broad. These are linear, lanceolate to spatulate, and may have a blunt or pointed tip. The leaves are tough, leathery, thick at the region of attachment, taper proximally, and have parallel venation (Fig. 13.5A). Several species have been described, which differ from each other in their epidermal and internal structure. Transfer specimens of thick cuticles of compression fossils show rectangular cells in rows. The cell wall is thick and straight, and the cells are parallel to the long axis of leaf. The haplocheilic stomata are arranged in longitudinal rows between the ribs formed by the vein in the lower epidermis. In a few species they are scattered on the upper epidermis (Stewart 1983). The stomata are elongated parallel to the veins, guard cells are in a pit surrounded by four to six subsidiary cells, of which two are polar.

The palisade layer is poorly differentiated and the spongy mesophyll shows large lacunae between the veins. Hypodermal strands of sclerotic tissue, and their extent of development and distribution, are diagnostic to distinguish the species. In some species the fibrous strands occur only above and below a vein, while in others they extend from the upper to lower epidermis, enclose the vein, and form I-shaped girders. The centripetal xylem in the veins is surrounded by a two-layered vein sheath. Most species also show centrifugal xylem (Fig. 13.5 B).

Several authors have made a detailed study of the leaf and branch trace of *Pennsylvanioxylon* and *Mesoxylon* (see Stewart 1983). The branching is axillary, and the leaf trace appears double at its point of inception at the periphery of the pith. The sympodia from which the leaf trace arises is mesarch in *Mesoxylon* and endarch in *Pennsylvanioxylon*. Each double leaf trace is flanked by two branch traces. As the leaf traces traverse the cortex of the stem and enter a leaf base, they dichotomize repeatedly to form 8–16 bundles. In *Mesoxylon* the leaf trace (within the leaf) shows both centripetal and centrifugal xylem, in *Pennsylvanixylon* exclusively centripetal xylem.

The two branch traces fuse after reaching the cortex, and form a single (branch) trace.

Fructifications

Because of morphological similarity, both pollen and ovulate fructifications are designated as *Cordaianthus*. The compound fructifications consist of a primary axis with secondary shoots/cones in the axils of modified leaves called bracts. The cones bear spirally arranged modified sterile (mostly) and fertile leaves, the scales. Close to the apex, a few fertile scales terminate in pollen sacs or ovules. The cones are monosporangiate. Figure 13.5A

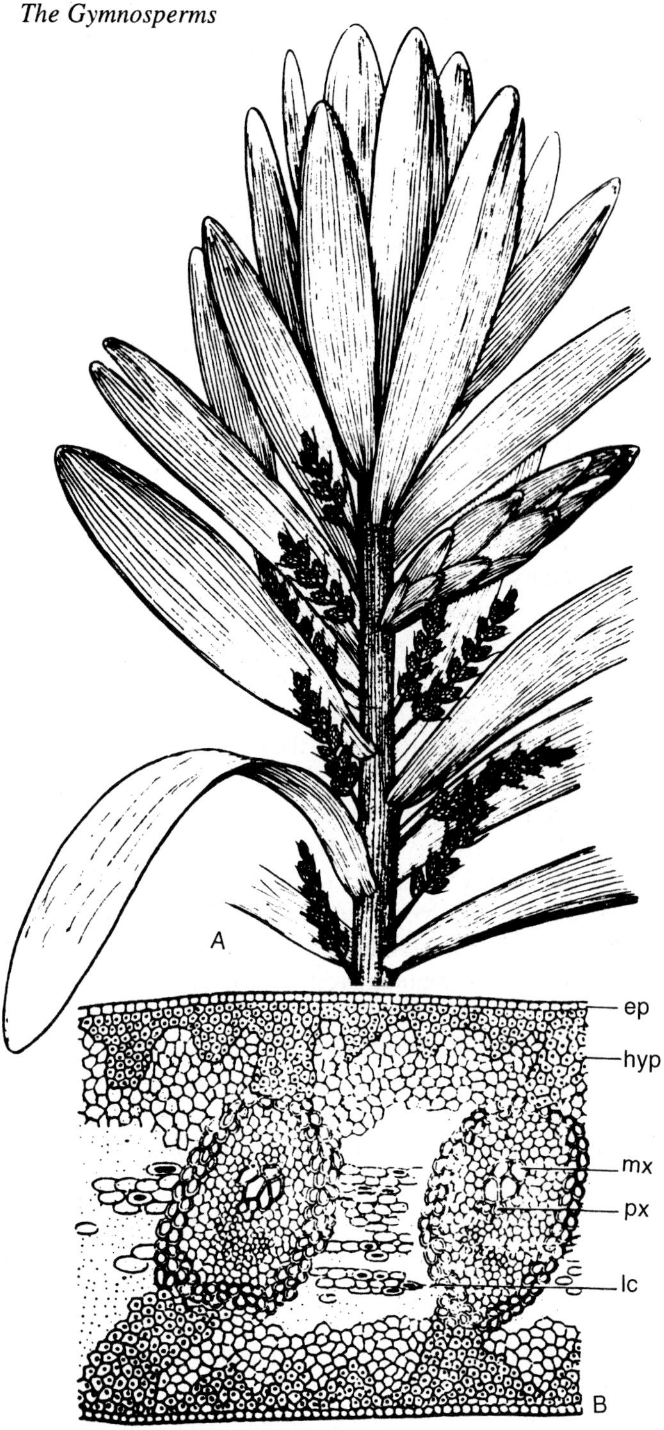

Fig. 13.5A, B. **A** Cordaitean branch bearing leaves, vegetative bud and fertile shoots. **B** *Cordaites* sp., vertical section leaf. *ep* epidermis. *hyp* hypodermis. *lc* transversely elongated cells connecting bundles, *mx* metaxylem. *px* protoxylem. (**A** After Grand 'Eury 1877, **B** after Renault 1879)

shows a branch bearing a vegetative bud (upper right), leaves and scattered fertile shoots.

Cordaianthus concinus. Besides compression-impression fossils several permineralized specimens have been discovered from the Middle Pennsylvanian coal ball material (Delevoryas 1953). A cone has, on a secondary shoot, ca. 25–40 scales, and only 5–10 distal scales are fertile (Fig. 13.6 A, B). Each fertile scale usually terminates in six microsporangia/ pollen sacs fused at the base (Fig. 13.6 C, D). They contain pollen grains of the Florinites type, monosaccate with reticulate ornamentation on the inner wall of the saccus. The latter is attached to the body of the pollen grain (corpus) on its distal and proximal surfaces (Millay and Taylor 1974).

Cordianthus pseudofluitans. Most of the species are compression-impression fossils from the Upper Carboniferous of England and Europe; very few permineralized specimens have been discovered. According to Florin (1939, 1950a, 1951), the morphology and anatomy of the ovulate fructification is similar to that of the male fructification. A cone and its subtending bracts are distichous on the primary axis. Each cone has ca. 16–20 spirally–arranged scales, and the distal four to six scales are fertile. Each scale may have a single reflexed ovule at the tip, or it may dichotomize so that there are two or more pendulous ovules at the tip of the branched scale (Fig. 13.7 A, B). The cordate ovules are platyspermic. In compression-impression fossils, the ovules are of the samaropsis type, while the structurally preserved permineralized ovules are assigned to *Miterospermum* (Lower-Upper Pennsylvanian) and *Cardiocarpus* (Middle Pennsylvanian to Permian). In transection, the ovules are biconvex, the primary plane of the ovule being parallel to the long axis, and the secondary plane at right angles to it. In *C. spicatus* the conspicuous testa consists of a sarcotesta; its outer zone comprises large thin-walled cells, and the inner small cells. The sclerotesta has thick-walled sclerotic cells and spine-like projections into the sarcotesta. The endotesta cells line the sclerotesta. In the primary plane, the sarcotesta of *Cardiocarpus* and *Miterospermum* expands into a flattened "wing" in some species (Fig. 13.7 C).

The nucellus is free from the integument, except at the base. Its distal portion, below the inner opening of the micropyle, is differentiated into a beak and well-developed pollen chamber (Fig. 13.7 D). Several workers (see Stewart 1983) have observed the florinites dispersed type of pollen in this chamber. A few specimens showed well-preserved gametophytes with at least two micropylar archegonia flanking a "tent pole' (as in some Pteridosperms and *Ginkgo*). In the mature seeds the nucellus appears thin and inconspicuous.

A single vascular strand enters the base of the ovule. In *Cardiocarpus*, it terminates in a plate of tracheids at the base of the nucellus. Two vascular

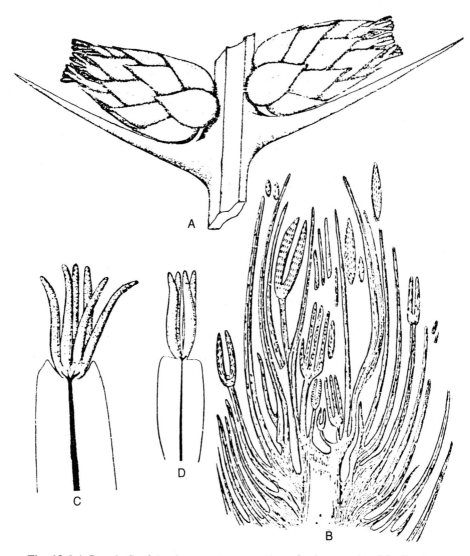

Fig. 13.6 A-D. **A** *Cordaianthus concinnus,* portion of primary axis of fertile shoot shows two male secondary shoots with spirally arranged sterile and fertile scales. **B, C** *C. penjoni,* **D** *C. saportanus,* **B** Longisection secondary shoot shows sterile and fertile scales with terminal pollen sacs. **C, D** Fertile scales. (**A** After Delevoryas 1953, **B** after Renault 1879, **C, D** after Florin 1951)

strands arise from the central vascular supply and extend laterally, in the primary plane, into the sarcotesta, and terminate almost at the apex of the ovule. In *Miterospermum compressum*, the two lateral strands pass through the sclerotesta into the sarcotesta where each strand divides (in the primary plane) to form several bundles (Fig. 13.7 D).

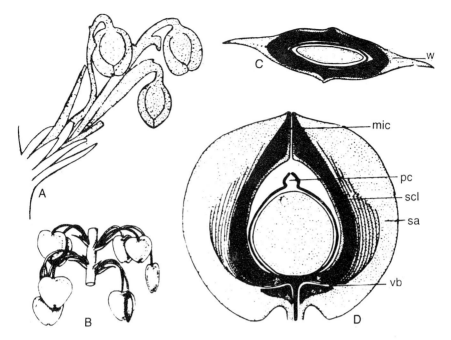

Fig. 13.7 A-D. A *Cordaianthus pseudofluitans*. **B** *Cardiocarpus cordei,* ovulate fructification. **C, D** *Mitrospermum compressum,* trans- (**C**) and longisection (**D**) of ovule. *mic* micropyle. *pc* pollen chamber. *sa* sarcotesta. *scl* sclerotesta. *vb* vascular bundle. *w* wing. (**A** After Florin 1951, **B** after Zimmermann 1959, **C, D** after Taylor and Stewart 1964)

Phylogenetic Considerations

Cordaitales is the oldest truly coniferophytic group of plants (Miller 1977). We do not precisely know how or from what ancestral group the Cordaitales evolved. An appraisal of the ontogeny, structure and reproductive biology of the most ancient seed ferns, Upper Carboniferous seed ferns, cordaites, and conifers shows differences between the last two groups and their supposed ancestors. The analysis is based principally on the investigations of *Cordaixylon dumusum, Mesoxylon priapi* (Cordaitales) and *Lebachia* and *Ortiseia* (Coniferales). The data indicate that the evolution of Cordaitales from the Lower Carboniferous seed ferns included several significant steps (Rothwell 1986) such as:

1. The transition from hydrasperman reproduction (the most primitive type of reproduction in gymnosperms), where the ovules had a complex pollen chamber which sealed the microgametophytes with radial, trilete exine structure, i.e. prepollen, to more modern type, where the integument plays an active role in pollination and post-pollination sealing of the pollen chambers.

2. The vegetative leaves changed from pinnately compound to strap-shaped simple leaves with entire margins.

3. Reduced number of ovules and microsporangia per microsporophyll.

4. Aggregation of sporophylls on compound shoots.

It is uncertain whether changes in anatomical features and cuticular anatomy (important in the identification of Cordaitales) were also instrumental in the evolution of Cordaitales. The evolution of conifers probably resulted from changes in vegetative (the conifer type of leaves along the entire branch system) and reproductive morphology (aggregation of sporophylls on compound shoots with derived appendage arrangement on secondary axes) among plants that already had the more modern type of gymnospermous ovules, as in Cordaitales and some Upper Carboniferous seed ferns (Rothwell 1986).

14. Voltziales

In the history of the plant world, the transition from the Carboniferous to the Permian has attracted much attention. Coexisting with the Cordaitales in the Late Paleozoic and persisting into the Mesozoic are the Voltziales, often called the transition conifers. They exhibit the reduction series and changes in symmetry of a cordaianthus-like fertile dwarf shoot (see Chap. 13) that lead to the ovuliferous scale of extant conifers. Florin (1939, 1944, 1945, 1950a, b) made an intensive study of the Voltziales. He compared the fertile dwarf shoot of *Cordaianthus* with the ovuliferous scale of modern conifers, and cited several intermediate stages within the Voltziales (Florin 1951). However, the Voltziales do not fill the many gaps between the cordaites and conifers (Stewart 1983).

Morphology

Externally, the Voltziales resemble extant taxa such as *Araucaria excelsa*, the Norfolk Island pine (Florin 1951). Older plants show monopodial stems with a regular arrangement of their (five or six) branches in a whorl (Fig. 14.1A). The leaves on the vegetative branches are acicular, while those on fertile branches have a bifurcate tip.

The best-known genera are *Lebachia* and *Ernestiodendron*. They are based on vegetative remains which were formerly included in *Walchia*, and are distinguishable from one another primarily by differences in their cuticle. They appear to lack calcium oxalate crystals in the cuticular layers. *Lebachia* and *Ernestiodendron* have simple unicelled hairs particularly on the underside of the leaves (unlike contemporary Cordaites or Voltziaceae). The haplocheilic stomata are amphistomatic (on both surfaces of the leaf) and have four to ten subsidiary cells. The leaves and bracts are supplied by a single vein. *Lebachia* leaves are "entire", except those on penultimate shoots and bracts of ovulate cones, which have bifurcate tips. They are decurrent along the branch (Fig. 14.1 B) and slightly spreading. The stomata are longitudinally oriented in two bands. The leaves of *Ernestiodendron* are borne at right angles to the branch (Fig. 14.1 C). The stomata are in isolated longitudinal rows. *Walchia* represents fossils which lack sufficient preservation of the cuticle for identification, either in *Lebachia* or *Ernestiodendron*.

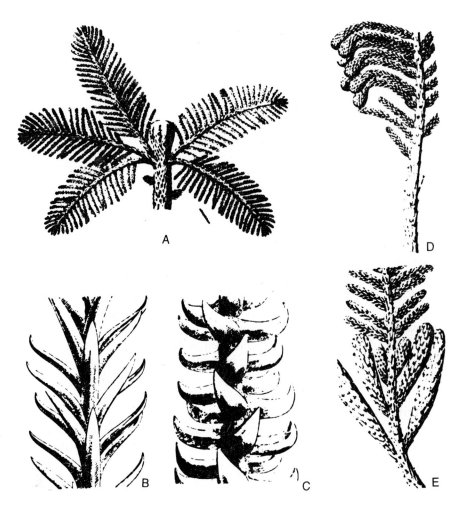

Fig. 14.1A-E. A, D, E *Lebachia piniformis*, **B** *Lebachia* sp., **C** *Ernestiodendron*.
A Whorled branches at the apex of shoot. **B, C** Leafy shoot, spirally-arranged leaves
(**B**), and leaves at right angels to the axis (**C**). **D, E** Branches with pendulous pollen
cones (**D**) and erect ovulate cones (**E**). (After Florin 1951)

Anatomy

The wood is not widely known in the Late Palaeozoic forms (Miller 1977).
Those on record show a eustele with endarch primary xylem (Florin 1951).
Secondary xylem has tracheids with araucaria type of pitting (one to three
rows of closely arranged alternating bordered pits). Occasionally, xylem
parenchyma is also present. The rays are uniseriate and resin ducts are
absent. Early or older Voltziales (see later) had fewer rows of pits on the
radial tracheid walls than *Cordaites*. A relatively large and irregularly ruptured
pith is present (Florin 1951). In addition to the above characteristics, Rothwell
(1982) observed in *Lebachia* twigs (from the Middle Pennsylvania of

Oklahoma) the origin of leaf traces from the eustele, and apparent resin ducts without epithelial cells in the pith and cortex of twigs and mesophyll of leaves.

Reproductive Organs

There is considerable variation in the structure of the seed cone within the Voltziales. The changes begin in the Palaeozoic and continue into the Mesozoic. On the basis of relative modification of the ovuliferous shoot and its subtending bract, Florin (1951) recognized two groups of Voltziales. These are placed in two families: (*a*) Lebachiaceae (Pennsylvanian and Early Permian)—the older Voltziales. (*b*) Voltziaceae (Late Permian and mainly Mesozoic)—the younger Voltziales.

Lebachiaceae

The Lebachiaceae have inflorescences with strobiloid nature and show relatively much less modification than that of the *Cordaianthus* (the strobiloid nature of the dwarf shoot is evident).

The plants are monoecious. Morphologically the fructifications appear like cones. They are monosporangiate, ellipsoidal to cylindrical borne at the tip of leafy branches. The male cones are pendulous and the female cones upright (Fig. 14.1 D, E).

The male cone of lebachias is radially symmetrical to ellipsoidal. There is a primary axis around which are attached spiral dorsiventral scales or microsporophylls. In these cones there are no bracts subtending a secondary shoot. Florin (1951) and others (see Stewart 1983) interpret these cones as simple, and each microsporophyll has a narrow stalk-like base and a distal upturned leaf-like portion (Fig. 14.2 A, B). There are two microsporangia on the abaxial side (Fig. 14.2 A), and the pollen grains are monosaccate or bisaccate (Fig. 14.2 C).

The compact ovulate cone has a primary axis which bears several spirally arranged bracts with bifurcate tips (Fig. 14.2 D, E). In the axil of each bract is a tangentially flattened secondary shoot with scales attached assymetrically (Mapes 1982, Rothwell 1982). In *Lebachia*, there is only one fertile scale, terminated by a single erect bilaterally symmetrical ovule (Fig. 14.2 E). In *Ernestiodendron* and *Walchiostrobus*, there are several ovule-bearing scales (unlike *Lebachia*). In the former, all the scales on the secondary shoot are fertile (Fig. 14.2 F); in *Walchiostrobus,* sterile scales are present on the proximal and fertile scales on the distal portion of the secondary shoot (Fig. 14.2 G). The ovules may be erect or reflexed (Fig. 14.2 F, G). Florin (1951) regarded the lebachia type of cone as primitive, and *Ernestiodendron* as more advanced.

The ovulate fructifications in *Lebachia* and *Cordaianthus* are compound, this is the most important character which indicates their relationship. The primary difference is that in *Cordaianthus* the secondary shoot-bract

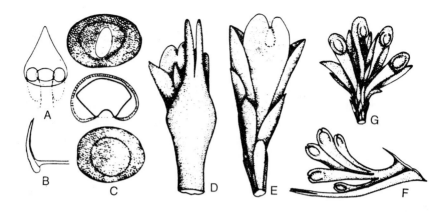

Fig. 14.2A-G. A, B *Lebachia hypnoides*, microsporphyll. **A** Two microsporangia. **B** Lateral view. **C** *Lebachia* sp., pollen grains; proximal (above) and distal (below) views; optical longisection in centre. **D, E** *L. piniformis*, secondary shoot. **D** Abaxial view, bract with bifurcate tip. **E** Adaxial view, spirally arranged scales and erect ovule. **F** *Ernestiodendron filiciforne*, bract and secondary shoot complex; all scales fertile. **G** *Walchiostrobus* sp., secondary shoot shows fertile and sterile scales and recurved ovules. (After Florin 1951)

complex is lateral and distichous on the primary axis, and spirally arranged in *Lebachia*.

Voltziaceae

Voltziaceae includes plants where the seed scale of the ovulate cone shows significant modification from the strobiloid condition of the Lebachiaceae, yet with parts sufficiently free from one another and from the bract to indicate derivation from a strobilus. They show reduction and fusion of fertile dwarf shoot elements, change in their symmetry from radial to bilateral (dorsiventral), and also fusion between the fertile shoot and its subtending bract. These changes begin in the Palaeozoic and continue into the Mesozoic (Miller 1977).

Several types of modifications of the bract-scale complex are known. Some of the genera may possibly represent intermediate evolutionary stages between the Lebachiaceae and modern conifers, while others show interesting variations, which probably became extinct without giving rise to an extensive lineage (Miller 1977).

In *Pseudovoltzia* (Permian) five sterile lobes, their bases fused to one another, occur axillary to each bract (Fig. 14.3 A, B). The middle and two marginal lobes are larger than the remaining two, and each bears a recurved ovule. This lobed unit is fused to the subtending bract for a short distance. Florin (1951) pointed out that the five lobes and three ovules represent eight distinct parts. Schweitzer (1963) studied petrified material of *Pseudovoltzia*, where the original three-dimensional arrangement is preserved. It shows a greater fusion of parts than was formerly known, and also

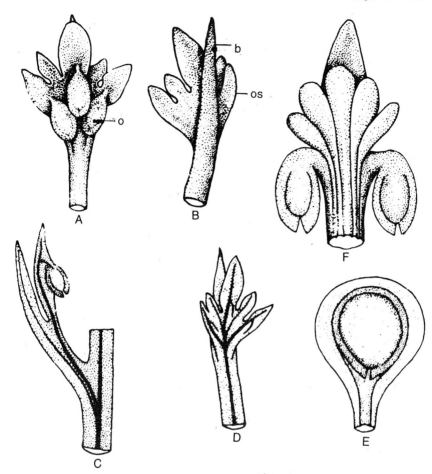

Fig. 14.3A-F. **A-D** *Pseudovoltzia liebeana*. **A** Ovuliferous scale shows three recurved ovules (*o*). **B** Abaxial view of bract-scale complex. *b* bract, *os* ovuliferous scale. **C, D** Vascularization of bract-scale complex (**C**) and ovuliferous scale (**D**). **E** *Ullmania bronii*, ovuliferous scale, adaxial surface shows single ovule. **F** *Glytolepis longibracteata*, fertile ovuliferous scale, note five sterile lobes and two recurved ovules. (**A-D** After Schweitzer 1963, **E, F** after Florin 1951)

supports Walton's (1928) view that the three larger lobes and their associated ovules represent megasporophylls and there are only five distinct parts to the unit. The vascular bundles to the bract and scale are not fused, the lobes and ovules are each vascularized by a single strand, five of which arise from the end of the scale bundle (Fig. 14.3 C, D). The three larger lobes and their associated ovules represent megasporophylls. *Pseudovoltzia* is, thus, reinterpreted as too advanced to be an intermediate between the Lebachiaceae and the Voltziaceae (Miller 1985). *Voltzia* also has five sterile elements in its cone scales, but there are only three ovules. The middle and two marginal lobes are larger than the remaining two lobes, as in *Pseudovoltzia*. The seed stalks are fused to their subtending sterile lobes and are free only at their tip.

Ullmania (Permian) has a single recurved ovule adaxial to a large orbicular scale (Fig. 14.3 E) subtended by a bract. In the structurally preserved material of *U. frementaria*, it is still uncertain if the seed has a separate attachment or is fused to the scale (Schweitzer 1963). The bract-scale complex is similar to that of *Araucaria* in general morphology and orientation of the single adaxial ovule.

Glyptolepis (Permian and Triassic) has elongate lax cones apparently borne in groups (Schweitzer 1963). The secondary shoot is planated and bilaterally symmetrical. It consists of five or six sterile lobes and two recurved ovules in lateral position on either side of the sterile scales (Fig. 14.3 F). The unit is axillary to a bract. In older species the bract is shorter than the sterile scales, while in younger species, it is longer (Florin 1951). The bracts and their secondary shoots are spirally arranged on the lax primary axis.

Voltziopsis (Early Triassic and possibly Permian) is known only from the Southern Hemisphere. Cylindrical seed cones terminate short branches and have ca. 25 bract-scale complexes. The narrow bracts have bifurcate apices and are nearly as long as the scale. The ovuliferous scale consists of a thick but flattened stalk which expands apically into five lobes (rarely six) that separate from the stalk at different levels. Each lobe bears a single recurved ovule which is free from the lobe but adnate to the seed stalk. Both complete cones and isolated bract-scale complexes are known, which suggests that the cones may have dissociated when mature (Townrow 1967).

Phylogenetic Considerations

The pollen-bearing structures in *Cordaianthus* are produced in sporangia which occur at the tip of microsporophylls (see Chap. 13). In Lebachiaceae and most other conifers, the sporangia are in the basal region of a laminar microsporophyll. These sporophylls (in *Cordaianthus* and Lebachiaceae) are arranged in a helix around an axis to form a strobilus, which is further organized to form compound strobili in *Cordaianthus*. The presence of simple or compound strobili is important to distinguish Cordaitaceae from Lebachiaceae and, according to Miller (1985), makes it difficult to accept Florin's (1951) hypothesis that the Lebachiaceae evolved from Cordaitaceae. Instead, it is more likely that the two groups share a common ancestor which was contemporary with or earlier than the earliest Cordaites. This may also explain the diversity seen in Pennsylvanian Lebachiaceae (Miller 1985).

As there is no precise information on how or from which ancestral group Cordaitales evolved, the major changes that occur in the evolution of compound female cones are interpreted from the existing information of the ovulate fructification of the upper Carboniferous-Permian Cordaitales and Voltziales.

The secondary shoot in the axil of a bract of *Cordaianthus* (see Chap. 13) is homologous with the bract and axillary ovuliferous scale of a conifer seed cone. *Ernestiodendron* and *Walchiostrobus* are like *Cordaianthus* in the radially symmetrical secondary shoots composed of spirally arranged sterile and fertile scales. The evolutionary trends apparent in Lebachiaceae, without implying an evolutionary series (Stewart 1983), which lead to the compound ovulate cone, are: a reduction in the number of ovules on the secondary shoot (*Lebachia*); the erect ovules become inverted with their micropyles directed towards the primary axis of the fructification (*Ernestiodendron* sp. and *Walchiostrobus* sp.), and a tendency towards bilateral symmetry by flattening of the secondary shoot and ovule (*Lebachia*).

In Voltziaceae the secondary shoot is planated and bilaterally symmetrical. In *Glyptolepis* and *Voltziopsis*, the fertile and sterile scales lie in a single plane and there is hardly any tangential fusion. The number of ovules is reduced to two. *Ullmannia* shows maximum tangential fusion of sterile scales to form a ovuliferous scale (cf. several conifers), and the single ovule is inverted. Permineralized specimens of *Lebachia* (Mapes 1982, Rothwell 1982) indicate how the shift occurred from the radial symmetry of the secondary shoot (Cordaitales and Lebachiaceae) to bilateral symmetry (Voltziaceae) followed by tangential fusion of scales to form an ovuliferous scale with adaxial ovules and their micropyle directed towards the primary axis, as in compound or ovulate cones of extant conifers. According to Stewart (1983), this is undoubted evidence that the ovuliferous scale of a conifer cone is a highly modified secondary shoot in the axil of a bract.

Late Permian conifers combine features of various Mesozoic taxa, and the Mesozoic and Cenozoic conifers those of many modern taxa. According to Miller (1982), the extant families may not have a common origin among the Voltziaceae. Some of them may have diverged directly from the Lebachiaceae, without intermediates from the Voltziaceae.

Certain characters of the conifers can be traced back to the Late Devonian. The growth habit appears to have already evolved (as seen in *Callixylon*), with 9-m-tall trunks and stumps 1.52 m in diameter. All features of coniferophytic secondary xylem, except the torus, had evolved by the end of the Devonian and are common to the progymnosperms.

The Coniferophytes were much more widespread and diverse during the Mesozoic than they are now. Also, modern conifer families appear to have originated somewhat earlier than was formerly presumed. This makes it difficult to visualize their evolution from the known Voltziales, where petrified seed cones show greater modification than was formerly known. According to Miller (1977), it would be better to search for precursors of modern families in Early Triassic and Palaeozoic sediments. While Florin's (1951) hypothesis remains valid, it may not be the only route that conifers followed in their evolution (see Miller 1985).

15. Coniferales

The conifers constitute more than three-fourths of the living gymnosperm flora, and cover vast areas of the temperate regions of both Northern and Southern Hemispheres.

The largest continuous, dense forest in the world is coniferous, and it encircles the earth just below the northern tundra of the Arctic and extends from Canada, Russia, Northern Europe to northern China and Japan (see Chap. 1). It began developing (ca. 10 000 years ago) when the glaciers retreated, and most of it still exists today. It extended southward along mountain ranges and merged with the deciduous forests of the south. The coniferous trees are ideally suited for the cold and snowy climate of the north. The dense stands deflect the wind, the snow slips off the down-sweeping branches of the trees, and the thick bark and waxy evergreen needles retain moisture for photosynthesis even in winter. The thick, rough bark of many conifers helps the tree to survive forest fires. There are 52 genera and ca. 566 species, which are classified into six families: Pinaceae, Taxodiaceae, Cupressaceae, Podocarpaceae, Araucariaceae and Cephalotaxaceae.

The plants are mostly trees or shrubs. Most trees are pyramidal or conical when young, round- or flat-topped when old. Many have columnar trunks free of branches to a considerable height. The tallest tree in the world is the Californian redwood—*Sequoia sempervirens* (Taxodiaceae), over a 100 m in height. The redwoods prefer the humid conditions of the Pacific coast, where fog is prevalent and the deep soil enables their roots to grow. *Sequoiadendron giganteum* (Taxodiaceae) is known as the giant sequoia, and grows on the western slopes of Sierra Nevada in California. Some of the trees are over 76 m in height and 9 m in diameter. The 'General Sherman' tree, the largest living example, measures 90 m in height and has a diameter of 33.5 m. The giant sequoias are more massive and heavier than the coastal redwood. Each tree (redwood or giant sequoia) yields a great deal of high-quality timber, and at one time it was felt that the trees would become extinct due to overcutting. Now, however, they are protected, and many of the areas where they grow have been declared National Parks. There are other giants like *Dacrydium cupressinum* (Podocarpaceae), which reaches a height of 55 m. *Taxodium mucronatum* is the "big tree of Tule" from southern Mexico and has a diameter of 17 m.

Massive buttresses support a few conifers, such as the *Taxodium distichum* (bald cypress) and some of the larger species of *Chaemaecyparis* and *Cryptomeria*. Such buttressed trees may have a circumference three times as great at ground level as at 2 or 3 m higher. Others may have no apparent swelling at the base, but have radiating roots which anchor the tree against strong winds. *Taxodium distichum* likes damp soil when growing in swampy habitats, and develops hollow vertical pneumataphores ca. 1 m high and 30 cm across.

Young trees of *Chaemaecyparis lawsoniana* (lawson cypress) and several other species have downwardly sweeping branches, the lower ones often rooting in the ground at their tips. *Dacrydium laxifolium* is the smallest conifer known. Mature specimens with cones are ca. 8 cm tall. It is a native of mountains of New Zealand. *Juniperous horizontalis* (Cupressaceae) is a completely prostrate form. Some trailing or creeping junipers, less than 30 cm tall, have been horticulturally developed as ground covers (see Chap. 23). *Podocarpus ustus* (Podocarpaceae), confined to New Caledonia, is the only conifer suspected to be a partial parasite on the roots of *Dacrydium taxoides*. The leaves are described as reddish bronze or purple (see Sporne 1965).

Some of the conifers have an exceptionally long life. In the Inyo National Forest of California, USA, there is a tree of *Pinus aristata* which is more than 4600 years old, and which even produces cones occasionally (P. Maheshwari and Konar 1971). The giant sequoias live perhaps as long as 3000 years, while the redwoods live for 2000 years. A cross section from the base of a *Cedrus deodara* (Pinaceae) tree—displayed in the timber museum of Forest Research Institute, Dehra Dun (India)—shows that this tree was a sapling in the 12th century and died early in the 20th century (P. Maheshwari and Biswas 1970).

Most conifers have needle-like leaves and are evergreen. In *Araucaria* the leaves may persist and remain green for as long as 15 years; the leaf base expands as the branch bearing it enlarges. A few conifers are deciduous, shedding their leaves or branches in autumn: *Larix decidua*, *Pseudolarix*, *Metasequoia* and *Taxodium distichum*. In *Phyllocladus* (Podocarpaceae) the foliage leaves are scaly and succeeded by phylloclades, which arise in their axils. The strobili are borne at the edge of the phylloclades near the base.

Anatomically, transfusion tissue is associated with veins in the leaves of all investigated conifers. The prominence and arrangement of this tissue varies in the genera. It may surround the veins (*Pinus*), form a cap adaxial to the xylem (*Araucaria*), occur as wing-like projections lateral to the vascular bundles (*Cephalotaxus*, *Podocarpus*), or be abundant lateral to the phloem. Transfusion tissue may consist entirely of tracheids, or have parenchyma interspersed with the tracheids (see Griffith 1971).

The roots of conifers have ectotrophic mycorrhiza which occurs naturally. The mycorrhiza is a symbiotic association between the root cells and fungi. It has been established that the symbiosis increases the uptake of phosphorous by the host plants for the mycelium. In Araucariaceae, the mycorrhiza is endotrophic, and Podocarpaceae also have root nodules (cf. Leguminaceae).

15.1 PINACEAE

Pinaceae is the largest and the most recent of the modern conifer families. The exact time of origin is not clear, but the family is evident by the Early Cretaceous. Seed cones of this Cretaceous assemblage show considerable diversity but have more features characteristic of *Pinus* than of any other modern taxa (Miller 1977). Anatomically also, the cones are distinctly *Pinus*-centred (which occurs only in the Early Cretaceous). *Pseudolarix*, and possibly *Picea*, may have evolved by the end of the Cretaceous, but reliable evidence of other modern taxa is absent during the Mesozoic.

The characteristic features of the family are: spirally arranged parts; two pollen sacs in a microsporophyll, pollen mostly bisaccate; ovuliferous scale free or slightly fused at the base of the bract scale, two ovules per scale, and seeds generally winged. There are over 200 spp. included in ten taxa: *Abies, Cathaya, Cedrus, Keteleeria, Larix, Picea, Pinus, Pseudolarix, Pseudotsuga* and *Tsuga*. All are trees, and the family is economically very important (see Chap. 23).

Pinus
The genus has ca. 80 valid species (Dallimore and Jackson 1966). They are divided into two natural subgenera with distinct characters:

(a) **Haploxylon** or soft pines—the scaly shoot at the base of the short shoot is deciduous, ray tracheids in secondary wood of stem and root have smooth wall; the needle (leaf) has a single vascular bundle.

(b) **Diploxylon** or hard pines—the scaly shoot present at the base of the short shoot is persistent; the ray tracheids have a corrugated wall, and the needle has two vascular bundles.

Morphology
Young pine trees are pyramidal with horizontal branches at regular intervales (Fig. 15.1.1). This symmetry is lost as the tree matures and the crown becomes round, flat or spreading.

There is a primary tap root with a large number of laterals called long roots. The primary root soon becomes arrested while the long roots continue to grow and bear clusters of dwarf roots. Some of these roots branch dichotomously, form coralloid masses, have an ectotrophic mycorrhiza (Fig. 15.1.2 A,B) and are termed mycorrhizal roots.

Fig. 15.1.1. *Pinus roxburghii*, plantation at Ranikhet (India). (After P. Maheshwari and Konar 1971).

The majority of the fungi forming ectotrophic mycorrhiza belong to the families Boletaceae and Agaricaceae (see P. Maheshwari and Konar 1971). In the soil these fungi occur as loose hyphal strands and can infect directly. A seedling with mycorrhiza acts as an inoculum for the spread of infection to the adjoining seedlings. Excessive mortality has been recorded when good-quality seedlings have been grown in soil other than forest soil. In *P. nigra* var. *calabrica*, stocks grown in nurseries without proper mycorrhizal conditions suffer on an average 45% mortality, while only 7% has been recorded in those raised in a soil with a mycorrhizal fungus like *Boletus bovinus* (Raymer 1947).

The intensity of mycorrhizal infection in soil depends on strong sunlight, adequate soil moisture, aerated and acid soils, and low levels of soil fertility (Bakshi and Kumar 1968). There is a positive correlation between the intensity of infection and increase in the mobilization of soil nitrogen. It has been suggested that the internal nutrient status, especially N_2, phosphorus and potassium is responsible for determining the intensity of infection. When their concentration is optimum in the roots, there is resistance to mycorrhizal infection.

The effect of ectomycorrhizal fungi on growth and phosphorus uptake of *P. sylvestris* seedlings at increasing phosphorus levels has been studied. Even a low intensity of infection by *Laccaria laccata* greatly stimulates growth of the seedlings, which seems to be more related to its capacity to produce growth substances than to its capacity to stimulate phosphorus uptake. The poor efficiency of *Hebeloma crystuliniforme* compared with *Laccaria laccata*,

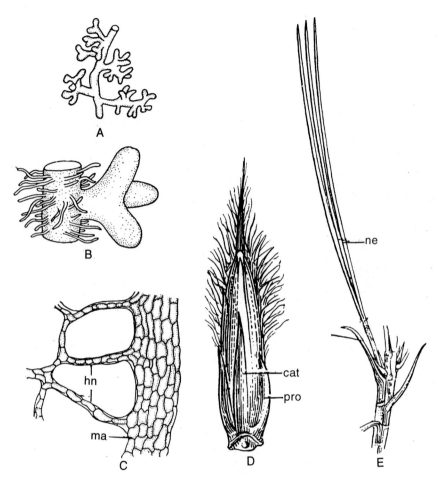

Fig. 15.1.2A-E. **A-C** *Pinus* sp., mycorrhiza. **C** Transection ectotrophic mycorrhiza shows mantle (*ma*) and Hartig's net (*hn*). **D E** *P. roxburghii* foliar spur. **D** Young, and **E** unfolded needles. *cat* cataphyll. *ne* needle. *pro* prophyll. (**A, B** After Hatch 1937, **C** after Hatch and Doak 1933, **D, E** after P. Maheshwari and Konar 1971)

at any level of phosphorus, could result from differences in diversion of carbohydrates from the host to the fungal structure.

Two types of branches are present: *long shoot* with unlimited growth; *dwarf shoot* with limited growth. The long shoots occur on the main stem as lateral buds in the axils of scale leaves, and terminate in an apical bud. The latter is enclosed by a number of bud scales closely surrounded by a thick mat of hairs. The lateral buds grow, nearly horizontally, to a certain length termed *nodal* growth. In *uninodal* pines a single internode, while in *multinodal* pines two or more internodes are formed in a year.

The dwarf shoot or foliar spur develops on a long shoot in the axil of a scale leaf (Fig. 15.1.2 E) and lacks a terminal bud. Each dwarf shoot initially has two opposite scales—prophylls followed by 5–13 cataphylls

(Fig. 15.1.2 D). Finally, needle-like foliage leaves (Fig. 15.1.2 E) develop on the spur shoot, in fascicles of one (*P. monophylla*), two (*P. sylvestris, P. merkusi*), three (*P. insularis, P. roxburghii, P. gerardiana*), four (*P. quadrifolia*) and five (*P. griffithii, P. wallichiana, P. armandi*). The number of needles present is constant for a species and is used for identification of different pines.

The plants are monoecious. The male and female cones are borne on different branches of the same tree. The male cones are modified dwarf shoots and appear in clusters (Fig. 15.1.3A) on the lower branches of the tree. The number of cones in a cluster varies from ca. 15 (*P. griffithii*) to 140 (*P. roxburghii*). Each cone arises in the axil of a scale leaf, replaces a dwarf shoot and is surrounded by several bracts. The female cones replace the terminal buds of the long shoot and are a modified long shoot (Fig. 15.1.3 B). In the beginning, the female cones are protected by an involucre of bracts. The cone has an average length of 1.5–2 cm and a diameter of 0.8–1 cm. The cone axis elongates (at the time of pollination) and the cone protrudes beyond the envelop of scales and is open to receive pollen grains. After pollination the cone closes. Originally, the cones are pale green but about the time of pollination the colour changes to reddish purple, and finally to glaucous green or purplish (*P. wallichiana*). The seeds are shed nearly 27 months from the time of initiation of the cone. The cones (at this time) are ca. 20–24 cm in length and 5.5–6 cm in diameter. In most species the hard and woody mature female cone opens to release the seeds (*P. roxburghii, P. wallichiana*), in others the seeds are released only after the cones fall to the ground and rot. In a few species (*P. flexilis*)

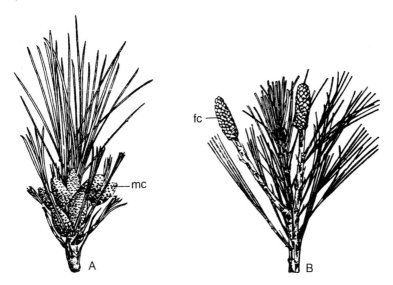

Fig. 15.1.3A, B. **A** *Pinus roxburghii*, long shoot bearing a cluster of male cones (*mc*). **B** *P. wallichiana*, long shoot bearing second year female cones (*fc*). (After Konar 1960)

the cones remain on the tree for several years and open only when scorched by forest fire.

Anatomy

Root.   In a mycorrhizal root the rootlet is enclosed by a fungal sheath, and the fungus is restricted to its cortex in a net of hyphae called Hartig's net (Fig. 15.1.2 C). The piliferous layer and the root hairs are replaced and covered completely by a mantle of mycelium. In *P. radiata* the invading fungus lyses the middle lamellae of the cortical cells, separates them by mechanical action, and occupies the intercellular spaces (Foster and Marks 1966). In the early stages of symbiosis, deep intracellular hyphae occur in the cortical cells, which are later digested by the host tissue due to its increasing resistance, and the fungus remains primarily in the intercellular regions.

The young long roots (in transection) have a circular outline (Fig. 15.1.4 A). The epidermis consists of isodiametric cells, and several cells become filled with tannin. The cortex is broad, parenchymatous and the cells frequently contain starch. The endodermis consists of suberized cells, usually impregnated with tannin which gives a brownish-orange colour. Casparian strips are indistinct. There is a six- or seven-layered pericycle, and many cells contain tannin. The stele is generally diarch or tetrarch, occasionally pentarch. A resin duct is associated with each protoxylem (Fig. 15.1.4 A). The protoxylem consists of mostly scalariform or scalariform-pitted tracheids, the metaxylem of only pitted tracheids. The phloem has parenchyma, sieve cells and tannin cells, and abundant starch (occasionally tannin) is present in the pith cells.

Secondary growth sets in even before the primary tissues are fully differentiated. The cambium differentiates from the parenchymatous cells beneath the phloem (Konar 1963a). It forms (by repeated periclinal divisions) secondary xylem towards the pith and secondary phloem towards the cortex. In the region of the resin ducts, the cambium cuts off only parenchymatous cells, which results in broad xylem rays (Fig. 15.1.4 B). Later, with the development of more secondary wood, the rays become reduced to a single cell. Secondary xylem tracheids have bordered pits on their tangential and lateral walls. In older roots tyloses block the tracheids. The rays are uni- or multiseriate, the latter associated with a resin duct. The secondary phloem consists of radially oriented rows of cells.

Simultaneously with the differentiation of the vascular cambium, a cork cambium is formed in the outer region of the pericycle. It cuts off cork cells externally and parenchyma internally. The cork cells may be highly suberized and cutinized (*P. roxburghii*), and some of them may be filled with tannin.

Stem. Shoot Apex.   The apex of the long shoot remains dormant from

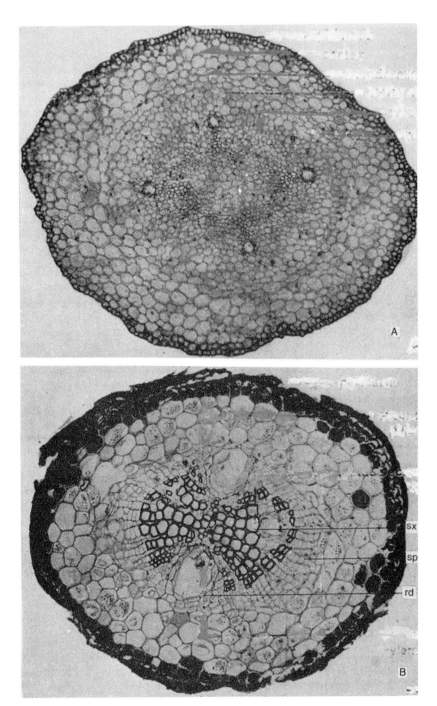

Fig. 15.1.4 A, B. *Pinus roxburghii*, transection root. **A** Young tetrarch root. **B** Old diarch root shows secondary growth. *rd* resin duct. *sp* secondary phloem. *sx* secondary xylem. (After Konar 1963a)

October to March. In *P. ponderosa, P. lambertiana* and *P. roxburghii*, the shoot apex is distinguishable into: (a) the region of apical initials, (b) the central mother cells, (c) the peripheral tissue zone and (d) rib meristem. During the dormant phase, the apex has a low parabolic dome averaging 72 μm in height and 277 μm in diameter. It resumes its activity in April, and by May attains a tall parabolic form. At this stage the needles produced in the axils of the scale leaves project beyond them. An active shoot apex is 175–300 μm in height and 305–500 μm in diameter. By the end of August, the phase of rapid growth ceases. The dwarf shoot apex is dome-shaped during the active period of growth and appears elongated and cone-like at maturity. There are four zones, highly telescoped, corresponding to those of the long shoot. The apical initials (by anticlinal divisions) contribute to the outer layer of the peripheral tissue zone, and by periclinal and oblique divisions to the rib meristem and the inner layers of the peripheral tissue zone.

A transection of a young stem, slightly below the apex, shows superficial ridges and furrows due to adpressing of the surrounding leaves. The epidermis is followed by a broad parenchymatous cortex, the vascular bundles are arranged in a ring, separated from one another by broad medullary rays. The resin ducts arise schizogenously, have an inner lining of epithelial cells and are arranged in a ring in the cortex. The pith is parenchymatous. Several cells of the cortex and pith contain tannin.

The interfascicular cambium arises in the region of the medullary rays and connects with its intrafascicular counterpart to complete the cambial zone, which is the entire area of undifferentiated dividing cells. The cambium forms two types of cells: (a) fusiform initials—these are elongated cells with tapering ends, and give rise to longitudinal or vertical systems of xylem and phloem; (b) ray initials—consist of isodiametric cells of relatively small size and give rise to the ray cells.

The vascular bundles increase in width due to tangential extension of the fascicular cambium. Further activity of the cambium produces a continuous zone of secondary xylem and phloem (Fig. 15.1.5. A). At places, the cambium cuts off parenchymatous cells which form the secondary medullary rays. These are mostly uniseriate, rich in cytoplasmic contents and have thick walls.

The phloem initial transforms directly into the sieve element (Srivastava 1963b). Occasionally, it may divide once transversely, the resultant cells develop into short sieve cells. The cambial derivative (destined to become a sieve cell) increases in cell volume accompanied by radial expansion. The secondary wall thickenings (characteristic of the family) are lamellate. The older cells have more lamellae in their walls than the younger ones. At this time, the cytoplasm has the usual complements of organelles. Rough ER is evident in the form of cisternae and vesicles and the ground substance is rich in free ribosomes. Occasionally, fine fibrillar material can be detected

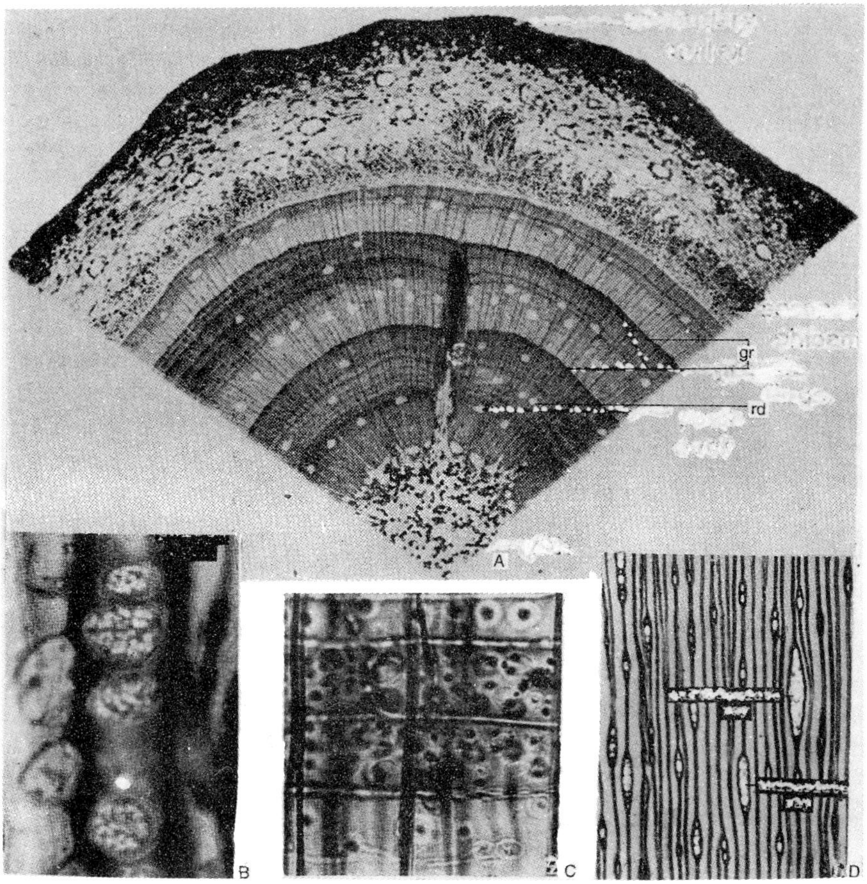

Fig. 15.1.5 A-D. *Pinus roxburghii*, stem. **A** Old stem in transection to show growth rings (*gr*) and resin ducts (*rd*). **B, C** Radial longisection. **B** Secondary phloem with lateral compound sieve plates. **C** Vascular ray parenchyma shows simple pits on radial walls. **D** Tangential longisection with uni- and multiseriate rays. (**A, C, D** After Konar 1963a, **B** after P. Maheshwari and Konar 1971)

in the vesicles of the rough ER. Mitochondria and dictyosomes are abundant and the cortex of the cell contains microtubules, which are usually parallel to the long axis of the cell.

The plastids also undergo changes. They accumulate a few electron-dense bodies in the stroma, some of these are 1.2 μm in older sieve elements (Srivastava and O'Brien 1966, Barnett 1974). These crystals, presumed to be proteinaceous, show weak refrigence under polarized light. Their origin and significance is not known, but they may develop from the osmiophilic intralamellar inclusions in the plastids of cambial initials. In *Pinus pinea*, Wooding (1966) has described, in the ray plastids, accumulation of granules comparable in staining to the starch grains.

During further differentiation, significant changes occur in the organelles and membrane system of the cell. The cisternae of rough ER vesiculate further, lose their ribosomes, and subsequently the vesicles disappear. Simultaneously, a new reticulum is formed, composed of large cisternae which mostly occur near the wall. Bouck and Cronshaw (1965) refer it as the sieve tube reticulum (STR) while Srivastava and O'Brien (1966) termed it sieve element reticulum (SER). While these changes take place in the ER, the ribosomes and dictyosomes (in the cytoplasm) disappear.

Evert and Alferi (1965) report necrotic nuclei in the mature (coniferous) sieve cells. Comparable stages of nuclei in the mature sieve elements have been identified in *Pinus strobus* (Muramanis and Evert 1966, Srivastava and O'Brien 1966), *P. pinea* (Wooding 1966, 1968) and *P. resinosa*. In the elements immediately adjacent to the crushed phloem, the vesicular masses, nuclear remnants, and mitochondria associated with longitudinally stranded material persist along the wall, and the lumen of sieve elements appears empty. The inclusion of the sieve element plastids disappears by the time it is mature, only granules, resembling starch in their staining behaviour, persist (Wooding 1966, Wooding and Northcote 1965). In *P. radiata* and *P. strobus* the crystalline inclusions change to fibrillar form.

A mature sieve element of pine is approximately of the same size as the cambial initial from which it originates. It has a thick lamellated secondary wall, separated from the cytoplasm by a persistent plasmalemma. The peripheral cytoplasm is composed of an elaborate system of smooth ER in the form of long cisternae. Somewhat abnormal mitochondria are often enmeshed between these cisternae. Plastids are present in various stages of breakdown, alongwith the starch grains and protein crystalloids. The ground substance is composed of a fine fibrillar material derived either from the stroma of degenerated plastids, or from the rough ER, or both. Degenerated nuclei are mostly present, but the vacuolar membranes, dictyosomes, and ribosomes are absent. Microtubules may be present at maturity but they are difficult to distinguish with certainty. Plasmalemma persists even in the advanced stages of sieve-element differentiation. Barnett (1974) has recorded stacks of hexagonally packed tubules in the lumen of sieve elements of *P. radiata*.

Initiation of the sieve area has been noticed soon after the cell wall formation has begun. The formation of sieve pores needs to be explained, whether the sieve areas arise from the pit fields in the fusion initials from which the sieve cells have been derived, or whether their origin is de novo. According to Srivastava and O'Brien (1966), some boring of the wall occurs because a cavity median nodule is formed where none existed before. Furthermore, strands from the two sides may join. The "pores" in the walls of dead sieve elements are much wider than the plasmodesmata. Barnett (1974) did not observe simple pits in the radial wall of the cambium in *P. radiata*. Gradually, the wall thickens in these regions. In an almost mature sieve area, the difference in the structure between the cell wall and the material of which the cell plate is made is more obvious. Usually, a sieve plate is homogeneous, stains evenly, while the remaining cell wall is lamellar in structure. This is the stage when communication channels begin to develop between the cells. Later, they widen enough for entire organelles to pass through them. The pores in the wall are lined by plasmalemma (Srivastava and O'Brien 1966), Wooding (1966) observed these as lined by callose (as revealed by fluroscence microscopy).

The sieve elements have compound sieve plates with either individual or small groups of (two or more, very closely placed) sieve areas (Fig. 15.1.5 B) on the radial walls and consist of numerous narrow channels (0.1–0.5 μm in diameter) lined with callose. These channels widen and coalesce to form a central lacuna in the middle of the wall. The sieve elements are in close contact with albuminous cells through a rather specialized compound pore. The sieve element towards the pore is similar to a half-sieve area, but the channels connecting the central lacuna with the albuminous cells are much narrower.

The xylem comprises tracheids which have bordered pits on their radial and tangential walls. Frequently, bars or rims of sanio are present between the pits. Every year, the activity of the cambium reaches its peak during spring. The xylem elements produced in this season are broader, thin-walled and somewhat polygonal. With the onset of summer, the activity slows down and the diameter of elements becomes appreciably smaller, they are thick-walled, narrow, squarish and compact. The summer wood passes rather abruptly into the spring wood, which results in the formation of growth rings (Fig. 15.1.5 A). In *P. halepensis* (see P. Maheshwari and Konar 1971) the number of circles in one ring of spring or autumn wood reflects the climatic conditions of a given year much better than the anatomical features of the tracheids. The constant presence of water table under the roots also leads to a very low increment in wood thickness.

Resin ducts are distributed all along the wood (Fig. 15.1.5 A) and reaches maximal number just before the onset of summer.

In the secondary xylem, the rays are differentiated into two distinct zones. In the central zone the cells are similar and have only simple pits on their walls (Fig. 15.1.5 C). On either side, one or two layers of cells form the marginal ray tracheids. They are thick-walled and rectangular with bordered pits on their radial and tangential walls. In a tangential section the medullary rays can be distinguished into two types: (a) uniseriate rays which are single-cell wide, the number (of cells) vary from 1–12; (b) multiseriate or fusiform rays which are always more than one cell wide and several cells in height, and have a resin duct in the centre (Fig. 15.1.5 D).

Simultaneously with the differentiation of the vascular cambium in the stelar region, a cork cambium arises in the first or second layer of the cortex. It divides periclinally and gives rise to the cork or bark on the periphery, and secondary cortex towards the centre. By intermittent anticlinal divisions it increases the number of radial rows and enables the periderm to keep pace with the increase in circumference of the stem. The scaly bark thickens gradually, cracks externally and ultimately wears away.

Compression Wood. It is formed along the under-surface of branches, on the trunks of leaning or fallen living trees. It apparently forms in response to geotropic stimuli and may also be due to the concentration of auxin, as shown in *P. strobus* by Wershing and Bailey (1942).

Dwarf Shoot. Anatomically, the dwarf shoot resembles a long shoot except for its narrow diameter. It has a small cortex, few resin ducts, and the vascular bundles are open and collateral. The meduallary rays are broad and parenchymatous. The tracheids are mostly scalariform, occasionally pitted. A large parenchymatous pith has many cells filled with tannin.

Leaf. A cross section of needle (Fig. 15.1.6 A) shows an epidermis of isodiametric cells with lignified walls. It is discontinuous in the region of the stomata, which are sunken (Fig. 15.1.6 B) and arranged in axial rows all along the surface of the leaves.

The hypodermis is two- or three-layered, of polygonal or rounded cells of uniform thickness.

The mesophyll, the bulk of the leaf, is delimited on its inner side by the endodermis (Fig. 15.1.6 A). It consists of closely fitting chlorenchymatous cells with a number of plate-like infoldings projecting into the cell cavity (Fig. 15.1.6 C). Two or more resin ducts are interspersed in the mesophyll tissue. Depending on their position, the resin ducts are classified into: *external* (touching the hypodermis), *internal* (touching the endodermis), *medial* (touching neither hypodermis nor epidermis), and *septal* (touching both hypodermis and endodermis to form a septum).

The stele occupies the centre of the leaf. The prominent endodermis (Fig. 15.1.6 D) consists of a single layer of barrel-shaped cells, followed by the transfusion tissue present all round the bundle with a larger amount on the abaxial side (Hu and Yao 1981). There are several tracheidal cells (Fig. 15.1.6 E) interspersed with parenchyma and albuminous cells.

The number of vascular bundles in a needle may be one (*P. sylvestris* var. *monophylla, P. wallichiana*), two or more (*P. roxburghii, P. contorta*). When two bundles are present, they are disposed at an angle to each other (Fig. 15.1.6 D). The mature leaf has a partly crushed protoxylem. The metaxylem has helically thickened tracheids with bordered pits. The radial rows of xylem elements are interspersed with rows of parenchymatous cells. The phloem parenchyma is more abundant than xylem parenchyma. Many of its cells contain starch, or proteins, or crystals.

Structure and ontogeny of the stomatal complex has been studied in

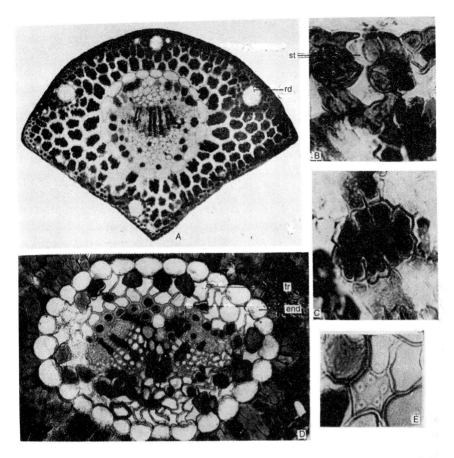

Fig. 15.1.6 A-E. A *Pinus gerardiana*, needle in transection. *rd* resin duct. **B-E**
P. roxburghii. **B** Epidermis shows sunken stoma (*st*). **C** Mesophyll cell with infoldings
of the wall. **D** Vascular bundles. *end* endodermis. *tr* transfusion tracheid. **E** Transfusion
tracheid. (After Konar 1963a)

P. strobus and *P. banksiana* (Johnson and Riding 1981). The stomatal complex
is eight-celled: two guard, two polar, two lateral and two hypodermal
subsidiary cells. The stomatal complex (in these two pines), therefore, cannot
be readily classified as haplocheilic because a polar subsidiary cell arises
from the same protodermal cell as the guard cell mother cell. A modification
of the classical concept of stomatal development is necessary to describe
stomata as eumesoperigynous (Johnson and Riding 1981).

Reproduction

Male Cone

A male cone consists of a number of spirally arranged microsporophylls
with upturned scaly apices (Fig. 15.1.7 A-C). Two microsporangia are

borne on the abaxial side (Fig. 15.1.7 B-C) and dehisce by a longitudinal slit.

Microsporangium. A transection of young microsporangium shows a mass of meristematic tissue surrounded by an epidermis. One, two or more cells of the hypodermal region, by repeated divisions, give rise to an archesporial tissue which has dense cytoplasm and prominent nuclei (Fig. 15.1.7 D). The peripheral cells of the archesporium undergo periclinal divisions and cut off the primary parietal layer towards the periphery (Fig. 15.1.7 E) and sporogenous cells on the inside. Further periclinal divisions in the parietal layer forms an outer three- or four-layered wall and the inner mass of sporogenous tissue (Fig. 15.1.7 F). The innermost layer of the wall cells develop into the tapetum (Fig. 15.1.7 G). Simultaneously, the epidermal cells undergo anticlinal divisions; most of the cells are filled with tannin except two rows of smaller cells, which form the line of dehiscence (Fig. 15.1.7 G). At the time of dehiscence of the microsporangium, the epidermis attains its maximal development with fibrous thickenings, and functions as the exothecium (Fig. 15.1.7 H). As the microspore mother cells initiate divisions, the epidermal cells lose their nuclei, become filled with a homogeneously staining substance which may extend (*P. strobus*) to the subepidermal layer of the wall. The outer wall of the epidermal cells gradually becomes cutinized.

The tapetum is of the secretory type in *P. sylvestris* (Willemse 1971b), *P. banksiana* (Dickinson and Bell 1972, 1976a) and *P. roxburghii* (see Moitra and Bhatnagar 1982). Young tapetal cells are richly cytoplasmic and multinucleate. They become very conspicuous during meiosis in the microspore mother cells, and degenerate soon after the spores are released from the tetrads. The tapetum is involved in the nourishing of the sporogenous cells/young microspores, and in the formation of exine on the spores (Dickinson 1970, 1971, Dickinson and Bell 1972, 1976a, b, Willemse 1971d).

The development of tapetum and sporogenous cells takes place simultaneously. There appears to be a correlation between the stage of the microspore mother cell (meiocyte) and the structure of tapetal cytoplasm. Accordingly, three phases (with subphases) of development of tapetal cell can be distinguished corresponding to pre-meiotic, meiotic and post-meiotic phases of microsporogenesis.

Pre-meiotic Changes. In *P. banksiana* the development of tapetum has been studied in detail (Dickinson and Bell 1976a). In a very young microsporangium, ultrastructurally the tapetal and sporogenous cells are alike. The membrane investing the plastids is difficult to distinguish and the contents of the organelles appear in continuity with the ground cytoplasm. Later, the envelope on the plastids becomes distinct, the number of dictyosomes and associated vesicles increases, and most of the starch is

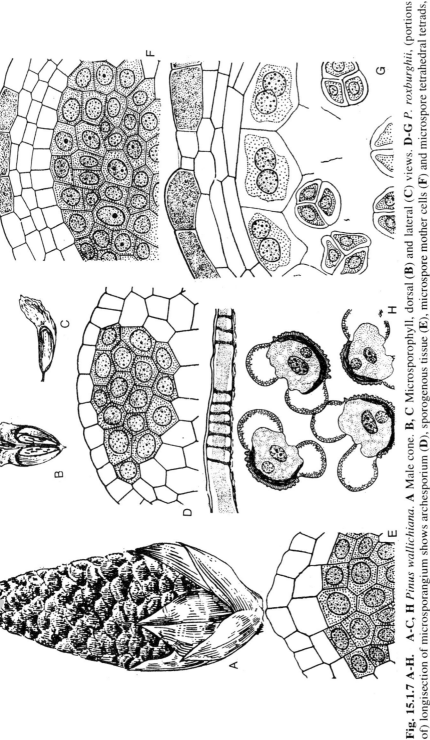

Fig. 15.1.7 A-H. A-C, H *Pinus wallichiana*. A Male cone. **B, C** Microsporophyll, dorsal (**B**) and lateral (**C**) views. **D-G** *P. roxburghii*, (portions of) longisection of microsporangium shows archesporium (**D**), sporogenous tissue (**E**), microspore mother cells (**F**) and microspore tetrahedral tetrads, and 2-nucleate tapetum (**G**). **H** Microsporangium (transection) shows fibrous thickenings in the epidermis, and 4-celled (2 prothallial cells have degenerated) pollen grains. (A-C, H After Konar and Ramchandani 1958, **D, E** after Konar 1960, **F, G** after P. Maheshwari and Konar 1971).

lost from the plastids (Fig. 15.1.8 A). These vesicles, which look "coated", may also surround a mass of fibrogranular, electron-opaque material. Small portions of rough endoplasmic reticulum (RER) appear. The protoplast of tapetal cells begins to contract from the fibrous PAS positive cell walls. Small quantities of a fibrous material accumulates between the contracting protoplasm and the original cell walls. From this stage onward, the tapetal and the sporogenous cells become distinct from each other. A large accumulation of RER and other associated coated vesicles become evident in the tapetal cells. The mitochondria in tapetal and sporogenous cells increase in size and frequency, and become more conspicuous (Fig. 15.1.8 A). The tapetal cells become radially compressed, and gather considerable RER at the end of this phase.

Meiotic Changes.   The tapetal cells increase in volume at the onset of meiosis. The RER continues to accumulate until the periphery of the tapetal cells is loaded with these membranes. The ribosomes are abundant in the form of polysomes only. Numerous large sudanophilic globules ca. 250 μm in diameter are visible. The initially strongly PAS-positive tapetal cell walls begin to loosen, with a concomitant increase in thickness. Osmiophilic globules arise outside the tapetal protoplast on middle lamellae between the outer tangential wall of the tapetal cells and the inner tangential wall of the innermost wall layer. These globules increase gradually from 0.1 μm to 1.0 μm in diameter. Occasionally, they adhere to the middle lamella between the radial walls of the tapetal cells. The globules continue to be deposited until a thin layer surrounds the pollen loculus. The tapetal cell walls show no change during the second half of meiosis.

Post-Meiotic Changes.   During the early tetrad stages, the electron density of the cytoplasm decreases. The tapetal cells become active again as the pollen wall formation begins. The Golgi bodies produce numerous vesicles, about 60% of each tapetal cell is filled with vesicles/inflated cisternae. The accumulation of RER continues. It is possible that sporopollenin precursors are stored or synthesized in these inflated cisternae and later passed on to the thecal fluid through these vesicles. The deposition of sporopollenin starts on the peritapetal membrane (described later) and orbicules. The peritapetal membrane completely invests the tapetum and developing pollen grains in *P. roxburghii* (see Moitra and Bhatnagar 1982). It is acetolysis-resistant and can be dissected as a bag containing pollen grains. During pollen wall formation, the tapetal cytoplasm again becomes osmiophilic and highly vacuolate. In *P. sylvestris*, Golgi bodies stop vesicle production (Willemse 1971b). At the tetrad stage, radial walls between some of the tapetal cells break down to form coenocytic cells. Tapetal cells degenerate soon after the release of microspores from the tetrad (Willemse 1971d, Dickinson and Bell 1976b), but the deposition of sporopollenin continues.

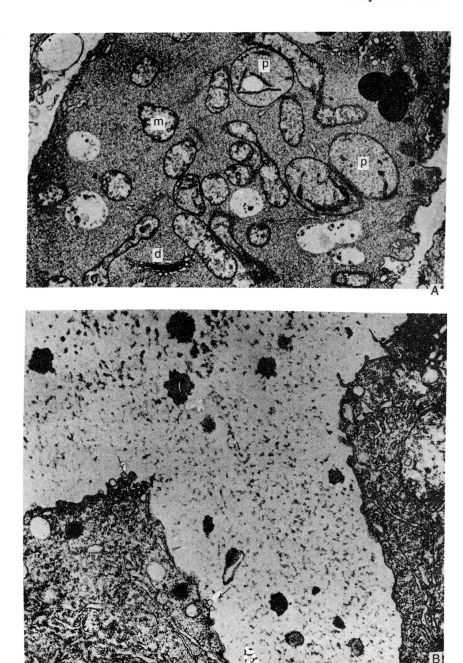

Fig. 15.1.8 A, B. *Pinus* sp., tapetum electron micrographs. **A** Tapetal protoplast in premeiotic stage shows conspicuous dictyosomes (*d*) and a starch grain in one of the plastids (*p*). *m* mitochondrion. **B** Tapetal protoplast during meiosis, note the deposition of sporopollenin on the lipid cores of orbicules and membrane-bound vesicles (arrows) discharged from the tapetal cytoplasm. (After Dickinson and Bell 1976a, b)

The cell organelles become swollen and show electron-transparent contents. After the young microspore stage, the tapetal cells disappear completely and the orbicules adhere to the peritapetal membrane.

A detailed developmental sequence of peritapetal membrane, at ultrastructural level, has been investigated in *P. banksiana* (Dickinson 1970, Dickinson and Bell 1972). The membrane develops between the innermost layer of the anther wall and the tapetum. Globules of saturated lipids (as shown by staining reaction) accumulate (at this site) and fuse to form a continuous outer electron-luscent layer, which coats the entire periphery of the loculus. A second layer of sporopollenin is deposited on the lipid substratum after the release of microspores from the tetrads (Fig. 15.1.8 B). This double-layered structure is called a peritapetal membrane. Deposition of sporopollenin is first detected as an accumulation of electron-opaque material upon the earlier-formed membrane. A large number of vesicles between the peritapetal membrane and tapetal protoplast persist even after the release of the tetrads from the callose wall.

The tapetal nuclei in younger sporangia show a higher Feulgen stainability than in older sporangia. In many tapetal cells the nuclei undergo regular mitotic divisions followed by inhibited cytokinesis which forms binucleate (rarely multinucleate) cells.

The tapetum generally has a higher RNA content than the microspore mother cell. There is a gradual increase of RNA synthesis in its cells from the time they are formed till the onset of meiosis in microspore mother cells. Thereafter, it declines sharply until final dissolution of tissue. In *P. banksiana* (Dickinson and Bell 1976a) the accumulation of proteins in the tapetal cells occurs from its initiation to the meiotic divisions in mother cells. This is associated with an increase in ribosomal population during the same phase.

In young microsporangia of *P. roxburghii*, there is a difference in staining pattern of the cell walls of sporogenous and tapetal cells and of the other wall layers. The cell walls of the wall layers are thicker and stain a brighter magenta by periodic acid-Schiff's (PAS) reaction, than the cell walls of tapetum and sporogenous cells. The epidermal and middle layer cell walls are largely cellulosic, while those of tapetal and sporogenous cells are mostly pectinaceous (see Moitra and Bhatnagar 1982).

There is a definite pattern of starch distribution in the microsporangium. Minute starch grains appear at the archesporial cell and are absorbed. They reappear in the sporogenous cells immediately prior to meiosis. In tapetal cells starch appears only after it appears in the sporogenous cells. During meiosis starch shows a very regulated pattern of distribution. At metaphase I the grains accumulate at the equatorial plate, and move to the opposite poles during anaphase I and telophase I, and are equally distributed at the end of meiosis.

Throughout their development, the sporogenous cells are rich in RNA which becomes maximal in meiocyte prior to meiosis. The young microspore is poor in RNA but the level rises again at the time of formation of the antheridial cell (see Moitra and Bhatnagar 1982). The archesporial and, later, the sporogenous cells become rich in proteins. The level is high in meiocytes but low during meiosis and, subsequently, rises during microspore maturation (*P. banksiana*—Dickinson and Bell 1976a, b).

Microsporogenesis. The differentiation and maturation of the sporogenous tissue in *Pinus* starts in the centre of the sporangium and proceeds centrifugally. Electron microscopic studies have revealed many more details of the basic course of development than is shown by light microscopy.

The meiocytes undergo a gradual stretching, thinning, and loss of PAS stainability before meiosis. With the dissolution of their walls, a thin layer of callose develops between the cell wall and the plasma membrane. It gradually thickens with the advance of meiosis.

Meiosis is asynchronous, and prophase I to telophase II are seen in adjoining meiocytes.

During meiosis, several changes occur in the cell organelles. EM studies reveal that the structure of nuclear pore in microspore mother cells (*P. sylvestris*) is very complex and remains constant throughout microsporogenesis. The annulus (made up of eight circular regions) and the centre of the nuclear pore are electron-dense (Willemse 1971a, b, c). At the beginning of meiosis, the evaginations (formed during premeiotic stages) of the nuclear membrane disappear; the large nucleus shows homogeneous nucleoli and the chromatin begins to shrink. The chromosomes comprise typical electron-transparent chromatin fibrils, covered by a membrane-like boundary. The nucleus is homogeneous till diplotene, but later it seems to produce granules which disperse in karyoplasm and mix with the cytoplasm after telophase I . The granules are the precursors of ribosomes and polysomes. The nucleolus disappears during diakinesis. The nuclear membrane shows invaginations during the young tetrad stages (Dickinson and Bell 1970b, Willemse 1971c), and transports chromatin-like and other materials into the cytoplasm. This peculiar phenomenon is connected with "primexine" formation (described later) in the spore tetrads.

During anaphase and telophase II, the Golgi bodies become active and produce a variety of vesicles which become arranged in the region of the future wall, and fuse with each other. Microtubules appear perpendicular to the fusing vesicles (Willemse 1971b). A very thin line appears in the direction of the cell plate in the centre of the fused vesicles. It is replaced by a fine fibrillar material, which is eventually taken over by the electron-transparent callose.

Following meiosis, wall formation is simultaneous, callose grows centripetally, separating the four nuclei, and tetrahedral and isobilateral tetrads are formed (Fig. 15.1.9 A-F). There are no plasmodesmatal connections

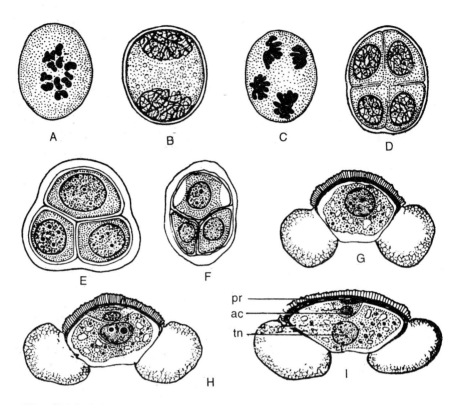

(**Fig. 15.1.9 A-I.** *Pinus roxburghii*. **A-H** Development of microspore and male gametophyte. **I** Pollen grain at shedding. *ac* antheridial cell, *pr* prothallial cell, *tn* tube nucleus. (After P. Maheshwari and Konar 1971).

either between the dividing mother cells, or the microspores in a tetrad. Callose has low permeability, isolates the mother cells and their products from the circumjacent influences, and the meiocytes follow an independent course of development (during meiosis).

The development of the pollen grain wall has been intensively investigated, mainly by using EM, and described in *P. sylvestris* (Willemse 1971c.) and *P. banksiana* (Dickinson 1971). The outermost layer, sexine, is laid down while the spores are still within the tetrad and callose covering. The Golgi bodies excrete small unit membranal vesicles (ca. 0.1 μm) with granular electron-dense contents, and coalesce during transportation towards the plasmalemma. The unit membrane of the vesicles fuses with the plasmalemma, secretes its cellulosic contents outside, but within the callose layer. It is the laying down of primexine. In *P. sylvestris* the granular content of these vesicles changes into fine fibrillar network. The sexine pattern is determined by placement of the vesicle content on the plasmalemma. A space forms between the plasmalemma and the callose wall, and is much enlarged in the region where subsequently air sacs develop. The plasma membrane shows many protrusions. Tectum and baculae are absent in the region between sacci where the germinal pore of the pollen develops later. In *P. banksiana* also sexine in not formed in the germinal pore region.

The next layer laid down is nexine I, formed by the deposition of sporopollenin on trilaminar tapes (at least five layers) present in the inner part of the primexine during the tetrad phase. In *P. banksiana* it is formed by the thickening of the plasma membrane and the appearance of a new membrane under it. This process is repeated until several layers develop. In *P. sylvestris* electron-dense tapes have been observed lying along the plasma membrane, and extending deep into the cytoplasm. Lipid granules form the primary source for nexine I and the tapes probably originate from plasma membrane or endoplasmic reticulum. The nexine surrounds the cytoplasm except at the regions where wings develop and is thicker on the side of germinal pore than on the other side.

The callose degradation begins after the formation of nexine I. The wings enlarge in volume due to an increase in the sexine layer. The sporopollenin in the thecal fluid and the tapetum are continuously deposited on the sexine.

The development of air sacs has been traced in *P. banksiana* using light, phase contrast, electron microscopy and histochemical techniques (Dickinson and Bell 1970a). They begin to form within the tetrad as the primexine is laid down between the plasma membrane and the callose layer. The sacci develop in an electron-transparent space containing PAS-positive fibrillar material. The callosic covering is conspicuous at this stage. The second phase of sacci development begins when microspores are released from the tetrad. The PAS-positive materials between sexine and nexine of the early pollen wall expands several times. The mechanism of expansion is not well understood. Lipid appears to be the main structural component of the wings as they collapse after lipid extraction. The wings are autofluorescent in *P. roxburghii* (see Moitra and Bhatnagar 1982).

The intine is the last layer to be laid down in *P. banksiana, P. sylvestris* and *P. roxburghii.* It is thinnest at the proximal end (the region of prothallial cells) and thickest at the distal end of the grain. In *P. sylvestris* the intine is lamellar. In *P. roxburghii,* a thin layer of callose (callosic outer intine) starts developing at the two sites where (later) the wings are formed. It spreads and covers the proximal (prothallial) end of the pollen grain. The inner intine is made up of cellulose and pectin (see Moitra and Bhatnagar 1982).

The nuclear envelope in a mature pollen grain consists of two membranes with occasional pores (ca. 0.1 μm in diameter). In *P. banksiana* and *P. sylvestris* finger-like invaginations (involving both layers of the nuclear envelope) extend into the nucleoplasm. They appear when the microspores are still within the tetrad. At the distal end of these invaginations, structures similar to nuclear pores have been observed. Fibrillar material, interpreted as chromatin, fills up the invaginations, which are discharged into cytoplasm, and the nuclear envelope becomes regular once again. In addition, in the outer nuclear membrane, evaginations are present which extend into the microspore cytoplasm. The latter becomes filled with vesicles (ca. 1 μm in diameter) surrounded by unit membrane. Nuclear pores occur at unvaginated portions only. Chromatin-like material is associated with invaginations in the nucleoplasm, and not with the evaginations in the cytoplasm. The invaginations, evaginations, and the formation of Golgi complexes are correlated with the production of the sexine layer.

Male Gametophyte. The uninucleate microspore/pollen grain is the first cell of the male gametophyte (Fig. 15.1.9 G). In all its divisions inside the microspore, the nucleus divides equally, but the cytoplasm becomes unequally distributed.

As the intine is being laid down, the divisions of microspore nucleus lead to the formation of two prothallial cells. Both the prothallial cells are ephemeral, and their remnants become embedded in the intine (Fig. 15.1.9 H, I). After the formation of prothallial cells, the antheridial initial divides, giving rise to a small antheridial cell, which remains attached to the intine at the site of prothallial cells, and a large vacuolate tube cell with a conspicuous nucleus (Fig. 15.1.9 I). The electron density and organelle distribution of the antheridial and tube cells are similar (Dexheimer 1969, 1970), but there is a cytochemical difference in the amount of DNA, RNA and proteins (see Moitra and Bhatnagar 1982).

From the end of April (at Simla, Mussoorie, Chakrata—Northern Himalaya, India, 2000–3000 m) to the beginning of June, *P. wallichiana* pollen grains are shed at the four-celled stage, i.e. two prothallials, antheridial and tube cell (Fig. 15.1.9 I). The pollen germination is arrested and further development takes place on the nucellar apex.

Female Cone

A young cone is small, elongated to spherical, and green in *P. roxburghii* and maroon in *P. wallichiana*. Each cone consists of 80–90 ovuliferous scales in the axil of bract scale, two together are termed seed-scale complex. They are arranged spirally on the cone axis (Fig. .15.1.10 A-C). Independent vascular traces supply the ovuliferous and bract scale (Fig. 15.1.10 C); in the former it is inversely oriented. The number of seed-scale complex varies with the species. Those present at the base and apex of the cone are sterile. An ovuliferous scale arises in the axil of a bract scale, which encloses it till the time of pollination. It bears two ovules on its dorsal surface, their micropyles face the cone axis. Later, the ovuliferous scale outgrows the bract scale (Fig. 15.1.10 C-E).

Ovule. The ovule is unitegmic. The integument is free from the nucellus

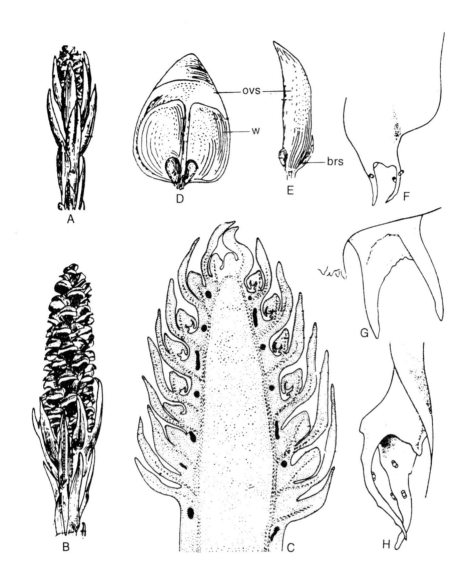

Fig. 15.1.10 A-H. A, B, D, E *Pinus wallichiana*. **A, B** Young female cone, **A** before and **B** at the time of pollination. **C** *P. roxburghii*, longisection young female cone. **D, E** Adaxial (**D**) and lateral view (**E**) of megasporophyll bearing ovules. *brs* bract scale. *ovs* ovuliferous scale. *w* wing. **F-H** Ovules in adaxial (**F**), abaxial (**G**) and lateral (**H**) views of cone scales. The micropylar canal is two-pronged, note pollen grains on the stigmatic prongs. (**A-E** After P. Maheshwari and Konar 1971, **F-H** after Doyle and O'leary 1935)

except at the chalazal end, and forms a symmetrical tube well beyond the level of the nucellus (Fig. 15.1.10 F-H). Adaxially, the edges of the integument extend into two long arms, which curve inwards before pollination, outwards during and curve back after pollination and finally dry up. There is no vascular supply to the ovule (Konar and Ramchandani 1958).

Megasporogenesis.   The archesporial cell becomes distinguishable while the female cone is still covered by the scale leaves. It differentiates at the broad apical end of the nucellus (Fig. 15.1.11A), and divides transversely to give rise to primary parietal and primary sporogenous cell. The former undergoes both vertical and transverse divisions, so that the sporogenous cell is pushed deep into the nucellus, and later functions as the megaspore mother cell (Fig. 15.1.11 B, C). Starch grains accumulate at the chalazal end of the megaspore mother cell (*P. sylvestris*). The latter undergoes meiosis I and produces a dyad (Fig. 15.1.11D, E). Only a triad (Fig. 15.1.11F) is formed if the upper dyad cell does not undergo meiosis II, or a tetrad if both the dyads undergo meiosis II. In a triad and linear tetrad, the upper dyad cell and the adjoining megaspores or the upper three megaspores in a tetrad usually degenerate. The chalazal megaspore functions so that the development of the gametophyte is monosporic.

Three to five layers of cells round the functional megaspore become densely cytoplasmic with prominent nuclei. This is the spongy or nutritive tissue. In the ovules, if the megaspore mother cell, or functional megaspore, fails to develop, the adjoining cells of the spongy tissue enlarge simulating a cellular gametophyte. The spongy tissue comprises a definite zone of physiologically active cells which are concerned in the nutrition of the young gametophyte, especially at the resting stage. The cells of this zone contain abundant starch.

Female Gametophyte (First Period of Growth).   The functional megaspore enlarges and shows a large vacuole even before the nuclear divisions commence. It forms only a few free nuclei and apparently remains inactive for 8–9 months (first period of rest).

Pollination.   Each tree produces an enormous number of pollen grains dispersed by wind. The surrounding area becomes clouded by the yellow powder (also known as sulphur shower). However, only a few pollen grains reach the pollen chamber and develop further.

A pollination drop is secreted (possibly from the nucellus), at the flared-out tip of the integument. The fluid contains sucrose, glucose and fructose (McWilliam 1958). The secretion starts a few days after the female cones emerge out of the scale leaves, and the ovuliferous scales separate sufficiently to permit the free entry of pollen. Under high humidity and cell turgor, the secretion begins around midnight, and by early morning the micropyle dries up. The arrival of the wind–borne pollen at the micropyle is purely a chance phenomenon. The pollen is caught in the pollination drop, grains stick to the two-pronged "stigmatic" micropylar canal (Fig. 15.1.10 F, H), and "migrate" to the nucellar tip.

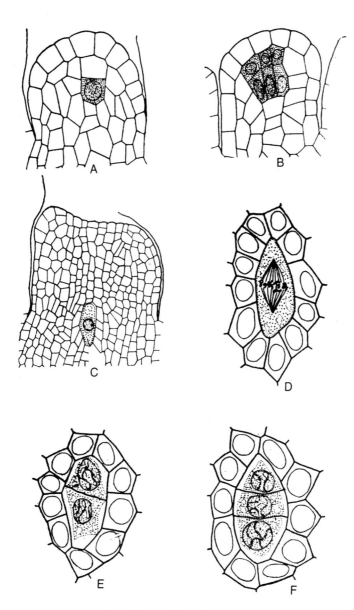

Fig. 15.1.11A-F. *Pinus roxburghii.* **A, B** Nucellus shows a primary parietal and a sporogenous cell in **A** and three parietal and three sporogenous cells in **B**. **C** Deep-seated megaspore mother cell. **D** Megaspore mother cell at metaphase. **E** A dyad. **F** Linear triad. (After P. Maheshwari and Konar 1971)

The pollen grain has a germinal furrow which closes in dry weather, but remains wide open in high humidity, or when in contact with the pollination drop. The pollen tube emerges through this furrow.

In *Pinus* the ovuliferous scale is "inverted" and the pollen grains are winged. Doyle and O'Leary (1935) and Doyle (1945) have studied the mechanism of pollination. According to Doyle, the wings orient the grain on the hanging pollination drop in such a way that the germinal pore/furrow of the grain faces the surface of the drop. This orientation is particularly necessary since the ovules are inverted. The pollen then floats upward (due to buoyancy caused by the air sacs) and reaches the nucellus with the germinal furrow faceing the nucellus. The pollen tube enters the nucellus without curving or twisting. McWilliam (1958) emphasizes that in *P. elliottii*, *P. nigra* and *P. wallichiana*, the stickiness of the neck and arm of the micropyle may be due to either a local secretion, or sugar residue resulting from the retreating micropylar fluid. This is an effective method for retaining the pollen at the site, and the prime mover of the grain is the active absorption of the fluid. McWilliams did not observe any preferred orientation of pollen on the nucellus.

At the time of pollination, the integument is four-to-five-layered, and soon after the pollen grains reach the nucellus, the micropyle closes due to a rapid division and enlargement of the cells of the inner layer of the integument. In *P. wallichiana* pollination takes place in May when the female cone contains a functional megaspore (Konar and Ramchandani 1958).

Post-Pollination Development of Male Gametophyte.   The pollen deposited on the nucellar apex germinates immediately and produces a short tube. Initially, the growth of the pollen tube is very slow as compared to the rapid development of the ovule. The apical part of the nucellus (penetrated by pollen tubes during the previous year) becomes thickened, and the cells lose their contents. The underlying cells are meristematic and rich in starch. Due to their growth, the tip of the nucellus with the pollen tubes is lifted above and away from the developing female gametophyte.

Towards the end of March or later, depending on the species (P. Maheshwari and Konar 1971), the pollen tube resumes active growth. In *P. sylvestris* the nucellar cells adjoining the pollen tube degenerate; the hemicellulose and pectin are affected first (Willemse 1968, Willemse and Linskens 1969). During this period, the antheridial cell in the pollen grain divides to form the stalk and body (spermatogenous) cell. However, there is much variation in the time of division, not only in different species but also in the same species. Willemse and Linskens (1969) observed that, following the division of antheridial cell (called generative cell by them), the stalk cell has fewer organelles as compared to the body (spermatogenous) cell. Also, the cytoplasm of the stalk cell is vacuolate, but that of the body cell is dense. The stalk and body cell migrate into the pollen tube. Soon after, the body cell divides into two male gametes. To begin with, both the male gametes are equal, but later, one of them enlarges considerably. These two unequal male gametes have a distinct cytoplasmic sheath around them. The male gametes are Feulgen-negative, stain faintly for RNA, and are rich in proteins (*P. roxburghii*). In *P. sylvestris* the cytoplasmic sheath contains plastids, mitochondria and ribosomes, and the ER remains coiled up like a watch spring (Willemse and Linskens 1969).

Female Gametophyte (Second Period of Growth). When the growth is resumed in the following year in spring, the free-nuclear divisions in the female gametophyte continue (Fig. 15.1.12 A-C). In *P. roxburghii* and *P. wallichiana,* when ca. 2500 nuclei have been formed, wall formation begins by alveoli (see Chap. 2) which grow centripetally (Fig. 15.1.12 D, E).

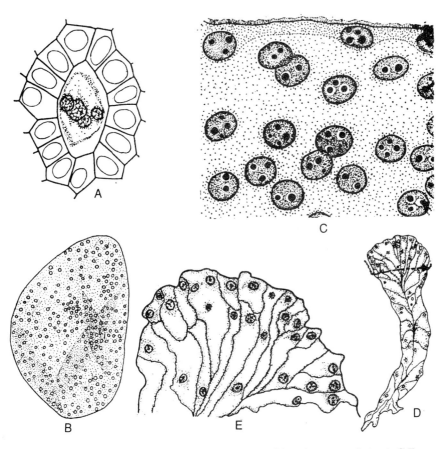

Fig. 15.1.12 A-E. *Pinus wallichiana,* development of female gametophyte. **A-C** Free-nuclear. **D** Gametophyte (longisection), note wall formation by alveoli. **E** Inset from **D**. (After Konar and Ramchandani 1958)

The additional tissues formed in the ovules after the resumption of growth are green. Both ovuliferous and bract scale become much enlarged and the end part of the ovuliferous scale is pushed up.

As the ovule increases in size, the gametophyte also enlarges and occupies the entire basal and central part of the nucellus. The densely cytoplasmic two- or three-celled thick spongy tissue surrounds the gametophyte. Due to its continuous growth, the nucellar cells become flattened. When wall formation is initiated in the free-nuclear gametophyte, the spongy tissue begins to disintegrate and is absorbed by the time the archegonia are ready for fertilization.

After the formation of the cellular gametophyte, the archegonia differentiate at the micropylar end. Occasionally, lateral and chalazal archegonia have also been observed. The archegonia occur singly, 1–7 in a gametophyte, the number varies with the species. A superficial cell of the gametophyte functions as the archegonial initial. It is larger and has a prominent nucleus. Its development (see Chap. 2) corresponds to the pattern in other gymnosperms (Fig. 15.1.13 A-F).

A mature archegonium has a multicellular neck, an egg cell, and a small ephemeral ventral canal cell. In *P. nigra*, the egg cell is 600 μm long and 300 μm wide, its nucleus is ca. 150 μm in diameter, and migrates to the centre of the cell.

Each archegonium has a single-layered jacket derived from the uninucleate cells with dense cytoplasm, adjoining the archegonial initial. In *P. palustris*, *P. gerardiana* and *P. pinea* the thick jacket wall has numerous pits on its inner wall, while in *P. wallichiana* and *P. roxburghii* the pits are poorly developed. The archegonium maintains contact with the circumjacent tissue through these pits. As the egg cell matures, the pit membrane between the egg and jacket cells breaksdown. Through this passage, the mitochondria, plastids, dictyosomes, portions of endoplasmic reticulum and even entire nuclei pass into the egg cytoplasm (Corti and Maugini 1964). The contents of the jacket cells are incorporated in the egg cell.

In *P. nigra* the ultrastructure of the egg cytoplasm has been investigated (Camefort 1959, 1960, 1962, 1965b, 1968). This study has invalidated the classical interpretation of the origin and nature of its inclusions. There are two types of inclusions:

a) Small inclusions, called granules by earlier workers. They are ca. 4 μm in diameter, and surrounded by a small ring or crescent form of vacuole.
b) Large inclusions, variously called paranuclei, vitellus, granules and proteid vacuoles. They are ca. 40 μm in diameter and structurally more complex (Fig. 15.1.14).

The small inclusions comprise cytoplasm and are partly enclosed in a simple membrane which is a portion of the envelope of the vacuole that caps the cytoplasmic nodule. These inclusions usually remain connected with the general cytoplasm by means of short peduncles. The enclosed cytoplasm is fairly granular with a scattered mass of osmiophilic granules, dictyosomes and mitochondria. The included cytoplasmic islet is situated within an electron-transparent ground substance which itself is separated from the general cytoplasm by double membrane.

The large inclusions arise by transformation of pre-existing plastids in

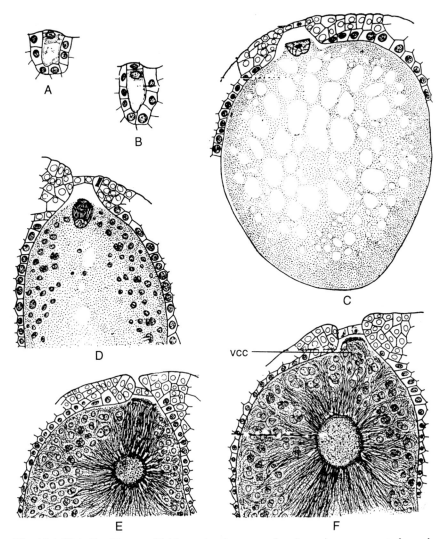

Fig. 15.1.13 A-F. *Pinus wallichiana*, development of archegonium. *vcc* ventral canal cell. (After Konar and Ramchandani 1958)

the young central cell. These are elongated and contain only a few lamellae. They become deformed while invaginating, and assume the form of a ring (in section) which encloses the islet of cytoplasm. This deformation is followed by others that completely change the structure of plastids to give them finally the general appearance of large inclusions. The plastid origin of these hypertrophied formations can always be distinguished by the presence of two layers around the various compartments. These layers separate the inclusions from the general cytoplasm.

Camefort (1962, 1965b) investigated the development, growth and maturation of the cytoplasm of the central cell/progamete in *P. nigra*. The

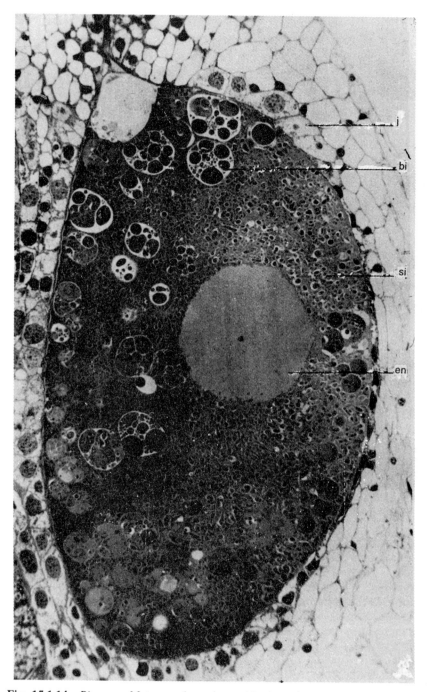

Fig. 15.1.14. *Pinus* sp. Mature archegonium with circumjacent gametophytic cel
(longisection). The dense egg cytoplasm contains numerous large (*bi*) and small (*s*
inclusions, while the contents of the egg nucleus (*en*) are relatively thin. *j* jacket cel
(After Camefort 1968)

"young" transparent cytoplasm contains mitochondria, plastids and dictyosomes, which are grouped at the periphery of the cell or near its nucleus. Numerous small vesicles tend to come together. Under the light microscope, the central cell shows a large number of vacuoles (foam stage; Fig. 15.1.14). The EM shows that these vacuoles are bounded by a simple membrane, which is often open, and the transparent content is continuous with the general cytoplasm and can enclose cytoplasmic organelles. During the growth of the cell, the vacuoles increase in number. The simple bounding membrane disappears and the contents become as dense as that of the general cytoplasm.

The cytoplasmic vesicles of the young central cell fuse and form layers of endoplasmic reticulum which caps the small inclusions. The plastids also become deformed before the vacuoles disappear. The general cytoplasm becomes increasingly dense. A mature central cell has fully differentiated protoplasm with abundant inclusions and a concentric layer of endoplasmic reticulum (as seen in cross section). The mitochondria are dispersed throughout the cell. They appear to have undergone structural regression (as shown by a reduction or disappearance of the tubules). The young central cell shows only a few mitochondria with short tubules. As the cell grows, the mitochondria become arranged in long, sinuous and branched chondrioconts (see Chesnoy and Thomas 1971). They appear constricted at this time, possibly fragment, which accounts for their increased number in the mature egg.

The nucleus of the egg cell (*P. sylvestris*, *P. uncinata*, *P. nigra*) shows chromosomes which consist of bundles of fibrils 100 Å in diameter (each 30 Å thick) made up of a light central line between two dark ones. These fibrils are parallel to one another and lie along the length of the chromosome. This particular arrangement appears to suit the generally inert female gametophyte before fertilization (Camefort 1964).

The nucleoli are formed of 50 Å thick fibrils. They may be distributed, tightly packed, in the dark zones of the nucleolus, or dispersed in the light zone. At the periphery of the nucleolus, the nucleolar substance is scattered either in the form of fibrils separating from each other, or as small muriform masses (0.1–0.03 μm in diameter) consisting of several granules of 170 to 200 Å. The latter indicates the beginning of a change from a fibrillar to a granular structure (Camefort 1964).

Fertilization

The pollen tube, just before fertilization, grows rapidly and reaches the neck of the archegonium (Fig. 15.1.15A, B). The tip has a dense cytoplasm, a large number of starch grains, tube nucleus, stalk cell and the two male gametes enveloped in a common cytoplasmic sheath. The pollen tube penetrates into the neck of the archegonium, and discharges its contents at the tip of the cytoplasm of the female gamete.

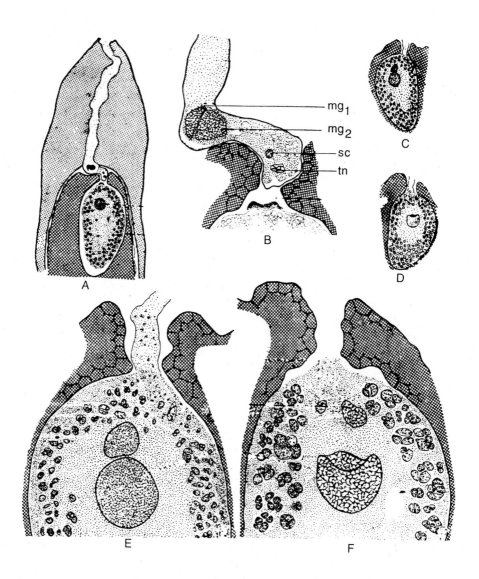

Fig. 15.1.15 A-F. *Pinus wallichiana.* **A** Longisection ovule shows pollen tube in the neck of the archegonium. **B** Pollen tube from **A**. mg_1, mg_2 male gametes. *sc* stalk cell. *tn* tube nucleus. **C-F** Stages in the fusion of one of the male gametes with egg. **E, F** Portion of archegonium from **C, D**. (After Konar and Ramchandani 1958)

One of the two male gametes moves towards the female gamete. The second male gamete, along with the stalk cell and tube nucleus, and the original male cytoplasmic sheath, persist at the apex of the egg and eventually degenerate. The functional male gamete is devoid of the cytoplasmic sheath as it moves down the archegonium to fuse with the female gamete (Willemse and Linskens 1969). The mitochondria and plastids brought by the pollen tube are morphologically different from those of the female gamete. They remain grouped in the upper part of the egg cytoplasm during the division of the zygote (Camefort 1965b, 1967).

The functional male gamete lodges in a depression on the female gamete (Fig. 15.1.15 C-F). As soon as contact is established, junctions between the nuclear membranes of the two gametic nuclei are established at several points. These areas of communication between the nucleoplasm gradually

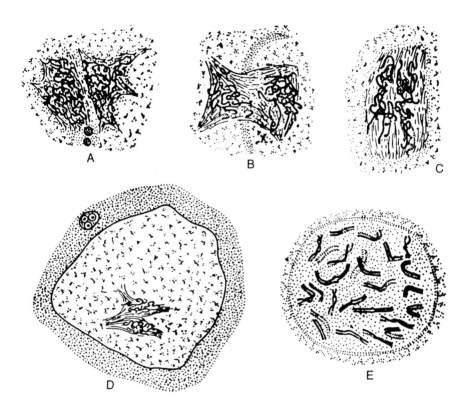

Fig. 15.1.16 A-E. *Pinus lambertiana.* **A** Male and female chromatin in separate groups in egg, each surrounded by spindle fibres to form a multipolar spindle. **B, C** Two chromosome groups on multipolar and diarch (**B**) and bipolar (**C**) spindles. **D** Zygote nucleus in mitosis, note separate male and female chromatin. **E** Polar view of transection through late metaphase shows diploid (24) chromosomes. (After Haupt 1941)

enlarge to form islets of cytoplasm surrounded by nuclear membrane. The latter slowly breaks down at the region of contact between the two nuclei and the two groups of chromosomes [which are initially quite distinct (Fig. 15.1.16 A)] converge and lie on a common spindle (Fig. 15.1.16 B). Finally, the chromosomes of the two gametes lose their identity (Fig. 15.1.16 C).

Embryogeny

Proembryo. At the time of mitosis of the zygote, the nuclear membrane separates the nucleoplasm from the surrounding cytoplasm (Fig. 15.1.17 A, B). Here and there the nuclear membrane is deeply invaginated into the nucleoplasm and collapses by the end of first mitosis. The nuclei of the coenocytic proembryo are formed within the nucleoplasm of the zygote (Camefort 1967) which progressively organizes into the cytoplasmic area of the embryo. It is essentially of nuclear origin and distinct from the cytoplasm of the egg (Fig. 15.1.17A-C) and is termed the neocytoplasm.

During the first embryonal mitotic division (*P. banksiana*), the two groups of chromosomes come to lie in one group at the equator of the spindle. In Fig. 15.1.16 D (*P. lambertiana*) the first embryonal mitosis shows separate male and female chromosomes. Each chromosome splits longitudinally, the daughter halves move to the poles. A transection through mitotic late metaphase shows (Fig. 15.1.16 E) polar view of diploid complement of 24 chromosomes (Haupt 1941). The two proembryonal nuclei divide once more and the four daughter nuclei migrate to the base of the proembryo and become arranged in one layer (Fig. 15.1.18 A, B). Initially, they are very small, as after every division there is a decrease in the size of the proembryonic nuclei; as they migrate to the base they enlarge considerably. All these nuclei divide with vertical spindles, resulting in eight nuclei which organize into eight cells arranged in two tiers of four cells each (Fig. 15.1.18 C-F). They are designated the primary embryonal tier (p^E) and primary (p^U) upper tier (Fig. 15.1.18 E, F). The cells of both the tiers divide transversely by internal division (nuclear division inside proembryo cells), so that four tiers of four cells each are formed. The lower two tiers comprise the embryonal tier (*E*) folllowed by disfunctional suspensor (*ds*) previously termed the rosette tier. The uppermost tier, *U*, is the upper derivative of p^U (Fig. 15.1.18 G, H).

After the proembryonal nuclei migrate to the base of the archegonium, the cytoplasm of the female gamete degenerates. The degeneration begins close to the neocytoplasm, extends gradually throughout the egg cell, and is completed by the time the proembryo has become cellular. A cytochemical study of the cytoplasm shows that there is a non-specific acid phosphatase activity in the endoplasmic reticulum formations enclosing the inclusions. This activity becomes intense in all the inclusions after the migration of the proembryonal nuclei to the basal pole of the archegonium (Camefort 1966, 1967).

Differentiation of Embryo. The upper four cells of the *E*-tier elongate and function as the embryonal suspensor (*Es*), and the lower four cells form the embryonal mass (Fig. 15.1.18 I). Subsequently, there are several layers of *Es*, which are termed Es_1, Es_2, Es_3 ..., formed by divisions in the *E*-tier. The *Es* cells elongate rapidly and thrust the terminal cells (embryonal cells) forward into the corrosion cavity (see Chap. 2). The latter enlarges and

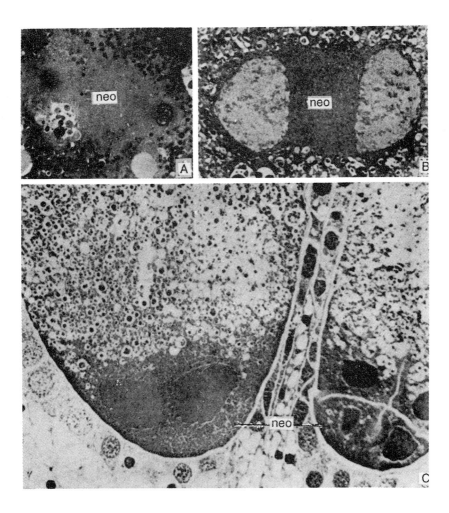

Fig. 15.1.17 A-C. Neocytoplasm differentiation in *Pinus*. **A** Zygote nucleus in telophase surrounded by archegonial cytoplasm. The nucleoplasm appears to form the ground substance of neocytoplasm (*neo*). **B, C** Progressive stages in the differentiation of neocytoplasm at two-, four- and eight- nucleate proembryo. The archegonial cytoplasm does not seem to be involved in the process. (After Camefort 1969)

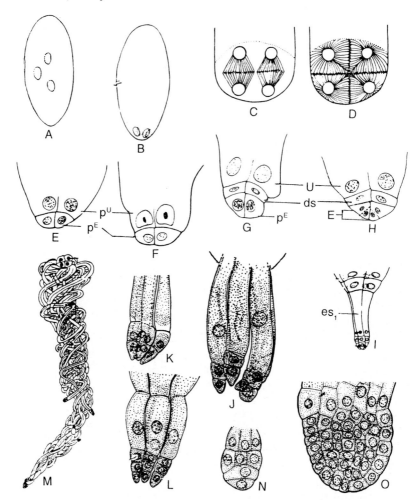

Fig. 15.1.18 A-O. Embryogeny. **A-D, I** *Pinus* sp., diagrammatic representation of the development of proembryo. **A, B** Four-nucleate proembryo, the nuclei are in the middle (**A**) and at the base (**B**) of the archegonium. **C, D** Wall formation at eight-nucleate stage. **E-H** *P. wallichiana*, lower portion of proembryo. **E** Eight-celled proembryo shows primary upper (*pU*) and primary embryonal (*pE*) tier. **F, G** Internal division in *pU* (**F**) and *pE* (**G**). *ds* disfunctional suspensor, *U* upper tier. **H** 16-celled proembryo. *E* Embryonal tier. **I** Formation of embryonal suspensor *es₁*. **J-O** *P. roxburghii*, development of embryo. (**A, B, I** After Buchholz 1929, **C, D** after Dogra 1967, **E-H** after Konar and Ramchandani 1958, **J-O** after P. Maheshwari and Konar 1971)

facilitates elongation of the *Es*. Later, as the elongation of *Es* exceeds the rate of enlargment of the corrosion cavity, the *Es* becomes coiled and twisted (Fig. 15.1.18 M). The elongation of the *Es* system pushes the growing embryonal mass into the central portion of the female gametophyte. As development proceeds further, the earlier suspensor system collapses, while newer *Es* are formed.

Polyembryony is common in *Pinus*. Additional embryos originate by simple polyembryony, i.e. from multiple zygotes, as well as due to cleavage polyembryony, i.e. a single zygote gives rise to multiple embryos by cleavage or splitting of the embryonal tier *E* into several embryonal units (Fig. 15.1.18 J-M). The separation into embryonal units occurs at Es_2 formation (see H. Singh 1978). These units remain unchanged during the earlier stages of Es elongation, but later they divide and form a multicellular mass. There is a period of embryonal selection due to competition between the four embryos from each zygote, and between the embryos from different zygotes. This phase of competition lasts for about 6 weeks from the time of fertilization (P. Maheshwari and Konar 1971). Generally, the deep-seated terminal embryo succeeds and develops further. The remaining embryos become arrested at different stages of development.

The embryonal mass develops from a single cell (Fig. 15.1.18 N, O). An apical cell with three cutting faces is formed during the earlier development, but it becomes indistinguishable soon afterwards.

The fully formed embryonal mass is a smooth paraboloid structure. It has a hemispherical apex at its distal end (Fig. 15.1.19 A, B), and a suspensor which is continuous with it at the proximal end. Further development at the proximal end gives rise to a well-developed root cap. It differentiates independently into a column and a peripheral region; the former does not contribute to the development of the latter (P. Maheshwari and Konar 1971).

As the root cap and root initials organize at the basal end, an incipient pith is recognizable in the hypocotyl between the epicotyl and the root initials. It is evident by the predominant transverse divisions, cell enlargement and vacuoliation. In *P. lambertiana* (Berlyn 1967) a pro-cambium differentiates at a lower level, following the formation of the pith. The cortex becomes demarcated soon after. The formation of pro-cambium is associated with the early phases of cotyledon formation though it is well developed before distinct cotyledonary primordia originate. Simultaneously, several long, uninucleate and multinucleolar secretory cells become distinguishable in the cortical region. Finally, 3–18 cotyledons and the shoot apex differentiate (Fig. 15.1.19 C, D). The cotyledons are traversed by the procambial strands, and show development of mesophyll. Generally, only one embryo matures, occasionally even two may reach maturity. The mature embryo has distinct epicotyl, root axis with remnants of the suspensors, and a hypocotyl-shoot axis bearing several cotyledons (Fig. 15.1.19 D).

The cells of the *ds* tier occasionally elongate and resemble suspensor cells. In some species it may undergo divisions to form a few to a mass of cells; they do not complete development, and very often abort.

Seed. The seed ripens within a few weeks without much change in the size of the embryo.

In the young ovule the integument is three-layered. It becomes six- or

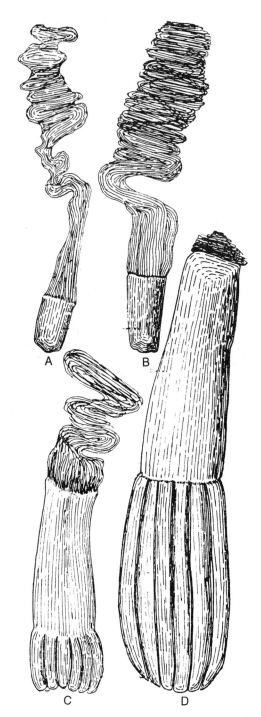

Fig. 15.1.19 A-D. *Pinus roxburghii*, progressive development of embryo. (After P. Maheshwari and Konar 1971)

seven-layered at the time of pollination (Fig. 15.1.20 A). There is an epidermis and hypodermis, the cells of the inner layers are richly cytoplasmic, have prominent nuclei and some of them are tanniniferous. The cells divide anticlinally and periclinally to form a 10–12-layered integument which differentiates into three distinct zones: outer fleshy, middle stony, and inner fleshy (Fig. 15.1.20 B, C).

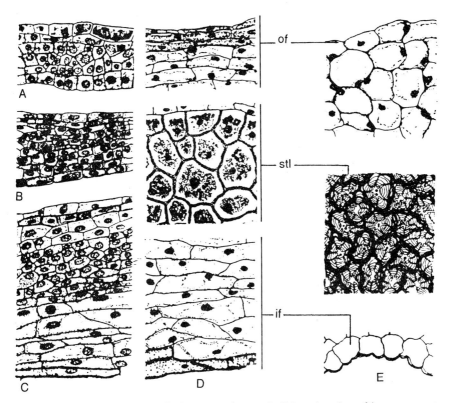

Fig. 15.1.20 A-E. *Pinus wallichiana*, seed coat. A-C Longisection of integument at the time of pollination (**A**), free-nuclear female gametophyte (**B**) and archegoinal stage (**C**). **D, E** Mature seed coat shows outer fleshy (*of*), middle stony (*stl*), and inner fleshy (*if*) layers. (After Konar and Ramchandani 1958)

The outer fleshy zone finally becomes seven- or eight-layered, the outer two or three layers contain a shiny granular material. At the time of seed-shedding, these cells become round with vacuolated cytoplasm and the nucleus shrinks (Fig. 15.1.20 E).

The middle stony layer differentiates when the ovule is at the archegonial stage. At maturity it comprises ca. 18–20 layers of pitted cells, the outermost layer has refractive granules. Lignification of the cells begins in the innermost layer and proceeds outward (Fig. 15.1.20 D, E).

The inner fleshy layer is seven or eight cells wide. The cells are thin-walled and elongated with scanty cytoplasm. At maturity only two or three layers persist and the rest are absorbed (Fig. 15.1.20 D, E).

In some species of *Pinus* (*P. roxburghii*, *P. insularis*, *P. balfouriana*, *P. strobus*), the seed is winged and the development of the wing is closely linked with the development of the seed coat. The latter extends into a wing which cannot be detached without injury. The wing is thin and papery or thick at the base.

Embryo development and the type of food reserve in endosperm has been studied by Hakansson (1956). Starch appears first around the apex of embryo when suspensor cells begin to elongate. There is no starch in the upper region around the empty achegonia while other parts of the gametophyte have scanty starch. Later, three zones can be demarcated in the endosperm (*P. sylvestris*): (a) the cells bordering the central cavity have no starch or only simple grains, (b) the middle zone has abundant starch, (c) the cells of the peripheral zone have dense contents (but no solid storage material). Later, the entire endosperm becomes packed uniformly with starch. The accumulated food reserves in the gametophyte are utilized at seed germination.

An ultrastructural study of the dry seed of *P. sylvestris* shows that the endosperm and embryo cells have the same components. Their different physiological and morphological role in the development of young seedlings needs further study (Simola 1974).

There is a cytochemical difference in the distribution of proteins and amyloplasts in shoot apex of the embryo in a dry seed of *P. banksiana* (Mia and Durzan 1974). The size and frequency of protein bodies and amyloplasts increase from the apical zone towards the flanks and subapical region.

Free sugars such as sucrose, stachyose, raffinose and 19 amino acids and 2 amides have been reported from the embryo and endosperm of *P. banksiana* (Durzan and Chalupa 1968). In the seeds of *P. thunbergii* Katsuta (1961) reported globulin, albumin and glutalin.

Hatano (1957) observed that pyruvic acid and alpha-keto acids are consumed during the first step of conversion of organic acid to amino acid.

The fat content in the seed of *P. roxburghii* is 31% of the total weight (Konar 1958). In *P. sylvestris* linoleic, oleic, palmitic and stearic acids and an unidentified component C have been reported (Nyman 1966).

Germination. The food stored in the seed breaks down during germination and is the only source of energy for the germinating embryo. Several changes take place after the seed has imbibed water. Mitochondria, dictyosomes and ER—at first absent (or present in small quantities)—appear and gradually increase (Durzan et al. 1971, Simola 1974). Small microbodies, interpreted as glyoxysomes, have been observed in *P. sylvestris* (Simola 1974), *P. ponderosa* (Ching 1970) and *P. pinea* (Lopez-Perez et al. 1974). These are the sites of fatty acid oxidation in oil-containing seeds.

Stachyose, rafinose and sucrose are the main sugar reserves in the seeds of *P. thunbergii* and *P. sylvestris* (see Konar and Moitra 1980). In the former, as germination proceeds, raffinose and stachyose are digested and sucrose, including glucose and fructose, appear. In *P. sylvestris*, small quantities of glucose and fructose, already present in dry seeds, rapidly increase during germination. In *P. strobus* the increase in the above sugars in the seedlings is due to photosynthesis. By using radioactive methods, it has been observed that of the total sugars formed sucrose is maximal followed by glucose and fructose together, while raffinose is minimal.

Almost all the seeds contain protein reserves as a source of N_2 for the young seedlings before they are able to absorb nitrogen by their roots. The breakdown of protein reserves in the endosperm and embryo and appearance of new proteins in other parts of the seedling takes place during germination. Durzan et al. (1971) and Hatano and Asakawa (1964) observed that reserve protein is high in basic amino acids. In the shoot apex of *P. radiata* (Riding and Gifford 1973) there is an initial breakdown and subsequent increase in protein content following seedling establishement. Fats and proteins are the major reserves in the mature seeds of *P. thunbergii* (Goo and Negisi 1952).

During germination, most of the protein reserve moves from the endosperm to the embryo, which shows a continuous increase in total nitrogen. With the germination of embryo in the seed, a sustained synthesis of proteins is indispensible and is accompanied by increased synthesis of RNA, which is triggered off within a few hours of imbibition of water.

For further details and information on seed germination, readers are referred to the excellent review by Konar and Moitra (1980).

The seed germinates 20–25 days after sowing. The radicle emerges and penetrates into the soil (Fig. 15.1.21A-D). The hypocotyl elongates, straightens, and carries above the ground the remnant of the seed alongwith the cotyledons. The cotyledons absorb nutrition from the gametophyte till it is exhausted, and then the seed coat drops off (Fig. 15.1.21E, F). The epicotyl, deep within the whorls of cotyledons, gives rise to the stem and leaves (Fig. 15.1.21 G, H). The cotyledons persist for a long time and are shed only after the juvenile leaves and the long shoot have grown considerably.

Chromosome Number

The karyotype analysis shows that *Pinus* has uniformly n = 12 and 2n = 24 chromosomes (Mehra 1988).

The chromosomes are numbered I-XII according to their decreasing size. Chromosome pair XII is heterobrachial in *P. gerardiana, P. grifithii, P. roxburghii* and *P. kesya*. In *P. kesya*, chromosome XI is also heterobrachial.

In *P. gerardiana* there is a secondary constriction in the proximal arm of the heterobrachial pair, as well as in isobrachial chromosome pair X, where it is located a little away from the centromere in one of the arms.

In *P. grifithii* and *P. roxburghii* (Fig. 15.1.22 B, C) in the haploid complement each of the six isobrachial chromosomes have a secondary constriction. In *P. roxburghii* the six homologous pairs with secondary

Fig. 15.1.21 A-H. *Pinus strobus.* **A-F** Germination of seed. **G, H** Seedling with cotyledonary and juvenile leaves. (After P. Maheshwari and Konar 1971)

constrictions are reported in the diploid complement (Fig. 15.1.22 A). In *P. kesiya*, each of the two isobrachial pairs have secondary constrictions.

Meiosis has been studied in detail in *P. roxburghii* and *P. patula* (Mehra 1988). In both the species, meiosis is normal. However, not infrequently the chromosome segments are left behind at the equatorial region as laggards (*P. roxburghii*), or are extruded (*P. patula*). Such irregularities can affect the chromosomal constitution as well as the morphology of pollen grains.

Polyploidy is not known in the wild population of any species.

Temporal Considerations

In *P. roxburghii* (growing at an altitude between 500 and 2500 m in the NW Himalayas, India) the male cones are initiated in September. Pollination takes place (pollen grains at the four-celled stage) in March. Pollen germinates

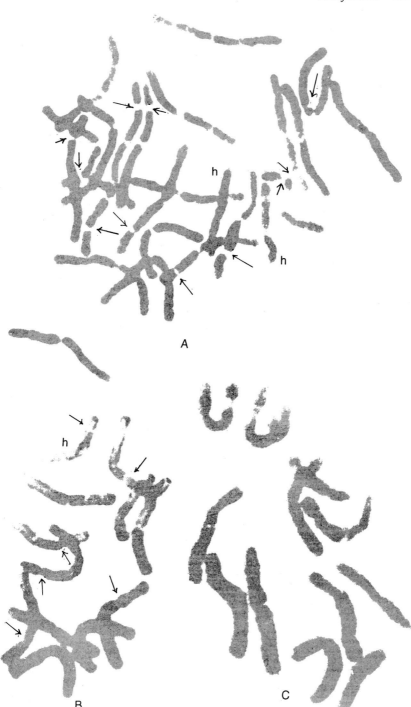

Fig. 15.1.22 A-C. *Pinus roxburghii.* **A** Root tip (squasn) shows 2n = 24 chromosomes.
B, C Endosperm (squash) shows n = 12. *h* heterobrachial chromosome; arrows indicate
position of secondary constriction. (After Mehra 1988)

on the nucellus immediately after shedding. There is a period of rest for approximately 10 months (from May to February—2nd year). Growth is resumed in March (2nd year) and fertilization occurs in April (2nd year). The female cones are initiated in February. Soon after pollination in March, the cone undergoes a period of rest from April to January for ca. 10 months. The female gametophyte is at the free-nuclear stage at this time. Growth is resumed from February (2nd year), and fertilization occurs in April. The embryo developes and matures by December (2nd year). The cones open and shed their seeds in April-May (3rd year). Usually, the cones dehisce in the third year; the undehisced cones may also contain mature seeds.

There are three species of *Pinus*—*P. pinea*, *P. leiophylla* and *P. torreyana*—which have a 3-year-type reproductive cycle (Dallimore and Jackson 1966, Francini 1958). These taxa undergo three winter rests, and the interval between pollination and fertilization is 2 years. The cones are initiated during autumn and the ovules overwinter (1st rest period). Pollination occurs during spring when the ovules show sporogenous cells. Megasporogenesis occurs during autumn and the ovule with a slightly enlarged megaspore overwinters (2nd rest period). The megaspore matures through the coming spring and summer. It undergoes free-nuclear divisions during autumn. A massive spongy tissue develops around the megaspore, and the young female gametophyte. The ovule overwinters again (3rd rest period) at the free-nuclear gametophytic stage. In the coming spring, the gametophyte develops more rapidly and fertilization takes place by June. The interval between pollination and fertilization is 2 years. The embryogeny is completed and the mature seeds are shed in autumn or early winter.

The reproductive cycle in tropical and temperate pines is conditioned by the environmental factors and, therefore, the 2- and 3-year cycle. The cyto- and histochemical changes during the rest period deserve further attention.

15.2 TAXODIACEAE

The characteristic features of the family Taxodiaceae are: monoecious trees; leaves narrow and linear (occasionally apparently two-ranked); vegetative buds without scales; young bract and ovuliferous scale nearly free from each other but fuse in the mature cone; each ovuliferous scale has 2–9 ovules; microsporophyll bears 2–9 pollen sacs; pollen grains without wings.

There are 10 genera and 15 spp.: *Athrotaxis* (3 spp), *Taxodium* (3 spp), *Cunninghamia* (2 spp.) *Cryptomeria*, *Glyptostrobus*, *Metasequoia*, *Sciadopitys*, *Sequoia*, *Sequoiadendron* and *Taiwania* are all monotypic.

There are several genera of seed cones, and vegetative remains are reported from the Jurassic. The Taxodiaceae were then a distinct line of evolution. All modern taxa of the Taxodiaceae, except *Sequoiadendron*, are known from the Mesozoic. The family shows increased diversity in the Cretaceous, a number of new organ genera as well as many more of the modern genera are represented. The living species probably existed as persistent relics of lineages that evolved nearly 100 m y ago, and may be regarded as modern representatives of lines of specialization (see Miller 1977).

Cryptomeria japonica

This monotypic genus includes ca. 25 varieties (Dallimore and Jackson 1966). It is a native of China and Japan. Japan owes much of the beauty of its groves to this taxon. The seeds were first brought to India in 1844 (see Troupe 1921), and cultivation was started in Darjeeling, which now has extensive forests of *C. japonica*. In the moist climate of the Eastern Himalayas, it grows very rapidly and yields excellent planted forests. It has also naturalized in the Western Himalayas and grows up to an altitude of 1800–2400 m.

Morphology

The plant is a tall, conical, much-branched evergreen tree, reaching a height of 50 m or more and a girth of 7–8 m (Mehra 1988). The bark is fibrous, reddish-brown, and peels off in long strips. The leaves are linear, awl-shaped and spirally arranged (Fig. 15.2.1 A, B). They are keeled on both the surfaces and have a somewhat spinous tip. The leaf bases are fused with the stem and form its outer covering. The branches are in whorls. The branchlets with the awl-shaped leaves are deciduous.

The plant is monoecious. Male and female cones are usually borne on different branchlets (Fig. 15.2.1 A), occasionally on the same branchlet (Fig. 15.2.1. B). The male cones arise in clusters near the tip of the young shoots, while the female cones are terminal on the branchlets (Fig. 15.2.1 A).

Anatomy

Stem. The wood is reddish brown and fragrant. A cross section shows clearly defined growth rings, variable in width and slightly wavy. Early and late wood are well differentiated, the tracheids are square, occasionally with smaller tracheids occurring between the larger ones. There are ca. 3600 tracheids/mm^2. The thickness of the wall is 1.5–3 μm in early wood and 4–6 μm in the late wood. The rays are relatively few, wide apart and traverse several growth rings. Their walls are horizontal and sparingly pitted. There is abundant wood parenchyma, chiefly in the late wood and at various distances from the growth ring limits. The parenchyma cells are thin-walled and can be readily distinguished from the thick-walled late tracheids. The resin ducts are absent (Greguss 1955).

Numerous pits occur on the tangential walls of tracheids (especially abundant in the late wood). The pits are 9–11 μm in diameter, commonly in a row, circular with poorly defined borders. Apertures in the early wood are circular or elliptic, in the late wood eye-shaped or slit-like. The tracheids

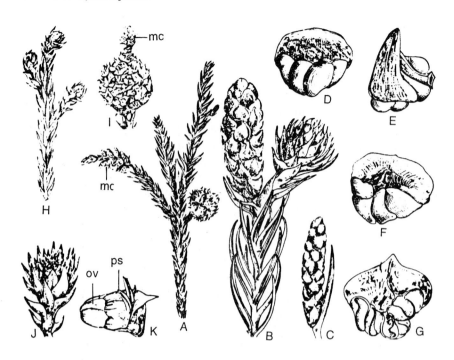

Fig. 15.2.1 A-K. *Cryptomeria japonica*, external morphology. **A, B** Male (*mc*) and female cones borne on different branchlets (**A**) or on the same branchlet (**B**). **C** Male cone with bract. **D, E F** Microsporophyll, abaxial, lateral and adaxial views. **G** Microsporophyll with dehisced sporangia. **H-K** Female cones. **H** Young cones. **I** Abnormal cone bearing a cluster of male cones at the tip. **J** Open cone scales at the time of pollination. **K** Bract and ovuliferous scales show the orientation of ovules (*ov*), and processes (*ps*) on the ovules. (After H. Singh and Chatterjee 1963)

are pointed. The rays are uniseriate with variable height (commonly 3–6 cells); tangential ray walls are thin and occasionally delicately verrucose. The wood parenchyma cells are fairly abundant with nodular cross-walls. The cells communicate with each other through simple pits, and with adjoining tracheids through half-bordered pits. The latter are smaller than the tangential pits, and their apertures differ in size and shape in the early and late wood. There are about 40–45 rays and 150–160 ray cells per mm^2.

A radial longisection shows uniseriate (occasionally two- to three-seriate) bordered pits. The pits are circular, and the apertures circular, elliptic, linear or slit-like. In slits, the pits are obliquely inclined or nearly vertical. The horizontal walls of the ray cells are relatively thick (1.5–3 μm) and smooth except for sporadic local thinnings; the tangential walls are smooth and thin (1–1.5 μm), while the radial walls show large, round simple pits with half-bordered complementary pits in the tracheids. The apertures vary in size and shape. Taxodioid cross-field pits are also present.

Leaf. The leaf is awl-shaped and amphistomatic. The stomata are haplocheilic. A vertical section (Fig. 15.2.2 A-C) shows: stratified cuticle of variable thickness; thick-walled epidermal cells with a much-reduced lumen containing tannin or other inclusions; well-developed hypodermis of thick-walled sclerenchymatous cells on one or both leaf surfaces.

The mesophyll is undifferentiated. It shows many elongated cells (especially on the adaxial side) which appear like chains/filaments of long narrow cells with large spaces between the rows. The outermost layer on the abaxial side is somewhat palisade-like, with radially elongate cells (Fig. 15.2.2 B, C). There is a single large resin duct with a two-layered wall beneath the vascular bundle (Fig. 15.2.2 A). A bundle sheath is not very distinct. In vertical section, the transfusion tissue appears as an inverted U-shaped arc clasping the vascular bundle with the tracheids and albuminous cells situated

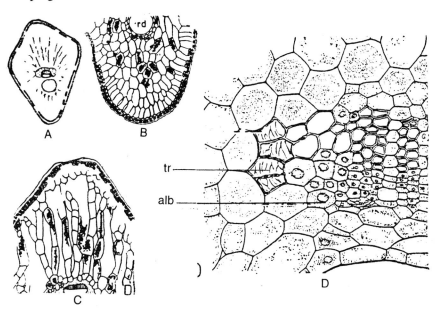

Fig. 15.2.2 A-D. *Cryptomeria japonica,* leaf. **A-D** Transection. **A** Outline diagram for **B** and **C**. **B, C** Internal structure. **D** Vascular bundle and transfusion tissue. *alb* albuminous cells. *tr* transfusion tracheid. (After Kausik and Bhattacharya 1977)

laterally to the xylem and phloem, respectively (Fig. 15.2.2 D). The transfusion parenchyma is distributed at random in the arc, especially on the adaxial side (Kausik and Bhattacharya 1977).

Reproduction

Male Cone

A male cone cluster comprises 10–18 cones, each borne in the axil of a bract (Fig. 15.2.1 C). Each cone bears 18–25 spirally arranged microsporophylls, the two lowermost microsporophylls are usually sterile. The stalk of each sporophyll is almost at right angles to the axis of the cone. It is a peltate structure which extends upward as a membranous flap and bears three or four microsporangia on its lower (abaxial) surface (Fig. 15.2.1 D-G).

Microsporangium. The hypodermal archesporial cells differentiate on the abaxial surface of the microsporophyll. They have dense cytoplasm and large nuclei, and are arranged in two layers (Fig. 15.2.3 A, B). The epidermal cells undergo anticlinal divisions; the archesporial cells divide in all planes to form a mass of cells.

The outermost layer of the archesporium divides periclinally to form an outer primary parietal layer, which divides again periclinally and gives rise to a three-layered wall. The innermost layer differentiates into a tapetum (Fig. 15.2.3 B, C, E). The tapetal cells become binucleate (Fig. 15.2.3 F) during reduction divisions of the microspore mother cells, and eventually degenerate with the maturation of microspores.

The two-layered hypodermal cells undergo rapid anticlinal divisions followed by tangential elongation to keep pace with the increasing sporogenous tissue. Finally, the hypodermal layers degenerate. The epidermis develops annular thickenings during the maturation of microspores (Fig. 15.2.3 E, G, H). In vertical section, the sporangia appear sessile and oval (Fig. 15.2.3 D). They dehisce by vertical slits which appear at the base, extend upward, and face the cone axis.

Microsporogenesis. Only some of the sporogenous cells function as microspore mother cells while the others degenerate. Cytokinesis follows meiotic divisions and tetrahedral, isobilateral and decussate tetrads are formed (Fig. 15.2.3 I-M). The callose wall around the microspores is consumed and the original wall of the mother cells breaks down, releasing them into the cavity of the microsporangium. The exine differentiates outwards to the intine, except where the latter projects in the form of a papilla (Fig. 15.2.3 I, N, O). The latter is occasionally balloon-like; sometimes an additional sac-like structure may occur beside the normal papilla (Fig. 15.2.3 P). The exine and the intine are of the same thickness; later the exine becomes thicker. Starch grains accumulate in the pollen grains, which show a high percentage of sterility. The pollen is shed at the uninucleate stage, which appears rare among the conifers; this condition is known only in *Cupressus* (Mehra and Malhotra 1947, Konar and Banerjee 1963) and *Taxus* (Dupler 1917, Sterling 1948a).

Male Gametophyte. After pollination, the pollen grains imbibe moisture and swell. The exine may persist in fragments or is entirely cast off. The

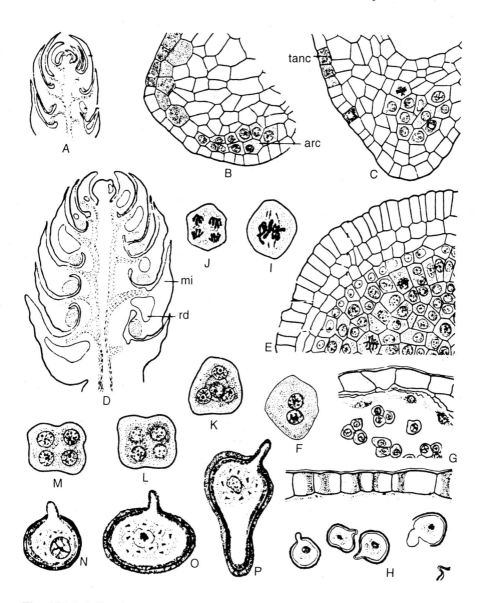

Fig. 15.2.3 A-P. *Cryptomeria japonica*, microsporangium and microsporogenesis. **A, D** Longisection of male cone. *mi* microsporophyll. *rd* resin duct. **B, C, E, G, H** Longisections of microsporophylls. **B** Archesporial cells (*arc*). **C** Sporogenous cells and wall layers. *tanc* tannin cell. **E** Later stage. **G** Epidermis, degenerated tapetum and microspore tetrads. **H** Exodermis with annular thickenings and microspores. **F, I-M** Whole-mounts. **F** Binucleate tapetal cell. **I, J** Microspore mother cells, meiosis I and II. **K-M** Tetrahedral, decussate and isobilateral microspore tetrads. **N, O** Uninucleate pollen grain. **P** Abnormal pollen grain, note starch grains in **N-P**. (After H. Singh and Chatterjee 1963)

nucleus moves to one side, and functions directly as the antheridial initial (prothallial cells are not formed in Taxodiaceae). The nucleus divides and forms a lenticular antheridial cell and a large tube cell (Fig. 15.2.4 A). At this stage, the starch grains are no longer discernible, the cytoplasm becomes less dense, and the pollen grains germinate. As the pollen tube enters and grows further, the nucellar cells disorganize (Fig. 15.2.4 B, C). Two to four pollen tubes have been observed in a nucellus; occasionally, the pollen tube branches (Fig. 15.2.4 D-F). The tube nucleus moves into pollen tube. The antheridial cell gives rise to the stalk and body cells. The nucleus of the body cell divides when the pollen tube has reached the archegonial complex. The two male gametes are equal and disc-like with dense cytoplasm and prominent nuclei (Fig. 15.2.4 G, H). The contents of the pollen tube are discharged in the archegonial chamber (H. Singh and Chatterjee 1963).

Female cone

The apex of the branchlets is generally used up during the differentiation of the terminal female cones (Fig. 15.2.1 H). Rarely, the axis of the cone may grow into a vegetative shoot which bears a cluster of male cones (Fig. 15.2.1 I).

The young female cone (ca. 4 mm in length and diameter) has a curved stalk and emerges from the rosette of leaves at the time of pollination. At this time, it has a flattened apex (Fig. 15.2.1 J), but subsequently becomes almost spherical. A mature cone (19 mm in length and 21 mm in diameter) is yellowish-brown, and comprises 26–30 spirally arranged scales; the upper nine or ten small scales are sterile. A mature fertile scale (ca. 9mm in length and 11 mm in width) is concavo-convex, and bears on the upper side three to five spinous processes at the distal end. The bract and the ovuliferous scales are fused except at the tip, where the bract appears as a recurved process. Each scale bears three or four adaxial ovules with the micropyle pointing away from the cone axis (Fig. 15.2.1 K).

Ovule.   After the demarcation of the integument, one or two hypodermal archesporial cells can be distinguished by their longer size and prominent nuclei (Fig. 15.2.5 A). These cells divide periclinally and form the outer primary parietal and the inner primary sporogenous cells. The epidermal as well as the parietal cells divide further and give rise to a massive nucellus, which thus has a dual origin (Fig. 15.2.5 B). The primary sporogenous cells divide and form a large group of sporogenous cells, some of which degenerate. At this stage the nucellar cells around the sprogenous tissue contain prominent nuclei (Fig. 15.2.5 C), and later contribute to the spongy tissue.

At its base, the integument is free from the nucellus (Fig. 15.2.5 E, G). The micropyle becomes wide open before pollination, and points upward. The lower part of the nucellus consists of smaller cells, while the upper has

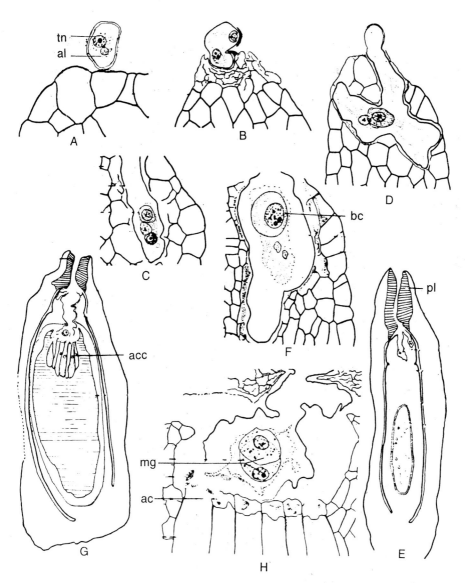

Fig. 15.2.4. A-H. *Cryptomeria japonica*, male gametophyte. **A** Two-celled pollen grain lying on the nucellus. *al* antheridial cell. *tn* tube nucleus. **B, C, F** Development of pollen tube in the nucellus. *bc* body cell. **D** Branched pollen tube. **E** Longisection of ovule with pollen tubes in the nucellus. *pl* plugging tissue. **G** Outline diagram for **H**. *acc* archegonial complex. **H** Pollen tube with male gametes (*mg*) in the archegonial chamber (*ac*). (After H. Singh and Chatterjee 1963)

larger cells. At the free-nuclear stage of the gametophyte, a constriction appears in the upper part of the nucellus; the underlying cells contain tannin. One of the sporogenous cells enlarges and functions as the megaspore mother cell (Fig. 15.2.5 D). Pollination takes place at this stage. Soon after,

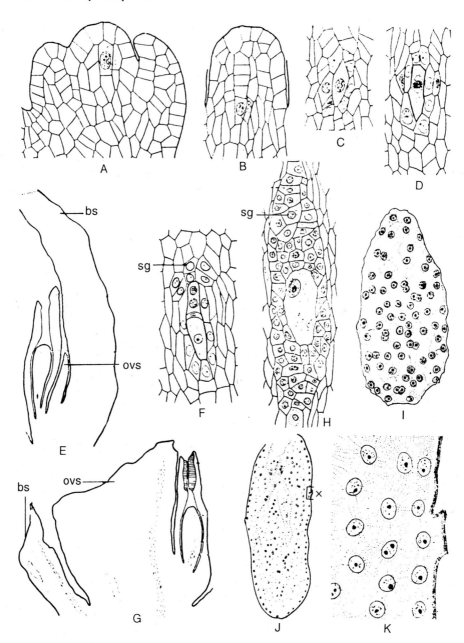

Fig. 15.2.5 A-K. *Cryptomeria japonica*, megasporangium and female gametophyte.
A Longisection of ovule, initiation of integument and archesporium. **B-D** Megaspore mother
cell. **E, G** Longisection of cone scale before (**E**) and after (**G**) pollination. *bs* bract scale.
ovs ovuliferous scale. **F, H** Spongy tissue (*sg*) and functional megaspore. **I-K** Free-nuclear
gametophyte. **K** inset *x* in **J** shows projection in the megaspore membrane. (After
H. Singh and Chatterjee 1963)

the epidermal and subepidermal cells lining the micropylar canal become richly cytoplasmic and elongate inward. They divide irregularly, the walls thicken at maturity, and the micropyle becomes closed (Fig. 15.2.5 G). One or two ovules (on a scale) may degenerate just before or after pollination.

Megasporogenesis. Meiosis I results in two dyads, of which only the lower undergoes meiosis II and a triad is formed. The chalazal megaspore functions, while the other megaspore and dyad degenerate (Fig. 15.2.5 F). With the enlargement of the megaspore, the circumjacent sporogenous and nucellar cells form the spongy tissue or tapetum. It is one to two cells thick laterally but six to eight cells thick above and below the megaspore (Fig. 15.2.5 C, D, F, H). The spongy tissue is very prominent and the densely cytoplasmic cells remain uninucleate. The nucellar cells disorganize in the immediate vicinity of the tapetum. The latter also collapses during the enlargement of the free-nuclear gametophyte.

Female Gametophyte. The functional megaspore enlarges, the cytoplasm becomes less dense, the nucleus also enlarges, migrates to one side of the cell, and undergoes division (Fig. 15.2.5 H). The gametophyte enlarges and a vacuole appears in the centre. The freee-nuclear divisions are synchronous and the nuclei lie in a peripheral layer of dense cytoplasm. The early gametophyte is ovate (Fig. 15.2.5 I). Later, it elongates, and at maturity shows a broad slightly convex micropylar and a tapering chalazal end. The megaspore membrane often has an uneven outline (Fig. 15.2.5 J, K).

A mature gametophyte shows rows of cells radiating from the centre to the periphery, which indicates centripetal wall formation. These uni- or binucleate cells are thin-walled and have scanty cytoplasm. With the differentiation of the micropylar archegonial complex, the adjoining gametophytic tissue grows upward and forms a prominent archegonial chamber (Fig. 15.2.4 G, H). The development of the embryo is accompanied by the deposition of starch grains in the gametophytic tissue.

A mature archegonial complex comprises 15–37 archegonia in a common jacket (Fig. 15.2.6 E). or the archegonia may be individually separated by the jacket cells and the gametophytic tissue. Usually, the archegonia develop side by side; occasionally they are superimposed. Lateral archegonia are frequent; they rarely differentiate at the chalazal end (Fig. 15.2.6 F, G). Sometimes, a gametophyte may show a chalazal archegonial complex in addition to the micropylar complex (Fig. 15.2.6 H, I). In the chalazal region, probably due to fusion, the archegonia assume an irregular shape. These archegonia do not become fertilized, their nuclei may divide but do not give rise to embryos (H. Singh and Chatterjee 1963).

The development of the archegonium follows the usual conifer plan (Fig. 15.2.6 A-C). A mature archegonium shows an ephemeral ventral canal nucleus and a large egg nucleus. Due to rapid elongation of the egg cell,

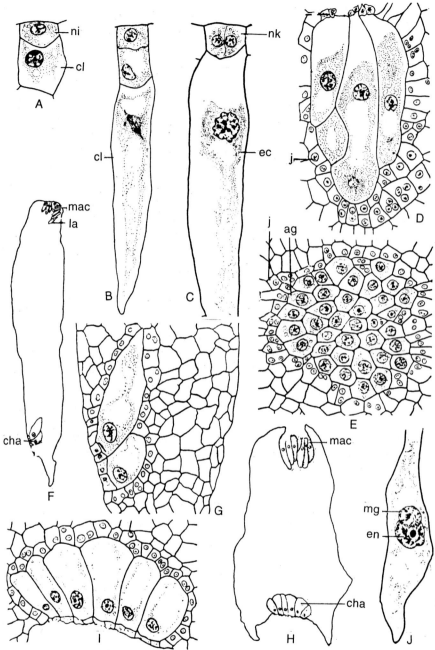

Fig. 15.2.6 A-J. *Cryptomeria japonica,* archegonial complex and fertilization. **A-C** Development of archegonium. *cl* central cell. *ec* egg cell. *ni* neck initial. *nk* neck. **D** Archegonial complex in tangential section; central archegonium shows a kinoplasmic body at the base. *j* jaclet cell. **E** Archegonial complex in transection. *ag* archegonium. **F** Longisection of gametophyte (outline), to show micropylar (*mac*), lateral (*la*) and chalazal (*cha*) archegonia. **G** Two archegonia (chalazal portion of **F**). **H** Outline diagram for **I. I** Chalazal archegoinal complex from **H. J** Fertilization. *mg* male gamete. *en* egg nucleus. (After H. Singh and Chatterjee 1963)

the cytoplasm becomes less dense and a large vacuole appears in the lower part; the egg nucleus is pushed up to the upper part of the archegonium (Fig. 15.2.6 C).

Electron microscopic studies show that in *Cryptomeria japonica* the egg cell is smaller than in the Abietaceae. It contains nodules of dense cytoplasm capped with a crescent-shaped vacuole, and is comparable with the "small inclusions" of the Abietaceae. The plastids undergo no appreciable change up to the maturation of the egg. These leucoplasts have few lamellae, and the ground substance is opaque to electrons. Their membrane is invariably supplemented with layers of endoplasmic reticulum, which lie parallel to it (see Chesnoy and Thomas 1971).

The jacket is usually one-layered laterally and two-layered at the base of the archegonial complex. The jacket cells contain dense cytoplasm and one to three prominent nuclei (Fig. 15.2.6 D). Some of the cells may contain rather large nuclei, probably due to the fusion of smaller nuclei.

Fertilization

The neck of the archegonium begins to disorganize before fertilization. The contents of the pollen tube are discharged into the pollen chamber, one or both of the male gametes enters the egg cell, and fertilization takes place in its centre. A depression appears in the upper part of the egg nucleus (Fig. 15.2.6 J), and the dense cytoplasm around the fusing nuclei contains starch grains. Vacuoles appear in the upper part of the zygote and the vacuoles in the lower part are no longer visible.

Embryogeny

Five to seven archegonia in a complex become fertilized, and the division of the zygote nucleus occurs at the base of the egg cell. Two further divisions follow, the resultant eight nuclei become embedded in dense cytoplasm and starch grains, and organize into the primary embryonal tier (*pE*) and primary upper tier (*pU*). The latter divides into a middle suspensor (*S*) and an upper open tier (*U*, Fig. 15.2.7 A-D). A mature proembryo has 13 or 14 cells arranged in three distinct tiers (*E, S, U*) and occupies nearly one-third of the length of the archegonium. The U tier eventually degenerates.

The cytoplasm in the upper part of the proembryo appears less dense and can be distinguished from denser cytoplasm in the lower part. Occasionally, one or two nuclei (probably derivatives of the second male gamete) have been observed in the upper part of the zygote; they may rarely persist up to the cellular stage of the proembryo (Fig. 15.2.7 D).

The suspensor tier *S* elongates and pushes the attached embryonal cells *E* beyond the archegonium. At the same time, the suspensor cells separate from each other and lead to cleavage polyembryony (H. Singh and Chatterjee 1963). Some of the separated suspensor cells do not bear any embryonal cell (the latter are fewer). The tip of such a suspensor cell becomes densely

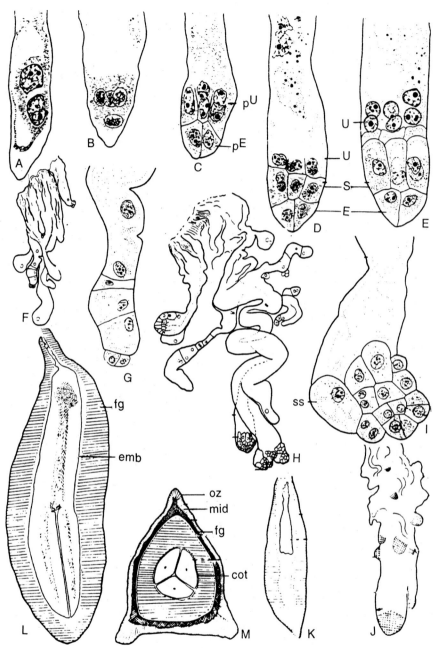

Fig. 15.2.7A-M. *Cryptomeria japonica*, embryogeny. **A-D** Development of proembryo.
E Embryonal tier. *pE* primary embryonal tier. *pU* primary upper tier. *S* suspensor tier.
U upper tier. *E* Elongated suspensor tier. **F** Young proembryo system, suspensor cells
with or without an embryonal cell at their tip. **G** Septate suspensor with embryonal cells
at the tip. **H** Outline diagram for **I. I** Secondary suspensor (*ss*) from **H. J** Mature embryo
system. **K, L** Longisection gametophyte, development of embryo. *emb* embryo. *fg* female
gametophyte. **M** Transection of mature seed. *cot* cotyledon. *mid* middle layer. *oz* outer
zone. (After H. Singh and Chatterjee 1963)

cytoplasmic, its nucleus enlarges and a small "embryonal cell" is cut off. Occasionally, a suspensor cell may become transversely septate (Fig. 15.2.7 E-G). According to Sugihara (1969), the suspensor cells elongate, but remain fused parallel to one another. Cleavage polyembryony takes place by elongation of embryonal tubes at the tip of the suspensor. Meanwhile, the cells of the gemetophyte adjoining the archegonial complex disorganize, and form a loose cavity. The elongation of the suspensors pushes the developing embryos (with the suspensors coiled around each other) into this cavity.

The embryonal cells divide to form an embryonal mass, whose upper cells elongate and produce secondary suspensors. The latter, eventually, become massive. The embryonal masses attached to the secondary suspensors (lying in the cavity of the gametophyte) show various stages of development. The embryonal mass lying deep in the cavity continues to grow; the other remains arrested (Fig. 15.2.7 H-J).

Differentiation of Embryo. The embryonal mass grows predominantly by transverse divisions. The differentiation of root initials is followed by the stem tip. The three cotyledonary primordia appear around the stem tip, and elongate towards the chalazal end of the seed. The gametophytic tissue below the shoot apex is not completely used up during the maturation of the embryo, and persists between the developing cotyledons. A provascular strand differentiates in each cotyledon. With the growth of the embryo, the suspensor is gradually absorbed. A mature embryo comprises three cotyledons, a well-organized root and shoot tip, and is surrounded by four or five layers of gametophytic cells filled with starch and fatty food reserves (Fig. 15.2.7 K-M).

Seed. The seed coat has the usual three zones : (a) an outer zone of thin-walled sarcotesta, (b) a middle zone of thick-walled sclerotesta, and (c) an inner zone of thin-walled endotesta (Fig. 15.2.7 M).

Germination. The seeds do not have a period of dormancy. The germination (%) is rather low, and is epigeal as in conifers. The three cotyledonary leaves are green, thick and linear, and show two resin canals, one near each margin of the leaf. The vascular bundles are collateral and tangentially elongated. The juvenile leaves are arranged in an opposite and decussate manner, and differ from the mature leaves.

Chromosome Number

The diploid chromosome number is 22 (Fig. 15.2.8 A, B); six chromosemes are heterobrachial and the rest isobrachial. Each of the two isobrachial chromosomes has a secondary constriction (Mehra 1988).

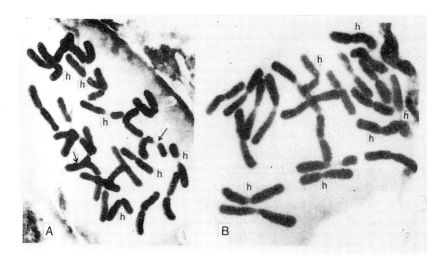

Fig. 15.2.8 A, B.   *Cryptomeria japonica*, cytology. **A, B** Root tip mitoses, 2n = 22; arrow shows secondary constriction. *h* heterobrachial chromosome. (After Mehra 1988)

Temporal Considerations

In Mussoorie (Western Himalayas), the life cycle is completed in 1.5 years (H. Singh and Chatterjee 1963). The male cones are initiated in June. The pollen is shed in February and male gametes are formed by May. The female cones appear in July—ca. 1 month after the male cones. It undergoes a rest period from September to January. Growth is resumed in the beginning of February of the following year and fertilization occurs in the middle of May. The embryo develops and seeds mature by October. The female cones dehisce from the beginning of December to the beginning of January.

15.3   CUPRESSACEAE

The trees or shrubs are characterized by shoots arranged in an opposite and decussate manner or in whorls; there is no differentiation into long and short shoots. The plants are monoecious. The scales in the female cone are arranged cross-wise. There is a pronounced fusion between ovuliferous and the bract scale. In some taxa, the cone scales overlap (*Thuja*), or are valvate (*Cupressus*) and may even be fleshy (*Juniperus*). The ovules are erect and 3–20 (rarely 1 or 2) per scale. The number of pollen sacs also varies from 3–6 (rarely 2).

The family comprises 19 genera; about 9 are confined to the Northern Hemisphere and the rest to the Southern Hemisphere. Eight taxa are monotypic. *Juniperus* is distributed in a continuous broad belt round the northern hemisphere, possibly due to its berry-like female cones which are relished and thereby distributed by birds. This taxon includes 70 spp. and is the second largest conifer genus.

The northern taxa are (species in parenthesis): *Arceuthos* (1), *Biota* (1), *Calocedrus* (3), *Chamaecyparis* (6), *Cupressus* (20), *Fokienia* (2), *Juniperus* (70), *Thuja* (4) and *Thujopsis* (1).

The southern taxa are: *Actinostrobus* (2), *Austrocedrus* (1), *Callitris* (15), *Diselma* (1), *Fitzroya* (1), *Libocedrus* (5), *Neocallitropsis* (1), *Papuacedrus* (3), *Pilgerodendron* (1), and *Widdringtonia* (5).

The Cupressaceae have existed for a considerable time. Their remains have been recognized in sediments of the Late Triassic (Lemoigne 1967), and again in those of the Lower to Middle Jurassic (Chaloner and Lorch 1960). Accordingly, most of the genera of the family can be viewed as relics in that they include only a few species with limited geographic distribution (see Miller 1977).

Biota orientalis

For a long time, the taxon *Biota* was included under *Thuja*. On the basis of morphological characters, Endlicher (1847) raised *Thuja orientalis* to a new genus, *Biota*. His views were accepted by Buchholz (1929) and Martin (1950), but opposed by Lindley (1853), Pilger (1926) and Pilger and Melchior (1954).

Biota orientalis is a monotypic genus. It is a native of the eastern part of Central Asia, and is cultivated as an ornamental in many parts of the world. This species has numerous cultivars.

Morphology

The plant is an evergreen tree reaching a height of ca. 15 m. It has an entirely different shape when young and displays considerable variations under cultivation. Wilson (1926) observed some plants in a park in China, that have been growing undisturbed for centuries. These were large trees with horizontal lateral branches, in contrast to the plants under cultivation with vertical lateral branches. It appears that in younger stages the lateral branches are vertical, but later become horizontal. Due to the slow-growing habit of the plant, and constant trimming of its branches to give it the desired shape, it almost never attains its natural appearance.

The bark is smooth, brownish, and separates from the older branches as papery scales. The terminal shoots are divided into a spray of branchlets (Fig. 15.3.1 A) covered by dark green, closely appressed, acute and decussate leaves. Each leaf is fused with the stem along one-third of its length. It has a long groove in the middle of its abaxial surface (Fig. 15.3.1B).

The plants are monoecious. The male and female cones occur on separate branch systems. However, both types of cones may sometimes be present on the same branch system (H. Singh and Oberoi 1962). The plants are either predominantly male with a few female cones, or vice versa. The sex

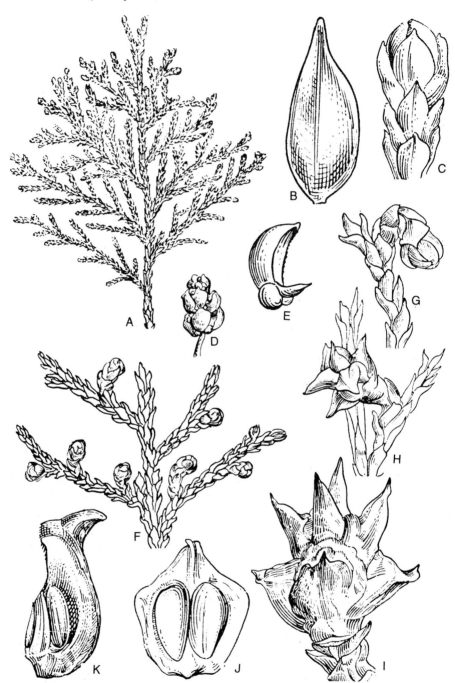

Fig. 15.3.1 A-K. *Biota orientalis.* **A** Twig with branchlets. **B** Leaf, abaxial surface shows longitudinal groove. **C** Young male cone enclosed by leaves. **D** Dehisced male cone. **E** Microsporophyll, lateral view. **F** Branch bearing male cones. **G, H** Female cones on curved branches. **I** Erect mature cone. **J, K** Lateral (**K**) and abaxial (**J**) views of scale-bearing ovules. (After H. Singh and Oberoi 1962)

expression (in this species) is quantitative, perhaps controlled by a multigenic factor (Mehra 1988).

Reproduction

Male Cone
The male cones are borne on the ultimate branchlets (Fig. 15.3.1 F); the young cones are completely covered by leaves (Fig. 15.3.1 C). Later, they emerge by the elongation of the basal portion of the cone axis. A male cone is 2–3 mm in diameter, and bears four to six pairs of microsporophylls arranged decussately (Fig. 15.3.1 D). At maturity the microsporophyll is small, round, leaf-like, and attached to the cone axis by a short stalk. It is slightly broad at the base, convex on its abaxial surface and has an inwardly curved crenate margin (Fig. 15.3.1 E). Three to five microsporangia are present on the abaxial surface near the base of each sporophyll. Young cones are yellowish green but turn pale yellow at maturity. The microsporophylls separate from each other and expose the sporangia (Fig. 15.3.1 D). After the pollen is shed, the cones turn brownish-yellow, dry up and fall off from the plant.

Microsporangium. A young sporophyll (in longisection) shows two to three hypodermal archesporial cells near the base of the abaxial surface. The epidermis and the archesporial cells divide anticlinally and their number increases (Fig. 15.3.2 A-C). The archesporial cells divide periclinally and form a primary parietal layer and a primary sporogenous layer. The cells of the latter give rise to a mass of sporogenous cells, by repeated divisions in all planes (Fig. 15.3.2 D). The primary parietal layer divides periclinally to form two layers, the layer in contact with the sporogenous tissue differentiate into the tapetum. A young sporangium consists of the epidermis, a single middle layer, a tapetum, and a mass of sporogenous cells (Fig. 15.3.2 F). The epidermis and the wall layer continue to divide anticlinally for a while, the sporangium becomes round or slightly elongated (Fig. 15.3.2 E). Later, with an increase in the sporogenous tissue, the anticlinal divisions become fewer. A mature sporangium shows a tangentially elongated epidermis with cutinized outer walls, and develops annular thickenings (Fig. 15.3.2 G). The cells of the middle layer degenerate as the mother cells prepare for reduction divisions. The tapetal cells become binucleate and collapse during maturation of the microspores.

Microsporogenesis. The microspore mother cells (mimc) undergo reduction divisions (Fig. 15.3.2 H-K). At anaphase I the cytoplasm of the mother cell retracts slightly and a callose wall is secreted around it. A cell plate may (Fig. 15.3.2 J) or may not (Fig. 15.3.2 I) be laid down between the two dyad nuclei, so that cytokinesis is of the successive or simultaneous type. Both tetrahedral and isobilateral tetrads are formed (Fig. 15.3.2 L-N).

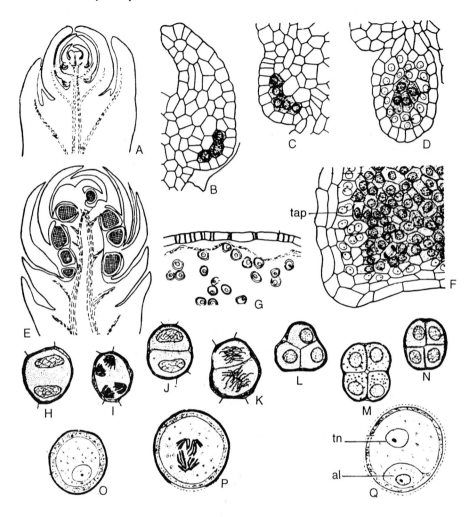

Fig. 15.3.2 A-Q. *Biota orientalis.* **A** Longisection very young male cone. **B, C** Longisection microsporophylls shows hypodermal archesporium. **D** Longisection young microsporangium with primary sporogenous and primary parietal layers. **E** Longisection male cone at microspore mother cell. **F** Young sporangium with wall layers and sporogenous tissue; some tapetal (*tap*) cells are binucleate. **G** Sporangial wall at microspore stage; epidermis shows thickenings. **H-N** meiosis I and II; the microspore tetrads are tetrahedral and isobilateral. **O** Microspore. **P** Microspore nucleus in division. **Q** Pollen grain at shedding stage. *al* antheridial cell. *tn* tube cell. (After H. Singh and Oberoi 1962)

Male Gametophyte.   A microspore is spherical, and shows a thin exine, a thick intine and a large nucleus (Fig. 15.3.2 O). Starch accumulates in the cytoplasm, and may even mask the nucleus. The latter shifts laterally and divides (Fig. 15.3.2 P) to form a small antheridial and a large tube cell (Fig. 15.3.2 Q). With further enlargement of the pollen grains, the intine becomes thinner and the starch grains begin to be digested. The pollen is shed at the two-celled stage. Some of the microsporangia, and sometimes all the sporangia on a plant, produce only sterile pollen (H. Singh and Oberoi 1962).

Female Cone

The female cones are borne on the terminal branchlets, the male cones slightly lower on the same branchlet. The young cones covered by leaves are borne on branches that curve downwards (Fig. 15.3.1 G). These branches become nearly upright when the cones emerge at the time of pollination. Each cone consists of three or four pairs of decussate scales, and bears a curved spine (Fig. 15.3.1 H, I, K). The scale is a fusion product of the bract and the so-called ovuliferous scale. It bears one to three orthotropous ovules near the base of its adaxial surface (Fig. 15.3.1 J, K). The uppermost pair of scales is usually sterile.

At the time of pollination, the tip of each scale curves backward to expose the ovules. With further growth, the scales become fleshy, come close to each other and the cone becomes compact (Fig. 15.3.1 I), and woody. A mature cone measures ca. 20–22 mm in length and ca. 18 mm in diameter. The cone scales separate from each other along the original margins and expose the seeds for dispersal. A seed is ca. 5 mm long and is covered by a three-angled stony seed coat.

Ovule. The young cone shows well-differentiated ovules on the adaxial surface of the scale (H. Singh and Oberoi 1962). The integument is three or four cells thick, tapers to one layer at the tip, and is slightly longer than the nucellus. One or two primary sporogenous cells differentiate in the nucellus at the level of insertion of the integuments (Fig. 15.3.3 A). The nucellar epidermis undergoes periclinal as well as anticlinal divisions. The ovule continues to increase in size, and the primary sporogenous cells undergo mitotic divisions to form the sporogenous tissue (Fig. 15.3.3 B).

Megasporogenesis. A single, more or less centrally placed mother cell (Fig. 15.3.3 D) undergoes reduction divisions to form a row of three cells. The uppermost cell is the undivided dyad; Tarchi (1969) observed callose around it. The chalazal megaspore functions (Fig. 15.3.3 E, F).

Female Gametophyte. The functional megaspore enlarges and a large central vacuole shifts the nucleus to one side. Repeated synchronous free-nuclear divisions occur, and due to the formation of a large central vacuole the nuclei become embedded in a peripheral layer of cytoplasm (Fig. 15.3.3 K). With an increase in the number of nuclei, they become smaller (Fig. 15.3.3 H-J). When the number of free-nuclei exceeds ca. 4000, centripetally advancing walls (Fig. 15.3.3 L, M) bring about cellularization of the gametophyte.

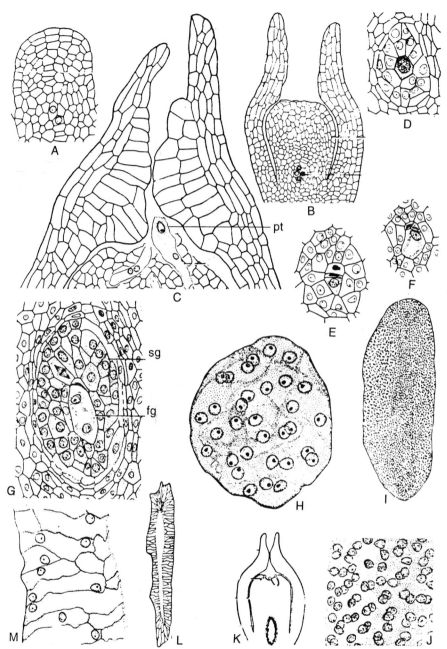

Fig. 15.3.3 A-M. *Biota orientalis.* **A, B** Longisection ovules with deep-seated sporogenous cells. **C** Longisection upper part of ovule after pollination; micropylar canal closed. *pt* pollen tube. **D** Megaspore mother cell surrounded by sporogenous tissue. **E, F** Megaspore triads, chalazal megaspore functional. **G** Two-nucleate gametophyte (*fg*) with well-developed spongy tissue (*sg*). **H, I** Free-nuclear gametophytes. **J** Inset from **I.** **K** Longisection ovule with deep-seated gametophyte. **L** Cellular gametophyte. **M** Inset from **L.** (After H. Singh and Oberoi 1962)

The megaspore membrane also becomes very conspicuous. The mature gametophyte elongates, tapers slightly at the chalazal end, and is flat or somewhat convex at the micropylar end.

A free-nuclear gametophyte is enclosed by two or three layers of a conspicuous spongy tissue derived from the non-functional sporogenous cells (Fig. 15.3.3 G). As the gametophyte enlarges, the nucellar cells outside the spongy tissue collapse. By the time walls are initiated in the gametophyte, the spongy tissue degenerates.

Archegonial Complex. A newly formed central cell along with a neck initial shows the megaspore membrane flush with the neck initial; the jacket cells are uninucleate (Fig. 15.3.4 A).

The neck initial divides twice vertically with the walls oriented at right angles to each other, and forms a neck of four cells lying in one plane (H. Singh and Oberoi 1962). Meanwhile, the central cell and its nucleus enlarge considerably and a vacuole appears just below the nucleus. About the time the pollen tube reaches the archegonial chamber, the nucleus of the central cell divides to form a small disc-shaped ventral canal nucleus (vcn) and a large egg nucleus (Fig. 15.3.4 C, D). Before the nucleus of the central cell divides, the organelles of the cytoplasm, particularly the mitochondria, collect below the nucleus and at the centre of the basal cytoplasm below the vacuole. During early stages, the cytoplasm around the vcn is vacuolar and distinct from that below it. As the body cell in the pollen tube divides to form two male gametes, the vcn begins to degenerate and is no longer recognizable by the time the male gamete enter the archegonium.

The egg is ca. 300 μm long and 50 μm across. The nuclear diameter is 35–40 μm and is situated in the upper part of the cytoplasm. The egg cell shows a vacuole representing the remnants of the vacuole earlier occupied by the greater part of the central cell.

The cytoplasm of the central cell has a large vacuole in the middle of the cell. The cytoplasmic organelles (Fig. 15.3.4 B) aggregate in its micropylar portion (below the nucleus), and the chalazal portion (below the central vacuole). The organelles comprise mitochondria and leucoplasts arranged radially around a mass of ribosomes and microtubules. It has been suggested that this area of organelle aggregation may also be the centre of organelle multiplication (Chesnoy 1971).

The filling up of the vacuole below the nucleus occurs by the breakdown of its tonoplast and the penetration of cytolasmic vesicles into the vacuole (Chesnoy 1971).

Electron microscopic observations have been made on the egg cytoplasm (Chesnoy 1971). The large number of mitochondria (present in the mature egg) are spherical and dispersed. In section, some of these appear as a ring enclosing a small core of cytoplasm. Only rarely do they reveal tubules,

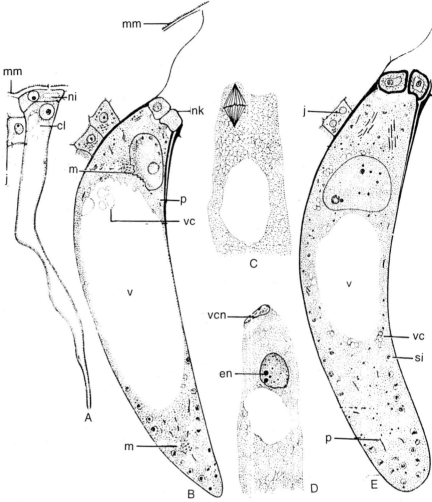

Fig. 15.3.4 A-E. B*iota orientalis.* **A** Longisection young central cell (*cl*). *ni* neck initial. *j* jacket cell. *mm* megaspore membrane. **B** immature archegonium; two asteroids lie in the cytoplasm of the large cell. *m* mitochondria. *nk* neck cell. *p* plastid. *v* vacuole. *vc* cytoplasmic vesicles. **C, D** Upper part of central cell. **C** Central cell nucleus in division. **D** Ventral canal nucleus (*vcn*) and egg nucleus (*en*). **E** Nearly mature archegonium with large egg nucleus, scattered mitochondria, the jacket cell is binucleate. *si* small inclusion. (**A, B, E** After Chesnoy 1971, **C, D** after H. Singh and Oberoi 1962)

which are very short and lie in a transparent matrix (Chesnoy 1969a, b). The leucoplasts are narrow, very long and often grouped into bundles (Fig. 15.3.4 E).

The archegonia are arranged in a single group or complex at the micropylar end of the gametophyte (Fig. 15.3.5 A). Each complex has 15–28 (usually 22) archegonia (Fig. 15.3.5 C). Occasionally, one or two supernumerary archegonia develop below the archegonial complex. In such an archegonium

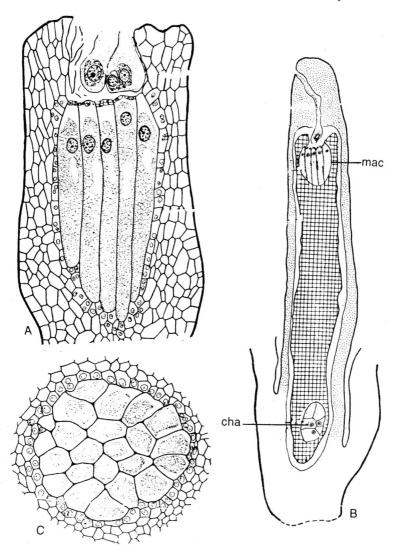

Fig. 15.3.5 A-C. *Biota orientalis.* **A, B** Longisection female gametophyte with micropylar (*mac*) and chalazal (*cha*) archegonial complex. **C** Archegonial complex in transection. (After H. Singh and Oberoi 1962)

the initial functions as the central cell and the neck cells are not formed. These complexes degenerate, probably due to lack of fertilization. Rarely, there is a chalazal archegonial complex with fewer archegonia (Fig. 15.3.5 B).

With the development of the archegonia, the adjacent tissue grows upward, which results in a depression, the archegonial chamber. The archegonial complex is surrounded by a common jacket layer. After fertilization, most of the jacket cells become binucleate; large polyploid nuclei may be formed by nuclear fusions.

Pollination

Pollination occurs by wind. At the time of pollination the tip of each scale curves backward so that the ovules are exposed. There is a sugary exudation at the micropyle (Fig. 15.3.6 A, B). The pollen grains swell when caught in the pollination drop (H. Singh and Oberoi 1962).

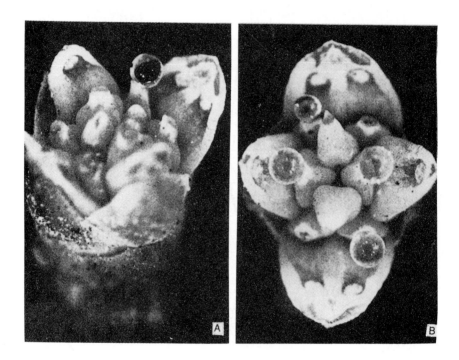

Fig. 15.3.6 A, B. *Biota orientalis.* **A, B** Female cones at pollination. (After H. Singh and Oberoi 1962)

After pollination, the cells of the integument (lining the micropyle) elongate radially and close the canal (Fig. 15.3.3 C), periclinal divisions may also take place. These cells become thick-walled and show simple pits during the maturation of the ovule.

Post-Pollination Development of Male Gametophyte.   After the pollen grains come to lie on the nucellus, the exine bursts and the intine extends and forms a small pollen tube containing the tube nucleus (Fig. 15.3.7 A, B). Meanwhile, the antheridial cell becomes free from the wall of the microspore, moves into the pollen tube (Fig. 15.3.7 B), and divides to form the stalk and body cells (on entering the nucellus). Later, a wall cannot be made out around the stalk cell and it persists as a nucleus.

Generally, three to five pollen tubes enter an ovule. They elongate and enlarge so that the adjoining nucellar cells are crushed; their contents always persist at the tip. Occasionally, a pollen tube may branch (Fig. 15.3.7 C). The growth of the pollen tube is accelerated by the differentiation of the archegonial initials.

The tube and stalk nuclei cannot be distinguished from each other and always lie close to the body cell in dense cytoplasm (Fig. 15.3.7 C). The spermatogenous (body) cell enlarges considerably (Fig. 15.3.7 E). On reaching the archegonial chamber, the tip of the pollen tube swells and brings about disintegration of the cells lining the archegonial chamber (Fig. 15.3.7 D, E). The spermatogenous (body) cell divides to form two equal male cells (Fig. 15.3.7 F-H). The stalk and tube nuclei begin to degenerate.

A mature male gamete is ca. 50 μm in diameter, its spherical nucleus is ca. 30–35 μm and shows a distinct nucleolus. It is separated from the cytoplasm of the pollen tube only by a plasmalemma. Electron microscope studies reveal that the cytoplasm is organized into three consecutive zones, which do not show uniform structure. The mitochondria, amyloplasts, groups of vesicles and abundant ribosomes collect in a deep zone ca. 3–15 μm in thickness (Fig. 15.3.8), separated from the nucleus by a narrow perinuclear, and from the plasmalemma by a marginal zone. The perinuclear zone is ca. 0.5 to 1 μm thick, free from organelles, and shows only fragments of reticulum arranged parallel to the nuclear zone. The marginal zone is 2–3 μm thick and contains only a few Golgi vesicles and dilated fragments of reticulum (Chesnoy 1969 a, b).

The EM studies of the male gametes have shown the ultrastructural characters of the organelles of the male cytoplasm, and have helped to locate these organelles after the union of the gemetes. The matrix of the mitochondria of the sperm cells is very dark and the cristae well developed. The amyloplasts contain several starch grains (Chesnoy 1969 a, b).

Fertilization

The pollen tube tip breaks down and discharges the male cells into the archegonial chamber. In a single chamber as many as 12 male cells have been observed. There is simultaneous fertilization of two adjoining egg cells by the two gametes from the same pollen tube. With the entrance of the pollen tube into the chamber, the neck cells degenerate, making a passage for the male cells to enter the archegonium. The male gamete (cell) penetrates into the cytoplasm of the egg; its organization is only slightly modified. The marginal cytoplasmic zone of the male gamete merges with the cytoplasm of the female gamete during penetration of the sperm cell into the egg. The male nucleus moves in the maternal cytoplasm surrounded by the perinuclear, deep zone of its own cytoplasm. It partially frees itself from the perinuclear, deep zone as it approaches the female nucleus. The latter has a depression in its upper part in which the male nucleus establishes itself. The nuclear membranes join together while the two nuclei turn at

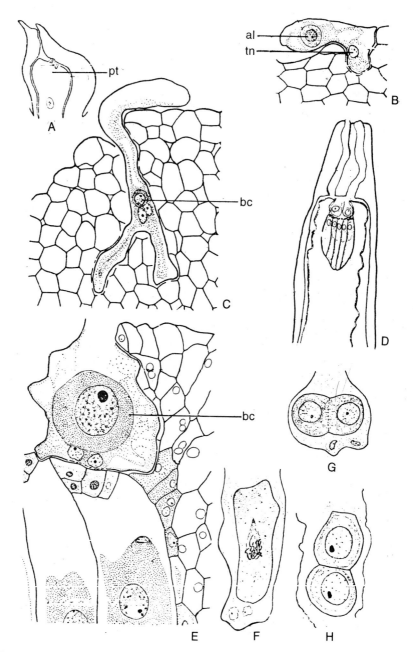

Fig. 15.3.7 A-H. *Biota orientalis.* **A** Longisection ovule shows germinated pollen on the nucellus. *pt* pollen tube. **B** Pollen tube from **A.** *al* antheridial cell. *tn* tube nucleus. **C** Branched pollen tube in nucellus shows tube and stalk nuclei and the body cell (*bc*). **D** Longisection (upper part) female gametophyte shows pollen tubes. **E** Part of pollen tube and female gametophyte from **D.** **F** Division in body cell. **G, H** Pollen tube with two equal male cells. (After H. Singh and Oberoi 1962)

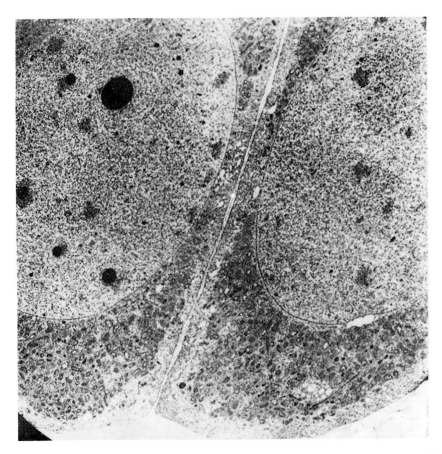

Fig. 15.3.8 *Biota orientalis.* Portions of two male cells, middle zone contains most of the organelles. (Courtesy Chesnoy, see H. Singh 1978).

180° (see Chesnoy and Thomas 1971). At the commencement of the nuclear encounter, the cytoplasm of the deep zone of the male gamete caps the two nuclei. It gradually surrounds them and isolates from the maternal cytoplasm. EM shows that the male cytoplasm retains all its cohesion during its descent through the egg cell.

Embryogeny

The division of the zygote takes place within the nucleoplasm, which is still bordered by the nuclear membrane that disappears only after the telophase. The mitotic figure is intranuclear and slightly oblique with respect to the long axis of the archegonium (Fig. 15.3.9 A). After the completion of mitosis, the cytoplasm of the male gamete, along with its organelles, penetrates into the original nucleoplasm and surrounds the two proembryonal nuclei. The two nuclei with the cytoplasm originating primarily from the deep zone of the male gamete migrate to the base of the archegonium and constitute

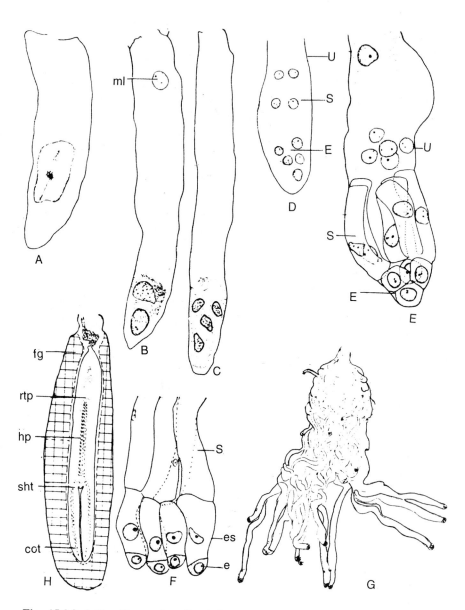

Fig. 15.3.9 A-H. *Biota orientalis.* **A** Intranuclear division of the zygote nucleus. **B, C** Two-and four-nucleate proembryos. *ml* male nucleus. **D** Three-tiered proembryo. *E* Embryonal tier. *S* suspensor tier. *U* upper tier. **E** Young embryo. **F** Elongation of embryonal suspensor (*es*). *e* embryonal cell. **G** Whole mount embryo shows mass of suspensor, embryonal suspensor and embryonal cells at the tip. **H** Longisection female gametophyte shows mature embryo. *cot* cotyledon. *fg* female gametophyte. *hp* hypocotyl. *rtp* root tip. *sht* shoot tip. (After H. Singh and Oberoi 1962)

the embryonic pole. The mitochondria, amyloplasts and cytoplasmic RNA are transmitted to the embryo. A few mitochondria of maternal origin (carried along by the male cytoplasm during its course through the egg cell) appear in the proembryonal cytoplasm. The plastids of maternal origin have not been obsrved in the proembryonic cytoplasm. EM studies confirm that the cytoplasm of the proembryo and its organelles are essentially of paternal origin (Chesnoy 1969 a, b).

Wall formation in the proembryo takes place at the eight nucleate stage. The lower one-quarter of the zygote contains the embryonal nuclei and is densely cytoplasmic (Fig. 15.3.9 B, C); the upper portion has scanty cytoplasm, which seems to lose contact with the cytoplasm of the lower portion.

Occasionally, one or two nuclei (probably derivatives of the second male nucleus) have been observed in the upper region of the proembryo (Fig. 15.3.9 B). These nuclei usually degenerate, but may persist up to the cellular stage of the proembryo (Fig. 15.3.9 E).

Wall formation in the proembryo results in the upper pU and the lower pE tier of four cells each. Two embryonal and five upper cells are not uncommon. The cells of the pU tier divide and give rise to a proembryo of 12–16 cells arranged in three distinct tiers of (a) embryonal (E), (b) suspensor (S) and (c) open tier (U), and occupies about one-third the space of the proembryo (Fig. 15.3.9 D).

Differentiation of Embryo. The suspensor tier elongates and pushes the embryonal tier beyond the archegonial complex (Fig. 15.3.9 E). As soon as the embryonal tier comes in contact with the gametophytic tissue, each of its cells divides to form the embryonal suspensor (Fig. 15.3.9 F), which elongates only after the primary suspensor tier ceases to function.

Cleavage occurs at this stage and the embryonal suspensors, carrying the small embryonal cell at its tip elongate (into the corrosion cavity). The suspensor system develops rapidly and becomes tightly coiled round (Fig. 15.3.9 G), in the limited corrosion cavity. The degenerated suspensor cells are pushed back by the developing suspensors. Occasionally, the suspensor cells become separated and bulge outward with prominent nuclei in the swollen portions. Rarely, the nucleus divides and as many as eight free nuclei may be formed in the bulge. Sometimes, these cells elongate considerably and simulate the suspensor. These can be distinguished by their swollen appearance and absence of an embryonal cell at the tip.

The embryonal cell remains inactive throughout the period of elongation of the suspensor cell. The first two divisions in the embryonal cells are vertical and at a right angle to one another. Further divisions take place in all planes, forming a small globular mass located at various levels in the cavity of the gametophyte. The leading embryo (lying farthest in the gametophyte) continues to grow, while the remaining embryos become arrested and gradually collapse. The proximal cells of the embryonal mass divide and elongate forming a massive secondary suspensor. The embryonal cells also increase by further divisions.

The root initials differentiate behind the apex of the embryonal mass, and this is followed by the organization of the stem tip. Simultaneously, two cotyledonary primordia appear and begin to elongate. The region between the shoot apex and the root initials also elongates considerably to form the hypocotyl. Each cotyledon is supplied by a provascular strand. As the embryo elongates, the suspensor collapses and its remnants persist until maturity (Fig. 15.3.9 H).

The mature embryo has two (rarely three) cotyledons with well-organized root and shoot tip. The embryo is surrounded by five or six layers of gametophytic tissue rich in starch.

Seed

Initially the integument comprises four or five layers of cells. These cells enlarge prior to pollination and undergo only anticlinal divisions. The cells of the outer epidermis and hypodermis contain tannin.

When the ovule contains a free-nuclear gametophyte, the integument is 10-12 layers thick. The cells lining the micropylar canal (after pollination) become thick-walled and show simple pits. Such thickening of the cells also extends downward (Fig. 15.3.10 A-C).

At maturity, the seed coat is 20–25 cells thick, comprising three distinct regions—an outer and inner zone of thin-walled cells, and an intervening zone of thick-walled cells (Fig. 15.3.10 G, H). The outer, thin-walled zone of the integument consists of three or four layers of elongated cells, the outer two layers are tanniniferous. In the mature seed this zone becomes disorganized and is no longer distinguishable when the cone "opens" to shed the seeds. The middle thick-walled zone consists of six or seven layers which are comparatively large and contain shrunken cytoplasm and small nuclei. The walls show numerous pits (Fig. 15.3.10 D, E). The inner zone consists of seven to ten layers of long, narrow cells which remain thin-walled. The cells contain small nuclei and scanty cytoplasm (Fig. 15.3.10 F). The cells of the inner epidermis contain tannin. The mature seed (in a cross section) is triangular (Fig. 15.3.10 H).

Germination. The seed germinates soon after shedding. The epigeal primary root branches only at a late stage. The juvenile leaves look very different from the mature leaves, and are arranged spirally on the stem. They are elongated and lanceolate, and are not fused with the stem for any appreciable length.

The epidermis of the cotyledons is covered by a poorly dveloped cuticle. The mesophyll consists of an undifferentiated mass of parenchymatous cells with abundant intercellular spaces. Each cotyledon has a single

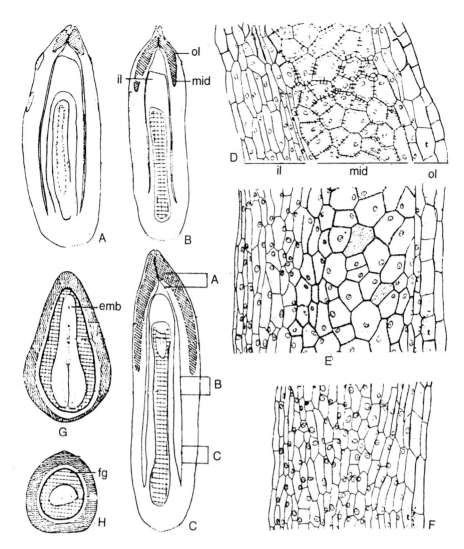

Fig. 15.3.10 A-H. *Biota orientalis.* **A-C** Longisection ovule at various stages of develoment. The thickening in the middle part of integument extends from micropylar to chalazal region. *il* inner zone. *mid* middle zone. *ol* outer zone. **D, E, F** Portions marked *A, B, C* in **C**. Three regions of integument. **E** Cell walls of middle zone have become thick. **F** All the cells are thin-walled. **G, H** Longi-and transection of mature seed. *emb* embryo. *fg* female gametophyte. (After H. Singh and Oberoi 1962)

(somewhat tangentially elongated) collateral and endarch vascular bundle. Resin ducts are not present in the cotyledons, although they are common in the foliage leaves.

Chromosome Number

The chromosome number, as determined from root tip squashes, is 2n = 22

(Fig. 15.3.11 A). Karyotypic analysis reveals three pairs heterobrachial and the remaining eight pairs isobrachial. One isobrachial pair showed a secondary constriction in one arm (Fig. 15.3.11 A, arrow). Two-chromosome configuration at mixed anaphase in endosperm squash is shown in Fig. 15.3.11 B, C (Mehra 1988).

Fig. 15.3.11 A-C.   *Biota orientalis.* **A** Root-tip mitosis, 2n = 22. h **B, C** Endosperm squashes show chromosomes at mixed anaphase I. (After Mehra 1988)

Temporal Considerations

In Delhi, the male cones are initiated at the end of May/begining of June, remain covered by foliage and emerge only after November; the growth of the cone is slow during this period. By early January, the microspore mother cells (mimc) differentiate. The tetrads and microspores develop by the end of January. The microspore nucleus divides in early February, and the pollen is shed in the second week.

Female cones appear in the beginning of July (a month later than the male cones). In early September the (small) branches bearing the female cones bend downward. The growth of the cones is very slow during this period. By the end of November, the megaspore mother cell (mgmc) differentiates. The female cone undergoes a period of rest during December and January. Reduction divisions in the mgmc take place in early February and simultaneously the cone emerges and "opens" to expose the ovules for pollination. From the second week of February for 20 days, a shiny conspicuous pollination drop appears at the micropyle, in the early hours of the morning, and dries up by about 10.00 h.

In early March, the pollen tubes are common on the nucellus; the female gametophyte is at the free-nuclear stage. Wall formation begins, archegonia develop early in April, and fertilization occurs by the middle of the month. By the end of May, embryonal masses show extensive suspensor. Cotyledons differentiate (mid-June), embryo matures (beginning of July), and dehiscence of the cones begins in the first week of August. Seeds germinate within 15–20 days without undergoing any period of dormancy.

15.4 Podocarpaceae

The family Podocarpaceae includes trees and shrubs, and the living representatives are predominantly distributed in the Southern Hemisphere. There are seven genera and ca. 150 sp. *Podocarpus*, with nearly 106 spp., is the largest genus of present-day conifers. The other taxa are: *Acmopyle, Microcachrys, Pherospherea (Microstrobus), Phyllocladus, Dacrydium* and *Saxegothea*.

The Podocarpaceae is of great interest because it shows a remarkably wide range of morphological characters. Most extant Podocarpaceae bear leaves and other parts in a helix, some are opposite. The leaf varies from sessile, decurrent and falcate with a single vein to large and flat, and contains many parallel veins.

The male cones may be simple (characteristic of conifers) or form a part of a compound structure. There is considerable variation in the development of male gametophyte, and a wide range of pollination mechanism.

In some spp. the ovules are borne in cones, while in others the female cone is so reduced that it forms a swollen fleshy receptacle with a single terminal ovule.

The Podocarpaceae is well represented in fossil record, the earliest occurrence is in the Early Triassic. The available evidence suggests that this family diverged from the Voltziales in the Late Palaeozoic and was already a separate line of evolution at the onset of the Mesozoic (see Miller 1977). However, *Podocarpus* is the only modern genus of the group with a record during the Mesozoic, having evolved by the Cretaceous.

Florin (1958) regarded the Podocarpaceae (along with Araucariaceae) as Southern Hemisphere conifers, and indicated that this had always been so. However, there are reports of this genus from the Northern Hemisphere. Krassilov (1974) reports remains of *Podocarpous* from the Late Cretaceous of Russia and states that the "northern" and "southern" conifers grew side by side in Mesozoic forests and at least *Podocarpus* persisted in North America until the Eocene (Dilcher 1969).

Podocarpus

It is distributed throughout the Southern Hemisphere, tropical and Southern Africa, Australasia to Japan. It is regarded as the dominant conifer in the Southern Hemisphere, as *Pinus* is in the Northern Hemisphere.

Podocarpus has been studied extensively, the various species show great diversity in structure and development. Pilger (1926) divided it into two subgenera: (a) *Stachycarpus* and (b) *Protopodocarpus*. The latter is further subdivided into four sections: *Nageia*, *Microcarpus*, *Dacrycarpus* and *Eupodocarpus*.

Morphology

The plants are mostly tall evergreen trees. *P. gracilior*, the African fern-pine, is 15 m high with a trunk of 1.8–2.9 m girth. The main trunk is crowned with innumerable branches. The leaves are ca. 8 cm in length and 4 mm in width.

Anatomy

Root.   There is a taproot system which shows numerous mycorrhizal tubercles on the main and lateral roots (Saxton 1930, Baylis et al. 1963, Konar and Oberoi 1969 a). The primary root is diarch (Fig. 15.4.1 A). The parenchymatous cortical tissue has large intercellular spaces. Most of the cells are filled with starch grains, and several contain tannin. Secondary growth and the development of bark occur at an early stage. A fungus infests the young roots, the cortical cells proliferate, becoming large and nodular. The root tubercles are endogenous and represent modified lateral roots. The large parenchymatous cells of the tubercles are filled with fungal hyphae and spores (Fig. 15.4.1 B, C).

Stem. Shoot Apex.   The lateral shoot apices are slightly smaller than the

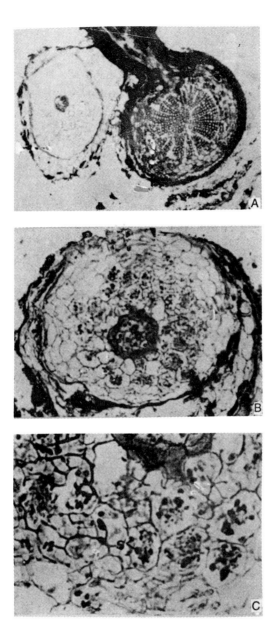

Fig. 15.4.1 A-C. *Podocarpus gracilior.* **A** Transection root shows secondary growth and attached tubercle. **B** Transection root tubercle with abundant fungal hyphae in cortical region. **C** Inset *C* from **B**. (After Konar and Oberoi 1969a)

terminal ones and slower in development in *P. gracilior* (Konar and Oberoi 1969a). Four cytohistological zones can be distinguished in the shoot apex (Pillai 1963, Konar and Oberoi 1969a): (a) apical initials, (b) subapical initials, (c) flanking zone, and (d) pith mother cells and pith. The shoot apex shows variations in activity, size, topography and histological organization.

A transection of young stem of *P. gracilior* displays a fluted outline due to the presence of broad leaf bases (Fig. 15.4.2 A). The epidermis is distinct and coverd by a thick cuticle. A young stem shows a narrow cortex of parenchymatous cells. In an older stem, three or four layers of the outer cortex and some isolated cells in the inner cortex contain tannin. Scattered sclereids occur in the inner layers of the cortex. In the cortex just outside the vascular bundles, there is a ring of large resin canals with a distinct epithelial layer, one canal opposite each phloem patch. Two or three leaf traces are also present in the cortex. A parenchymatous pith contains numerous sclereids and resin cells (Fig. 15.4.2 A, B). A cambial ring appears in the outer cortex, cutting off cork cells, which enclose many lenticels.

The vascular bundles remain separate from each other, just below the shoot apex (Fig. 15.4.2 A). Secondary growth occurs at a very early stage. The secondary phloem forms a narrow ring and consists of sieve cells, parenchyma, fibres and rays. The sieve cells (in transection) are rectangular and arranged in radial rows. Phloem rays are uniseriate and the tannin-filled cells of phloem parenchyma are grouped in tangential bands. The phloem fibres occur in a single ring.

A transection of wood shows distinct growth rings of variable width (Fig. 15.4.2 C). The xylem consists of tracheids, parenchyma and rays. The tracheids are rectangular (Fig. 15.4.2 D) and have uniseriate bordered pits (Fig. 15.4.2 E, H), on both their radial and tangential walls. The rays are homogeneous and consist of parenchymatous cells with thin, smooth horizontal walls. They are uniseriate (Fig. 15.4.2 F, G), and two to ten cells high, commonly three or four. One pit is present per cross field. The xylem paenchyma is diffuse, abundant, and becomes prominent in the older wood as it becomes resinous.

Leaf.   The leaves of *Podocarpus* are typically bifacial with a well-defined mid-rib: isobilateral in *P. gracilior* (Konar and Oberoi 1969a) and bifacial in *P. brevifolia*. The internal structure shows a sharp distinction into a compact palisade tissue on the adaxial and loosely arranged spongy cells on the abaxial side. There are three resin ducts below the vascular bundle, the central duct being slightly longer than the lateral ones (Kausik and Bhattacharya 1977; Fig. 15.4.3 A). The bundle sheath is not distinct. The vascular bundle is large and flattened dorsiventrally, with files of parenchyma between the rows of xylem and phloem.

In *P. gracilior* (isobilateral) the palisade is present on both sides. It is amphistomatic with sunken stomata. The epidermis is covered by a prominent cuticle, thicker at the margin and thinner over the stomata. The hypodermis consists of small groups of sclerenchymatous cells below the upper and lower epidermis except at the edges and above and below the vascular

Fig. 15.4.2 A-H. *Podocarpus gracilior.* **A-D** transection, **E, H** radial longisection, **F, G** Tangential longisection. **A** young stem with a ring of vascular bundles and peripheral resin ducts. **B** Mature stem with secondary growth. **C** Secondary xylem with growth rings. **D** Secondary xylem from **B. E** Wood with ray cells. **F, G** Wood with uniseriate rays. **H** Tracheids with bordered pits. (After Konar and Oberoi 1969a)

bundles. The palisade usually comprises two layers. Some isolated fibres occur around the vascular bundle and are dispersed throughout the adjacent mesophyll. A single median resin canal occurs on the abaxial side of the vascular bundle.

The transfusion tissue is lateral with a wing-like extension on the two sides of the vascular bundle, and extends outward into the mesophyll (Fig. 15.4.3 A). In *P. gracilior*, it comprises short spiral tracheids parallel

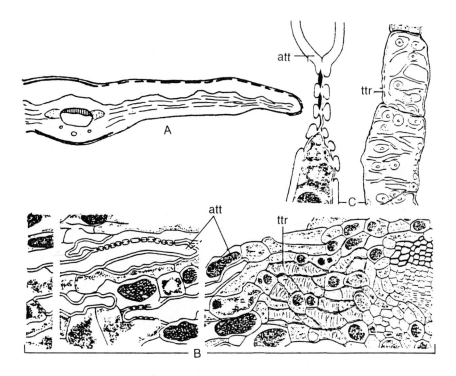

Fig. 15.4.3 A-C.   *Podocarpus brevifolius.* **A, B** Vertical section leaf. **A** Outline diagram to show transfusion tissue on either side of vascular bundle and three resin ducts below the bundle. **B** Transfusion tissue (*ttr*), accessory transfusion tissue (*att*), and vascular bundle. **C** Accessory transfusion tracheid and transfusion tracheid. (After Kausik and Bhattacharya 1977).

to the surface of the leaf. In *P. brevifolia*, the tracheids and a few parenchyma cells occur lateral to the xylem, and the albuminous cells in the phloem.

An additional histological feature of considerable importance is the accessory transfusion tissue in the mesophyll, between the palisade and the spongy cells extending from the mid-rib to the two margins of the leaf. In *P. brevifolia* this tissue is well developed. It consists of two distinct types of cells: (a) thin-walled parenchyma containing large quantities of tannin and other darkly staining inclusions (Fig. 15.4.3 B), and (b) thick-walled tracheids with groups of bordered pits (Fig. 15.4.3 B, C). These cells are elongated at right angles to the vascular bundle and form a net-like structure (in paradermal sections) and tiers of cells (in transections).

Reproduction
Podocarpus is dioecious. Most of the plants produce male cones and the seed set on the female plants is very low. In *P. gracilior* this is probably due to failure of pollination (Konar and Oberoi 1969b).

Male Cone
The male cones of *Podocarpus* conform to the general coniferous types and do not exhibit much diversity. In *P. gracilior* the male cones—widely spread along the short lateral branches—are borne on small pedicels which occur singly (rarely two together) in the axils of foliage leaves (Fig. 15.4.4 A, B). There are one to three short-stalked cones borne on the pedicel (Fig. 15.4.4 C-E). When there are three cones, the middle one is the longer of the two. A mature cone is ca. 2 cm long and 3–4 mm in diameter. Numerous sporophylls are spirally arranged on the cone axis. Each sporophyll has a short narrow stalk, and a large, triangular, upwardly projecting terminal region. Two ovoid sporangia occur on the abaxial surface, one on either side of the stalk near the cone axis (Fig. 15.4.4 F-J). The two sporangia are laterally fused. The line of dehiscence of the sporangium is transverse (Fig. 15.4.4 J), as in other podocarps (Stiles 1912). The cone axis elongates at maturity, separates the sporophylls and exposes the sporangia (Fig 15.4.4 D).

Microsporangium. The microsporangia (*P. gracilior*) initiate as small hump-like swellings on the abaxial surface of the sporophyll, near the base. With further growth, the sporangium appears as a round sac-like structure, and finally elongates parallel to the stalk of the sporophyll (Fig. 15.4.5 A-D). Three or four hypodermal archesporial cells differentiate, and divide anticlinally and periclinally to form a group of sporogenous cells surrounded by two or three wall layers (Fig. 15.4.5 E-G). As the divisions continue, the sporogenous tissue increases enormously and the wall becomes five- to seven-layered. The cells of the outermost and some of the inner wall layers become resinous. The innermost wall layer surrounding the sporogenous tissue differentiates into a tapetum. The tapetum is of the secretory type (*P. macrophyllous*). The cells are densely cytoplasmic and some of them become binucleate at maturity (Fig. 15.4.5 H). The tapetum is absorbed at the time of the reduction divisions in the mother cells. As the microspores mature, the cells of the middle layers become flattened and crushed (Fig. 15.4.5 I, J). At maturity, the epidermal cells enlarge, stretch, their radial and inner tangential walls become thick and develope upwardly directed fibrous thickening (Fig. 15.4.6 C-E). A row of wall cells, running transversely, remains small and unthickened, and dehiscence occurs along these cells with the epidermis alone functioning as the wall of the sporangium (Fig. 15.4.6 A-H).

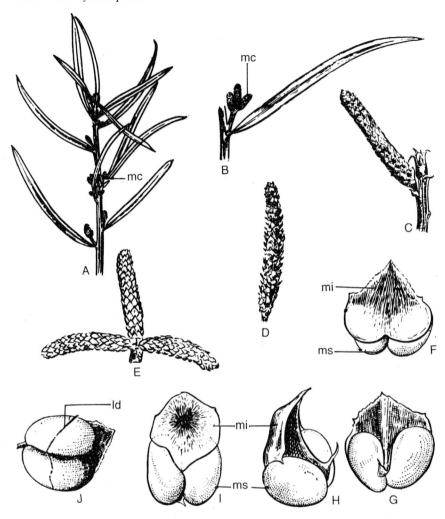

Fig. 15.4.4 A-J. *Podocarpus gracilior.* **A** Twig bearing male cones (*mc*) in the axils of leaves. **B, C** Male cones. **D** Dehisced cone. **E** Three cones in a cluster. **F-J** Microsporophyll (*mi*) and microsporangia (*ms*) in various views. **J** Ventral view of sporangium shows line of dehiscence (*ld*). (After Konar and Oberoi 1969b)

In *P. macrophyllous* the nuclei of the sporogenous cells are large with scattered chromatin. The outer membrane of the nuclear envelope shows frequent invaginations, which may reach to the plasmalemma (at some places). Inside these evaginations, there are several small vesicles with dark contents, their nature is not known. I.K. Vasil and Aldrich (1970) presume that this may constitute a line of transport. During the period of growth, large multilayered complexes of endoplasmic reticulum (ER) are scattered throughout the cytoplasm. Direct connections between these multilayered complexes and nuclear envelope or plasmalemma have not been observed (I.K. Vasil and Aldrich 1970).

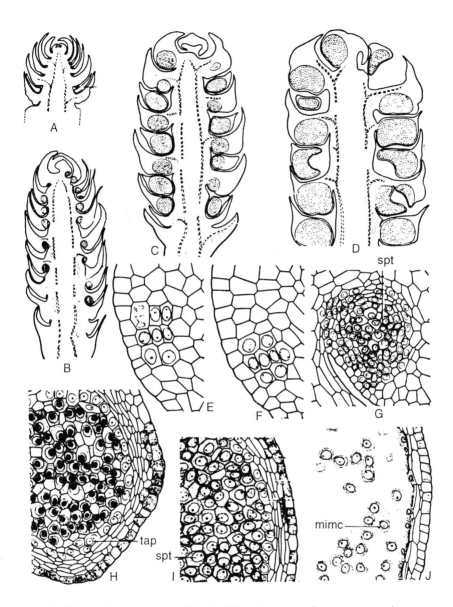

Fig. 15.4.5 A-J. *Podocarpus gracilior.* **A-D** Longisection male cone, progressive stages of microsporangial development. **E-I** Development of microsporangium. *spt* sporogenous tissue. *tap* tapetum. **J** Portion of sporangium with microspore mother cells (*mimc*). (After Konar and Oberoi 1969b)

There is a difference in staining pattern between the cell walls of sporogenous and tapetal cells, and that of other wall layers, in *P. macrophyllous.* The cell walls of the wall layers are thicker and stain a brighter magenta by periodic acid-Schiff's (PAS) reaction than the walls of tapetum and sporogenous cells. There are numerous vesicles in the cytoplasm

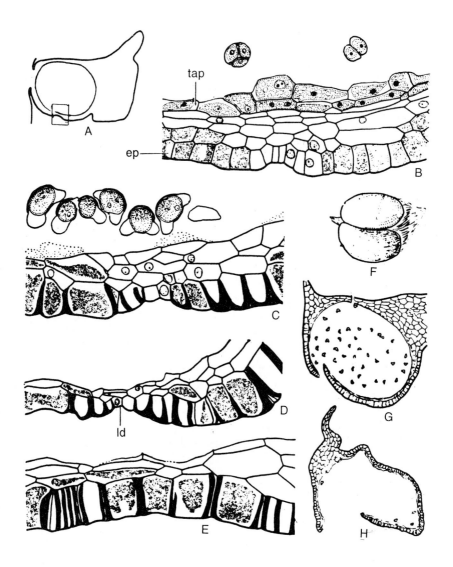

Fig. 15.4.6 A-H. *Podocarpus gracilior.* **A** Microsporophyll longisection (outline diagram). **B** Inset from **A**, epidermis (*ep*), tapetum (*tap*) and microspore tetrad. **C-E** Progressive development of wall and line of dehiscence (*ld*); note pollen grains in **C** and thickenings in epidermal cells in **C-E**. **F** Microsporophyll (ventral view). **G, H** Cross section dehisced sporangium. (After Konar and Oberoi 1969b)

of both sporogenous and tapetal cells, which probably carry hydrolytic enzymes. The walls of collapsed tapetal and sporogenous cells (prior to meiosis) show a basic similarity in their chemical nature (I.K. Vasil and Aldrich 1970).

In *P. macrophyllous* the sporogenous cells are rich in RNA throughout their development. The RNA level is maximal in the mother cells prior to meiosis. A loss of cytoplasmic basophilia has been observed especially during meiosis II. The young microspore is poor in RNA, but the level rises again on the formation of the antheridial cell.

Microsporgenesis. The microspore mother cells (mimc) undergo reduction division (Fig. 15.4.7 A-C). The cytoplasm of mimc contracts, the wall dissolves leaving the cytoplasmic masses with their prominent nuclei free in the sporangium (Fig. 15.4.5 J). In *P. macrophyllous*, prior to meiosis, the mimc undergo a gradual stretching, and loss of PAS stainability. With the dissolution of their walls, a thin osmiophilic layer of callose develops between the cell wall and the plasmamembrane. This layer fluoresces weakly by the aniline blue ultraviolet (UV) fluroscence test. It has been named a callose-like wall by I.K. Vasil and Aldrich (1970).

Meiosis is asynchronous. Plasma channels are not formed between adjacent sporogenous cells. In a microsporangium the meiotic stages in various mimc range from prophase I to telophase II. A callose wall has not been reported during microspore mother cell meiosis.

Cytokinesis is simultaneous, and isobilateral and tetrahedral tetrads are formed (Fig. 15.4.7 D, E). Several abnormalities occur during microsporogenesis. Frequently, incomplete cytokinesis leads to double pollen grains (Fig. 15.4.7 L). Wing formation is initiated in the microspores at the tetrad stage. Rapid development takes place as soon as the microspores separate from each other. The microspores increase in size, the exine at the proximal end becomes thick and cutinized. The large nucleus is centrally placed and the cytoplasm has a few starch grains (Fig. 15.4.7 F).

The development of the pollen grain wall has been described in *P. macrophyllus* by I.K. Vasil and Aldrich (1970). The outermost layer, the sexine, is laid down while the spores are still within the tetrad and callose covering. The Golgi bodies in the young microspore are large and very active. Small-unit membranous vesicles (ca. 0.1μm) with granular electron-dense contents are excreted by the Golgi bodies. These vesicles coalesce during their transport towards the plasmalemma. A space is formed between the plasmalemma and callose wall by contraction of the former. The space enlarges in the region where subsequently air sacs develop. The plasmamembrane shows many undulations or protrusions. The unit membrane of these vesicles fuses with the plasmalemma. The cellulosic contents of these vesicles is secreted outside the plasmalemma but inside the callose layer, and represents the initiation of the primexine. At this stage, long stretches of distinctly trilaminar lamellae or tapes are seen in the primexine. Lateral evaginations appear to give rise to these trilaminar structures. A deposition of sporopollenins on the outer surface of the vesicle contents (lying below the callose layer) is the first sign of tectum of the sexine. Deposits of sporopollenin also continue along the side of these vesicle

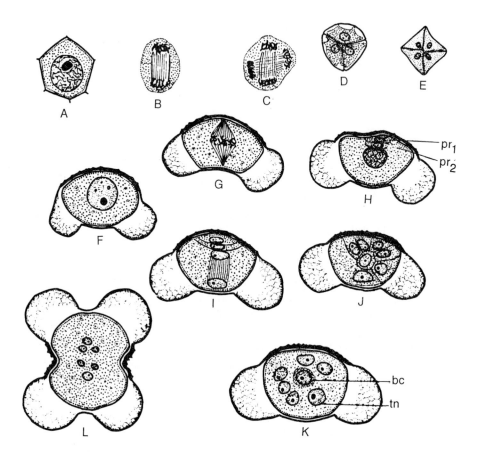

Fig 15.4.7 A-L. *Podocarpus gracilior.* **A-C** Meiosis in microspore mother cell. **D, E** Isobilateral and tetrahedral tetrad. **F** Uninucleate pollen grain. **G** Division of nucleus. **H** Pollen grain with two prothallial cells (*pr1, pr2*) and central cell nucleus. **I** Division of central cell nucleus (to form a tube nucleus and a centrally placed generative cell). **J** Further divisions in prothallial cells result in a tissue; due to displacement the prothallial cells do not appear in tiers. **K** Pollen grain at shedding stage; sterile nuclei lie free in the cytoplasm. *bc* body cell. *tn* tube nucleus. **L** Double pollen grain. (After Konar and Oberoi 1969b)

contents (sides of trilaminar tapes) to form the baculae. The fine fibrillar material of Golgi vesicles condenses against the callose wall and its protrusions.

The nexine I is the next layer to be laid down. It is formed by the deposition of sporopollenin on trilaminar tapes in the inner part of the primexine during the tetrad phase. At least five such layers of tapes are visible. The outermost layer has a richer deposition of sporopollenin than the inner foot layers. Lipid granules form the primary source for nexine I, and the tapes probably originate from the plasmamembrane or endoplasmic reticulum. The nexine surrounds the cytoplasm except at the regions where wings develop.

Further development takes place when the microspores are liberated from the tetrads. Small granular particles of sporopollenin present between plasmalemma and nexine I coalesce to form nexine II. A third layer, nexine III, is also present in *P. macrophyllous*. This layer is formed after the microspore nucleus has given rise to supernumerary prothallial cells (see below). The development of nexine III is similar to that of nexine II, and no trilaminar membranes or tapes are thus involved. This is the most electron–dense layer of the pollen wall.

The intine is the last layer to be laid down. Numerous Golgi vesicles take part in its development and empty their contents across the plasmalemma. It is homogenous and without any lamellations of fibrous character.

The unit member, which plays an important role in pollen wall formation, develops from the plasmalemma.

The nuclear envelope in a mature pollen grain consists of two membranes with occasional pores of 0.1 μm in diameter. Finger-like invaginations, involving both layers of the nuclear envelope, appear when the microspores are still within the tetrad, and extend into the nucleoplasm in *P. macrophyllous* (Aldrich and I.K. Vasil 1970), At the proximal end of these invaginations are structures comparable to nuclear pores. Fibrillar material interpreted as achromatin (I.K. Vasil and Aldrich 1970) fills the invaginations, and is discharged into the cytoplasm, the nuclear envelope then becoming regular once again. In addition to invaginations, there are evaginations of the outer nuclear membrane extending into the microspore cytoplasm. These are filled with vesicles ca. 1 μm in diameter, surrounded by unit membranes. Nuclear pores occur at non-evaginated portions only. A continuity between the outer nuclear envelope and the plasmalemma has also been observed, indicating nucleocytoplasmic transfer or exchange (Aldrich and I.K. Vasil 1970). These membranes can serve as sites for the synthesis of various substances in preparation for meiosis. They may also contribute enzymes for the breakdown of the mother cell wall and subsequent formation of the callose layer. The space between the two nuclear membranes may act as a channel for transport of exine precursors. Sporopollenin-like deposition occurs on Golgi cisternae and elements of endoplasmic reticulum within microspore mother cells during microsporogenesis (see Moitra and Bhatnagar 1982).

Male Gametophyte. At the proximal end, a mature microspore nucleus cuts off successively two lenticular primary prothallial cells (Fig. 15.4.7 G, H), which undergo further divisions to form three to four cells (Fig. 15.4.7 J). Simultaneously, or prior to the second division of these cells, the nucleus of the central cell divides to form a large tube nucleus towards the distal end and a centrally placed antheridial or generative cell towards the proximal end (Fig. 15.4.7 I). The generative cell divides to form a stalk and a body cell. It is difficult to distinguish the stalk cell from the prothallial tissue, whose cells also contain prominent nuclei (Fig. 15.4.7 J). At maturity, the wall of all the cells (except the body cell) dissolve and the nuclei lie free in the general cytoplasm of the pollen grain (Fig. 15.4.7 K). The cytoplasm contains abundant starch grains. At the time of shedding, a pollen grain shows a maximum of six free nuclei, in addition to the body cell (Fig. 15.4.7 K).

Female Cones

The female strobili in *P. gracilior* is axillary and occurs at the distal end of a vegetative shoot. The fertile branch can be distinguished from a vegetative branch only after pollination when the ovule emerges from the subtending bracts. Each fertile branch is ca. 2.5 cm long and bears five to ten leafy bracts (Fig. 15.4.8 A). The uppermost bract (rarely two) is fertile, and bears in its axil a single unitegmic anatropous ovule (Fig. 15.4.8 B). At the time of pollination it is nearly ovoid. The fertile bract does not grow further, and remains on the side away from the micropyle. The ovuliferous scale, or epimatium, encircles the ovule completely and is free from it at the micropylar

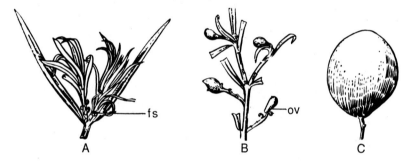

Fig. 15.4.8 A-C. *Podocarpus gracilior.* **A, B** Twig bearing female strobili (*fs*). *ov* ovule. **C** Mature seed. (After Konar and Oberoi 1969b)

end. The young ovule is externally covered by a white, shiny waxy deposition. The mature seed (Fig. 15.4.8 C) is ca. 2.5 cm in length and 2 cm in diameter. It is covered over by a thick fleshy seed coat which appears yellowish green at mautrity. The inner part of the seed coat is hard, stony, sharply pointed at the micropylar end, and has numerous small pits.

Ovule.   In *P. gracilior,* the earliest stage showed an undifferentiated hump which is carried a little above the shoot apex by a short stalk, a short projected nucellus in the middle, and the single integument above it. At this stage, the ovule is more or less hemianatropous. Simultaneously, the epimatium (another integument-like structure) appears on the side of the ovule (Fig. 15.4.9 A). The ovule grows rapidly, becomes anatropous, and has a wide micropylar opening. The epimatium is completely fused with the ovule except at the micropylar end, and is slightly shorter than the integument (Fig. 15.4.9 E).

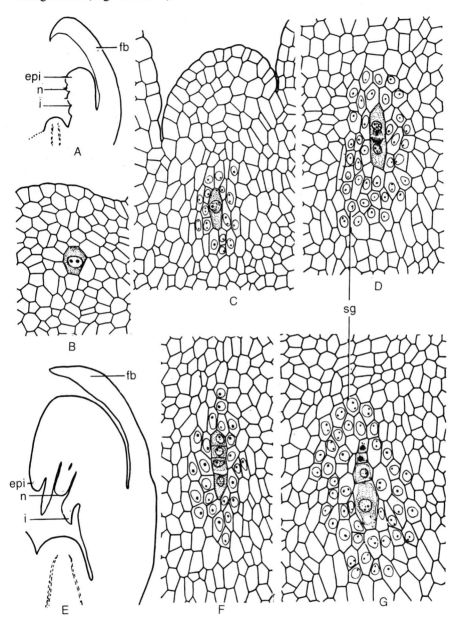

Fig. 15.4.9 A-G. *Podocarpus gracilior.* **A, E** Longisection young ovule (outline diagram). *epi* epimatium. *i* integument. *n* nucellus. **B** Nucellus from **A**, megaspore mother cell. **C, D, F, G** Formation of linear megaspore tetrad; note the circumjacent spongy tissue (*sg*). (After Konar and Oberoi 1969b)

Megasporogenesis.   A megaspore mother cell (mgmc) originates from a hypodermal archesporial cell and becomes deeply embedded due to the periclinal divisions of the cells above it (Fig. 15.4.9 B, C). A tapetum is present in *P. andinus, P. nivalis, P. falcatus* and *P. gracilior.* The ovule (*P. gracilior*) becomes somewhat exposed, and, for about a week, a pollination drop is secreted at its micropylar end. As the ovule grows, the part of the integument lying above the nucellus elongates to form a long narrow micropylar canal, which may be occasionally curved. The cells lining the upper part of the integumentary canal elongate inwardly, develop thickenings in the form of striations, and close the micropyle after pollination. Later, these cells lose their contents and look like tracheidal cells.

The reduction division in the mgmc (Fig. 15.4.9 D) results in a row of three cells (uppermost dyad and two megaspores) in *P. andinus* (Looby and Doyle 1944a), *P. nivalis* (Boyle and Doyle 1953), and a linear tetrad in *P. gracilior* (Fig. 15.4.9 F; Konar and Oberoi 1969b), *P. spicatus, P. ferrugineus* (Sinnott 1913) and *P. falcatus* (Osborn 1960). The chalazal megaspore is functional (Fig. 15.4.9 G).

Female Gametophyte.   The functional megaspore enlarges and becomes vacuolate (Fig. 15.4.10 A). Free-nuclear divisions follow and the nuclei are distributed peripherally (Fig. 15.4.10 B-D). Centripetal wall formation begins after several hundred free nuclei are formed. In *P. andinus*, alveolation occurs with an open inner face (Looby and Doyle 1944a). The gametophytic cells are arranged in files converging towards the centre; they are thin-walled and contain scanty cytoplasm. The two-layered megaspore membrane remains comparatively thin in the micropylar region of the gametophyte. The latter is without any definite shape due to its early contact with the pollen tube. The prothallial tissue may grow around the pollen tube.

The archegonial initial is a superficial cell (Fig. 15.4.11 A). It has scanty cytoplasm and a nucleus which grows in size. The development is very rapid (Fig. 15.4.11 B-D, G) and a mature archegonium has a neck which consists of four to six cells (*P. gracilior, P. nivalis*) or 10–15 cells (*P. andinus*) arranged in one (rarely two) tier. The neck cells degenerate or become inconspicuous soon after their formation. There is a small ventral canal nucleus and an egg nucleus, the latter shifts to a central position and is surrounded by a thick sheath of cytoplasm (Fig. 15.4.11 G). Gradually, the cytoplasm of the egg becomes thick and has few vacuoles and darkly staining masses appear, especially near the egg nucleus.

A mature archegonium is covered by a thin layer of the megaspore membrane which sometimes breaks down due to the impact of the pollen tube (*P. gracilior*).

A conspicuous jacket layer is present, its cells become multinucleate at maturity. Occasionally, before fertilization, the nuclei may escape into the archegonium. Groups of archegonia surrounded by a common jacket are also frequent.

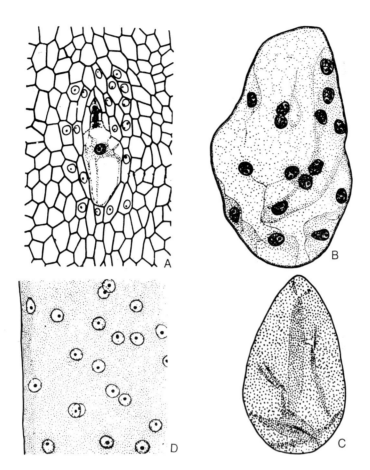

Fig. 15.4.10 A-D. *Podocarpus gracilior.* **A** Longisection central part of nucellus with functional megaspore, and three degenerated megaspores. **B, C** Free-nuclear gametophyte (whole-mount). **D** Portion of free-nuclear gametophyte. (After Konar and Oberoi 1969b)

There are two, rarely three, archegonia in *P. andinus* and *P. nivalis*. In *P. gracilior*, the number varies from 10–12; occasionally, as many as 22 have been counted (Konar and Oberoi 1969b). The shape of the archegonium depends on its position on the prothallus, and may be round, oval or elongate (Fig. 15.4.11 E, F). The formation of archegonia is markedly influenced by the presence of the pollen tube, and shows various stages of development in the same prothallus. Immature archegonia are present along the path of the pollen tube. The archegonia situated close to the spermatogenous (body) cell show a more advanced stage of development than those away from it.

Pollination

At the free-nuclear stage of the gametophyte, the ovules (*P. gracilior*) secrete a pollination drop in the early hours of the day, continuously for nearly a

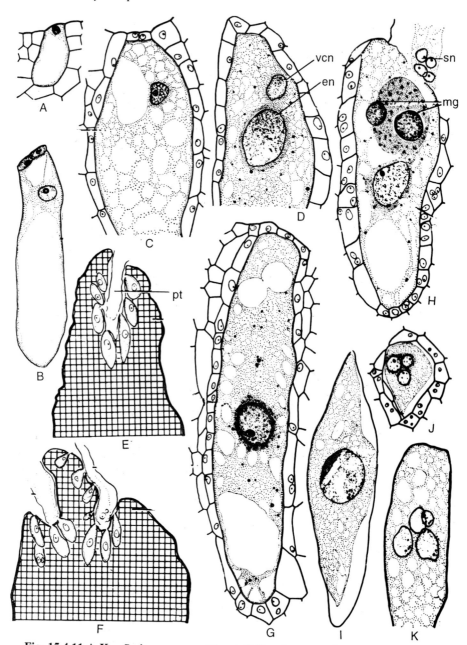

Fig. 15.4.11 A-K.   *Podocarpus gracilior.* **A-D** Development of archegonium. *en* egg nucleus. *vcn* ventral canal nucleus. **E, F** Outline diagrams of micropylar portion of gametophyte at fertilization, irregular arrangement of archegonia. *pt* pollen tube. **G** Mature archegonium shows thick cytoplasmic sheath around the egg nucleus. **H** Two male gametes (*mg*) in archegonium. *sn* sterile nuclei. **I** Fusion of male and female nuclei. **J, K** Archegonia show several nuclei in the egg cells formed by division of the unfertilized egg nucleus. (After Konar and Oberoi 1969b)

week. The pollen grains are caught in the pollination drop and sucked into the micropyle. Konar and Oberoi (1969b) also observed a large number of ovules pollinated with the pollen grains of *Pinus* (in the micropyle), besides the pollen of *Podocarpus* (Fig. 15.4.12 A, B). There was a pine forest in the neighbourhood. The pollen grains of pine abort after some time.

Post-pollination Development of Male Gametophyte. On reaching the nucellar surface, the pollen grain germinates immediately (Fig. 15.4.12 A, B). The intine bursts through the exine, a small pollen tube forms at the proximal end of the pollen grain and penetrates the nucellus but remains unbranched (Konar and Oberoi 1969b). The tube nucleus is the largest among the sterile nuclei and is the first to enter the pollen tube. It is followed by the rest of the sterile nuclei. The spermatogenous (body) cell is the last to enter the pollen tube (Fig. 15.4.12 B). In all species of *Podocarpus*, the spermatogenous cell lags behind in the pollen grain.

In *P. andinus* the spermatogenous cell overwinters in the grain, while the sterile nuclei pass into the short pollen tube. Later, the spermatogenous cell overtakes the sterile nuclei in the pollen tube and comes to lie ahead of them. In its early stages of growth, the pollen tube remains narrow and unbranched in the nucellus (*P. andinus, P. gracilior*). Later, its growth through the nucellus is rapid, and the pollen tube comes in contact with the (enlarged) female gametophyte, which is still at the free-nuclear stage (*P. gracilior, P. falcatus*). The sterile nuclei, as well as the spermatogenos cell, increase in size (Fig. 15.4.12 C-E).

On reaching the inner surface of the nucellar cap, the pollen tube expands into a more or less cylindrical vesicle (bulla) and expands maximally in the space between the nucellar cap and the female gametophyte (Konar and Oberoi 1969b). The bulla gives off lateral branches and the narrow tubes may themselves branch again (Fig. 15.4.13 A-E) and ramify on the inner surface of the nucellar dome and outer surface of the female gametophyte. The branched nature has been clearly documented from dissections by Konar and Oberoi (1969b). A lateral burrowing of the branches is a normal feature in *P. gracilior* and *P. falcatus*. The nuclear contents of the pollen tube lie within the bulla just above the female gametophyte in *P. nivalis* (Boyle and Doyle 1953), *P. gracilior* (Konar and Oberoi 1969b, Mehra 1991) and other members of the family Podocarpaceae.

Johri (1992) discusses the haustorial role of pollen tubes in several angiosperms and gymnosperms. In *P. gracilior* the pollen tube branches but does not grow intercellularly in the nucellar and gametophytic tissue. There is also no evidence of the collapse of the adjoining epidermal cells. Therefore, a haustorial function is out of the question (see also Konar and Oberoi 1969b). Mehra (1991), who also studied *P. gracilior*, concludes, on the basis of denser cytoplasm at the tip of tubes (dissected from macerated previously fixed material), that the pollen tube branches have a haustorial role. This presumption is not correct (see Johri 1992).

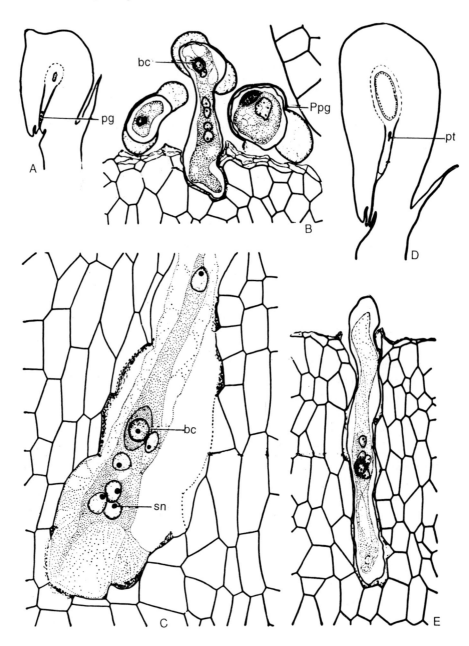

Fig. 15.4.12 A-E.   *Podocarpus gracilior.* A Longisection ovule (outline diagram) with pollen grains (*pg*) in micropylar canal and nucellus. **B** Nucellar tip from **A,** the tip of one of the three pollen grains has grown up to the nucellus. *bc* body cell. *Ppg Pinus* pollen grain. **C** Further growth of pollen tube in nucellus. *sn* sterile nuclei. **D** longisection ovule (outline diagram). *pt* pollen tube. **E** Pollen tube from **D**. (After Konar and Oberoi 1969b)

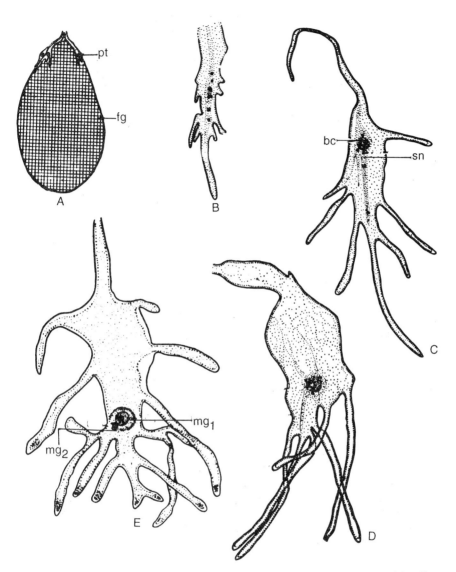

Fig. 15.4.13 A-E. *Podocarpus gracilior.* **A** Cellular female gametophyte (*fg*) with pollen tubes (*pt*). **B-E** Profuse branching of pollen tubes from bulla (whole-mount). *bc* body cell. *mg*1 functional, *mg*2 nonfunctional male gamete. *sn* sterile nuclei. (**A-D** After Konar and Oberoi 1969b; **E** after Mehra 1991)

The functional pollen tube, on reaching the female gametophyte, contains an enlarged spermatogenous cell (Fig. 15.4.14 A). The nucleus is eccentrically placed in the dense mass of cytoplasm (Fig. 15.4.14 B). Two or three small sterile nuclei surround the spermatogenous cell, the remaining nuclei lie away from it. The prothallial nuclei, tube nucleus, and the stalk nucleus reach the female gametophyte in a healthy state. By the time cell formation begins in the free-nuclear female gametophyte, the spermatogenous nucleus divides. In *P. andinus*, two unequal male gametes are formed (Looby and Doyle 1944a). In *P. nivalis*, one gamete remains small and is pushed out of the functional male cytoplasm. In *P. gracilior*, at first the two gametes appear to be equal but later become unequal (Fig. 15.4.14 C-F). Gradually, the smaller gamete is partly or wholly extruded, while the larger gamete comes to lie in the centre of the cytoplasm. The lattter shows a thick nuclear membrane, and a dark, densely staining homogenous nucleoplasm with no nucleolus. Later, irregular densely staining bodies appear in the cytoplasm of the male gametes and granular bodies appear within its nucleus (Fig. 15.4.14 E, F). Frequently, the non-functional male gamete enlarges slightly, shows a homogenous cytoplasm, and becomes demarcated from the functional gamete. The functional male gamete escapes from the cytoplasm, and enters the archegonium laterally, or through the neck cells. Along with the functional gamete, several sterile nuclei may also enter into the archegonium. More commonly, however, they degenerate outside the archegonia.

Fertilization

In *P. gracilior*, the pollen tube displaces the neck cells laterally, demolishing a part of the archegonial wall, and the contents of the tube are discharged into the archegonium. The functional male gamete with the cytoplasm enters the archegonium. The non-functional male gamete and some degenerated sterile nuclei lie outside the fertilized archegonium. Sometimes, both the male gametes and the sterile nuclei have been observed in the archegonium (Fig. 15.4.11 H). The egg nucleus of an unfertilized archegonium may divide to form a few nuclei (Fig. 15.4.11 J, K), which eventually degenerate. The fusion of the male gamete and egg nucleus occurs in the upper part of the archegonium (Fig. 15.4.11 I). In *P. andinus*, a unique phenomenon has been described (Looby and Doyle 1944b): the archegonial cytoplasm is discharged betwen the neck cells and pushes the male gametes backward. This is followed by a contraction of the cytoplasm and the functional male gamete back into the archegonium. According to Looby and Doyle (1944b), this is a normal feature and not due to any pressure on the turgid archegonia during collection and fixation.

Embryogeny

Podocarpus is an exceptional taxon among the conifers, as its embryogeny conforms to several distinct types (see Roy Chowdhury 1962). The basic plan of development is, however, uniform and consists of: (a) nuclear division of the zygote followed by four to five free-nuclear divisions to form 16–32 nuclei, (b) arrangement of nuclei in two tiers followed by wall formation, (c) the cells of both the tiers divide, (d) the upper tier (pU) forms an open tier (U) and the lower a suspensor tier (S), and (e) the division in

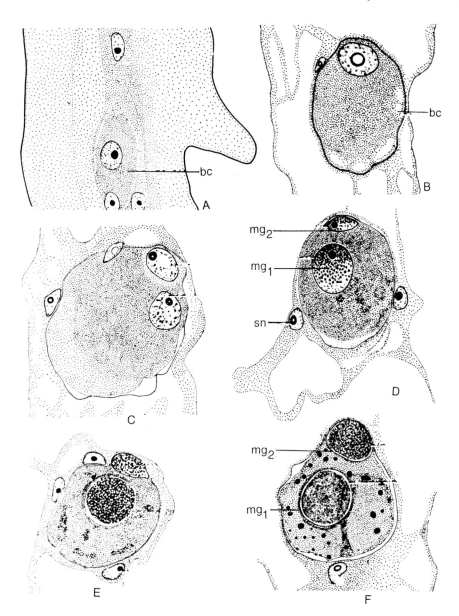

Fig. 15.4.14 A-F. *Podocarpus gracilior.* **A** Part of pollen tube with body cell (*bc*) and some sterile nuclei. **B** Body cell just before divisions. **C** Body cell with nearly equal eccentric male nuclei. **D-F** Maturation of functional male gamete (*mg*1, larger) and extrusion of the non-functional male gamete (*mg*2, smaller). *sn* sterile nucleus. (After Konar and Oberoi 1969b)

the lower tier (p*E*) is not followed by a wall, so that its cells (*E*) become binucleate. The binucleate condition of the cells of tier *E* is characteristic of the family Podocarpaceae.

In *P. gracilior,* the zygote is surrounded by dense non-vacuolate protoplasm. The nucleus moves to the lower part of the archegonium and divides (Fig. 15.4.15 A). These nuclei undergo four free-nuclear divisions, forming 32 nuclei (Fig. 15.4.15 B-E). This is also characteristic of the

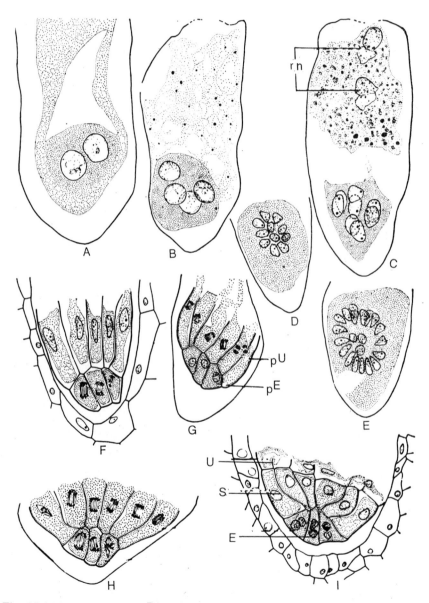

Fig. 15.4.15 A-I. *Podocarpus gracilior.* **A-C** 2-, 4-, and 8-nucleate proembryo. *rn* relict nuclei. **D, E** Transection 32-nucleate proembryo through upper region. **F-H** Two-tiered proembryo with nuclei dividing in one or both tiers. *pE* primary embryonal tier, *pU* primary upper tier. **I** Three-tiered proembryo, the embryonal cells (*E*) are binucleate. *U* upper tier. *S* suspensor tier. (After Konar and Oberoi 1969b)

primitive species of the sections *Stachycarpus, Afrocarpus, Sundacarpus* and *Nageia*. In the advanced sections, *Eupodocarpus* and *Dacrycarpus*, only 16 free nuclei are formed.

In *P. falcatus (Afrocarpus)*, the proembryo shows 16 and 32 nuclei, 16 nuclei are more frequent (Osborn 1960).

During the free-nuclear phase, some of the proembryonal nuclei escape into the upper part of the archegonium, persist till a much later stage of embryogeny, and eventually degenerate. Looby and Doyle (1944b) consider these as relict nuclei. They are more frequent in the sections of *Podocarpus* where five free mitoses occur.

During progressive divisions, the proembryonal nuclei decrease in size. Wall formation results in two tiers. The lower embryonal tier (p*E*) consists of 9–12 cells and the upper open tier (p*U*) of 15–22 cells. The embryonal cells are usually aranged in a single tier, so that each cell is in direct contact with the cells of the open tier, or the main group of p*E* cells surrounds 3–4 cells in the center. The group of inner cells are comparatively smaller and remain in direct contact with the suspensors; some of the cells lying below the inner cells have no direct contact with the cells of the open tier. All the nuclei in p*U* and p*E* tier undergo an internal division (Fig. 15.4.15 F-H). In p*U*, wall formation results in an upper tier (*U*) whose cells remain open, and a middle suspensor tier (*S*). In the p*E* tier the divisions are not followed by walls and the cells of this tier (*E*) become binucleate (Fig. 15.4.15 I).

There is variation in the ratio of cells in the *S* and *E* tiers among different species of *Podocarpus*. The total number in these tiers is determined by the number of free mitoses and the number of relict nuclei. However, the number of *E* cells in a species or a group of related species shows a narrower range of variation than the corresponding number of *S* cell. In *Afrocarpus, Stachycarpus*, and *Nageia* sections the average number of *E* cells is ca. ten, in *Dacrycarpus* it is four, while the most advanced *Eupodocarpus* shows one or two *E* cells.

A terminal cell in the embryonal group is reported in *P. spicatus, P. ferrugineus* (Buchholz 1936) and *P. andinus* (Looby and Doyle 1944b). This cell probably facilitates penetration of the embryo into the starchy endosperm. In *P. gracilior* the distal end of some of the *E* cells develops thick cellulose caps (Konar and Oberoi 1969b). In *P. nivalis* an apical enucleate portion is cut off in each embryonal cell (Boyle and Doyle 1954).

Differentiation of Embryo. After the embryonal cells become binucleate, the tier *S* elongates (Fig. 15.4.16 A, B). The nuclei of the tier *U* escape from the cells, form a group and eventually degenerate. The development from the time of fertilization to the elongation of suspensor takes 3–4 days. The gametophytic cells around the developing proembryo become multinucleate and filled with starch. The embryonal cells are pushed through these cells due to the elongation of the *S* tier. During elongation, many suspensor cells become detached, so that only a few cells remain in contact with the embryonal cells. The open end of such detached suspensor cells frequently proliferates to form small groups of cells (Konar and Oberoi 1969b). Such proliferation is common to other species (see Roy Chowdhury 1962). Occasionally, the cells of the *U* tier form groups of cells simulating embryonal masses.

The period of suspensor elongation is ca. 3 weeks. During this time the embryonal cells are in a "resting" stage, appear collapsed, and the cytoplasm stains deeply. Due to continued elongation, the suspensor becomes highly coiled and lies in the cavity formed by the breaking up and degeneration of the gametophytic tissue. After the suspensor tier ceases to elongate, the embryonal cells become active and the nuclei and nucleoli become prominent. The two nuclei of the embryonal cell divide simultaneoulsy (Fig. 15.4.16 C, D), followed by wall formation to form uninucleate cells. These cells are in groups of four, referred to as the embryonal tetrad (Looby and Doyle 1944b), and are a characteristic feature of the family (see Doyle 1954). The divisions in different cells of the embryonal group are not synchronous; some of the smaller inner cells may fail to divide and degenerate (Fig. 15.4.16 C, D).

Rapid divisions in the cells of the embryonal tetrad result in distinct lobing (Fig. 15.4.16 E), each lobe corresponds to a separate embryonal tetrad and remains distinct during later development. Finally, the lobes separate from each other due to the elongation of their upper cells, which function as secondary suspensor. Each embryonal tetrad separated by cleavage is a potential embryo (*P. gracilior*).

Cleavage polyembryony in *Podocarpus* appears to be related to the arrangement of cells in the tier *E*. In *P. andinus* (Looby and Doyle 1944b), *P. spicatus* and *P. ferrugineus* (Buchholz 1936), the *E* cells are arranged in superimposed tiers, and there is no cleavage of the embryonal group. In other species, such as *P. amarus*, *P. usambarensis* (Buchholz 1941) and *P. gracilior* (Konar and Oberoi 1969b), and in other advanced sections where the *E* cells are arranged in one horizontal tier, each embryonal cell represents a potential embryo.

In *P. gracilior,* most of the embryonal groups may become suppressed at a very early stage, so that ultimately there is only a single mass of embyonal cells at the tip of the massive secondary suspensor. For some time, meristematic activity and divisions in all planes in the lower half of the embryonal mass adds to the over-all size, while the other half continues to contribute to the massive secondary suspensors.

During early ontogeny, a dermatogen-like layer differentiates around the proximal end of the embryonal mass, and undergoes periclinal divisions. By the time the cotyledons are initiated, a stem tip and a root tip can also be clearly discerned. In the central part of the embryo, between the root and shoot tip, the cells are elongated and the procambial strand passes into

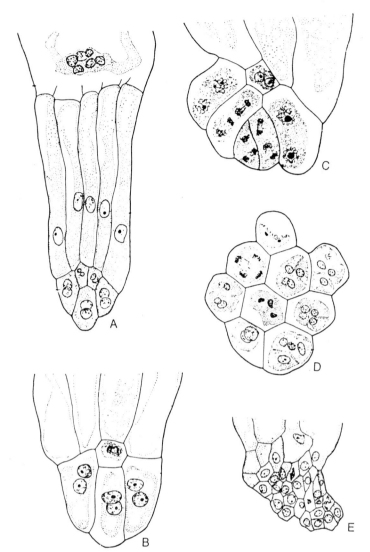

Fig. 15.4.16 A-E. *Podocarpus gracilior.* **A, B** Three-tiered proembryo at the beginning of suspensor elongation. **C, D** Longi-and transection of embryonal tier, some of the embryonal cells have degernerated. **E** Lobing of embryonal tier after formation of embryonal tetrad. (After Konar and Oberoi 1969b)

each cotyledon. Before any further organization, the embryo elongates considerably. The mature embryo shows a well-marked stem and root apex. In some mature seeds, the embryo has a well-developed plumular leaf lateral to the shoot apex.

The fleshy seeds of *P. andinus* and *P. ferrugineus* fall to the ground before the embryo matures (cf. *Cycas, Ginkgo* and *Gnetum*, where further development of the embryo takes place only after the seeds are shed). In *P. andinus*, the oldest embryo showed initial stages of cotyledonary development (see Roy Chowdhury 1962).

Seed

In addition to the integument, the seed coat includes a thick layer of epimatium. The tissues of the integument and epimatium are difficult to demarcate, as they are completely fused with each other except for a short distance at the micropylar end.

At the time of pollination, the integument consists of several layers of thin-walled cells, some of which, in the hypodermal region, contain tannin. At the free-nuclear stage of the gametophyte, some cells of the integument at the level of the nucellar beak contain tannin and resinous material. As the ovule enlarges the tanniniferous layer extends towards the micropyle and around the nucellus (Fig. 15.4.17 A-D). Simultaneously, the cells around the female gametophyte and interior to the tannin layer become sclerenchymatous (Fig. 15.4.17 A-D). This thickening extends towards the micropyle to the tanniniferous layer. During and after the cellular stage of the gametophyte, the seed coat differentiates into three zones (Fig. 15.4.17 E, F): (a) an inner zone of sclerenchymatous cells around the nucellus; (b) a middle two-/three-layered zone of tannin-filled cells, and (c) an outer several-layered zone of parenchymatous cells. Stomata are present on the epidermis. During the maturation of archegonia and early embryogeny, two to three layers below the epidermis become completely resinous. A few scattered tracheids are also present in the outer parenchymatous region. The hard, stony sclerenchymatous zone becomes exposed after the seed is shed and the two outer zones disintegrate.

The mature seeds of *P. henkelii* at shedding have a moisture content of ca. 62%; when incubated on a moist substrate they show slow and sporadic germination (Palmer and Pitman 1972). The removal of epicuticular wax, the epidermis, or the entire epimatium, leads to rapid water uptake and germination (Noel and van Staden 1975). The epimatium acts as a barrier to germination by restricting water uptake, and also permits a rapid and uncontrolled water loss, which can lead to loss of viability (Noel and van Staden 1975, Dodd and van Staden 1981). The seeds are at a double disadvantage by their inability to imbibe water readily and control dessication, which probably contributes to the limited plant regeneration in nature (von Breitenback 1965).

Short-term stroage of seeds (up to 18 months) is possible by holding seeds of high moisture content at 4 °C (Dodd and van Staden 1981). Over a 16-week period of storage, the seeds remain metabolically active and further embryonic growth occurs. Dodd et al. (1989) investigated the biochemical changes and fine structure of the female gametophyte and embryo of such stored and scarified seeds, during the course of germination. After a 3-day incubation at 25 °C, the cells of the root tip (at the electron microscope level) show increased vacuolation, numerous amyloplasts and lipid mobilization. By day 6, there is measurable embryonic growth and ultrastructural evidence of synthetic activity by the presence of abundant

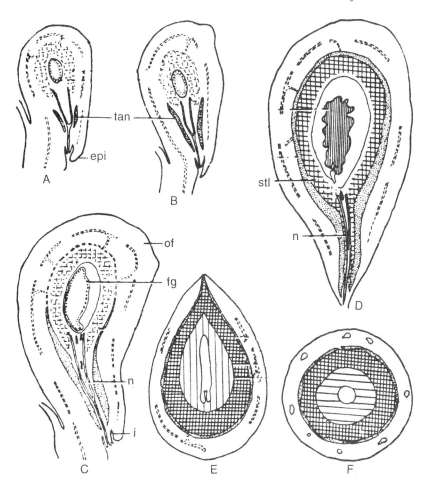

Fig. 15.4.17 A-F. *Prodocarpus gracilior.* **A-D** Longisection ovule (outline diagrams). *epi* epimatium. *fg* female gametophyte. *i* integument. *n* nucellus. *of* outer fleshy layer. *stl* stony layer. *tan* tanniniferous zone. **A-C** Free-nuclear gametophyte. **D** Cellular gametophyte at fertilization. **E** Longisection of mature seed with persistent female gametophyte, and dicotyledonous embryo. **F** Transection mature seed through hypocotyledonary region of embryo. (After Konar and Oberoi 1969b)

endoplasmic reticulum (ER), ribosomes, and dictyosomes. Fine-structural changes occur in the female gametophyte and suggest mobilization of reserve food. By 9 days of incubation, biochemical studies indicate a gradual decline in lipid and protein in the female gametophyte and embryo. At day 6, a decline in the embryonic starch levels contrasts with the increase in the female gametophyte. Between days 3 and 6, minor changes occur in the sugar level of the female gametophyte in contrast to the increase in embryonic tissues. These changes coincide with the first phase of germination. The retention of high moisture and the evidence of metabolic activity suggest that, following scarification, the transition between maturation and germination is characterized by a continuation of earlier synthetic activity and reserve inter-conversions.

Germination. There is no dormancy period, and young seedlings develop within 3 months (*P. gracilior*). The germination is epigeal (Fig. 15.4.18 A-E). The radicle emerges after the stony layer cracks laterally. The stony layer is cast off at an early stage and the hypocotyl elongates carrying the cotyledons covered with remnants of the female gametophyte (endosperm). There are two, rarely three, cotyledons, which persist for a long time. The juvenile leaves are short and arranged in loose whorls (Fig. 15.4.18 F-H).

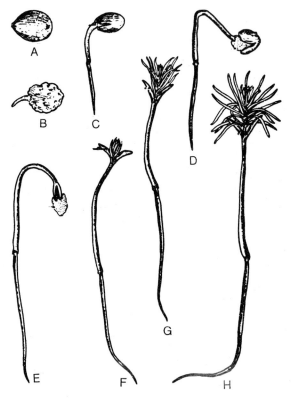

Fig. 15.4.18 A-H. *Podocarpus gracilior.* **A** Mature seed. **B-E** Germination of seed. **F-H** Later stages of seedling with cluster of leaves. (After Konar and Oberoi 1969b)

Chromosome Number

The haploid number, n = 17, has been determined from endosperm squashes in *P. neriifolius* (Mehra 1988). Three chromosomes are isobrachial with median primary constriction (marked M) and six chromosomes with terminal or near-terminal primary constriction (Fig. 15.4.19 A).

In *P. macrophyllus* the haploid number is n = 19. Three chromosomes are isobrachial, ten heterobrachial, and six have terminal or near-terminal primary constriction (Fig. 15.4.19 B).

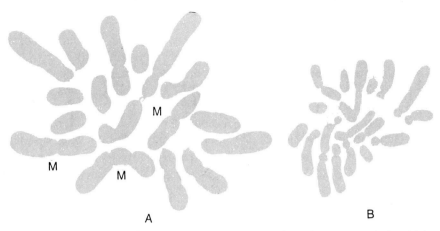

Fig. 15.4.19 A, B. A *Podocarpus neriifolius* (n = 17), three chromosomes isobrachial with median primary constriction (marked M), eight chromosomes whith subterminal primary constriction, and six chromosomes with near-terminal primary constriction. **B** *P. macrophyllus* (n = 19), three chromosomes isobrachial, ten heterobrachial, six with terminal primary constriction. (After Mehra 1988)

Temporal Considerations

Two types of reproductive cycles have been recognized in *Podocarpus* (Doyle 1954: a 2-year type, as in *P. ardinus* (Looby and Doyle 1944b), *P. falcatus* (Osborn 1960) and *P. gracilior* (Konar and Oberoi 1969b), while in the other type, the cycle is completed in 1 year, as in *P. nivalis* (Boyle and Doyle 1954).

15.5 ARAUCARIACEAE

The family Araucariaceae comprises two extant genera, *Agathis* and *Araucaria*. These are tall evergreen trees reaching a height of 60 m or more. *Agathis* has 13 spp. (Whitmore 1980a) and *Araucaria* 18 spp. (de Laubenfels 1972). They are confined to the Southern Hemisphere, with a very restricted distribution (see Chap. 1).

The Araucariaceae has a long and extensive fossil record. The remains of plants from the Southern Hemisphere are also recorded from the Northern Hemisphere. The fossil record indicates that this family was considerably larger, more diverse and widespread during the Mesozoic. Unquestionable araucarian remains extend into the Jurassic; there were numerous taxa in the Northern Hemisphere at that time. Petrified seed cones from the Middle Jurassic of Argentina indicate that *Araucaria* had evolved by that time (Calder 1953). While there are several reports of araucarian remains in the Triassic (Bock 1954, 1969), these are mostly impression fossils with dubious affinities to the family (Stockey 1982). The members of Araucariaceae first appear with certainity in the Late Triassic, seem to reach a peak of diversity during the Jurassic and decline towards the end of the Cretaceous (Miller 1977).

Araucaria

The name is adopted after Arauco, a province of Chile, the native habitat of *A. araucana* (monkey puzzle tree). According to Endlicher (1842), *Araucaria* comprised two sections: *Eutacta* and *Columbea*. Wilde and Eames (1952) divided *Araucaria* into four sections based on extant species.

a) Section *Eutacta* Endlicher. Leaves reduced, thick, often keeled, imbricate and erect. Male cones solitary, terminal, female cones on long peduncles. Ovulate cone scales thinly winged, indehiscent, the seed retained on the scale at shedding. Vascular supply of bract-scale unit single at source. Germination epigeal. Cotyledons subsessile, freed from seed at germination. Seedling not fleshy; 15 spp., including: *A. biramulata* Buchholz, *A. columnaris* (Forster) Hooker, *A. cunninghamii* Aiton ex Lambert, *A. heterophylla* (Salisbury) Franco, *A. memorosa* de Laubenfels.

b) Section *Columbea* Endlicher Emend.Wilde and Eames (1952). Leaves large, generally thin, flat. Male cones axillary, usually two to several on foliaceous branches. Ovulate cone scales nut-like, wings absent, indehiscent, the seed retained on the scale at shedding; vascular supply of bract-scale unit single at source. Germination hypogeal. Cotyledons long-stalked at germination, retained in seed coats, stalks not fused. Seedling fleshy, without a long subterranean dormant period; two spp. *A. araucana* (Molina) K. Koch, and *A. angustifolia* (Bertoloni) O. Kuntze.

c) Section *Intermedia.* White (1947). Leaves large, generally thin, flat, spreading, sometimes slightly imbricate; juvenile leaves needle-like, flat, small. Male and female cones axillary; ovulate cone scales samara-like with broad thin wings, indehiscent; seed retained on the scale at shedding. Germination epigeal. Cotyledons subsessile, freed from seed coats at germination. Seedling not fleshy. Three spp: *A. hunsteinii* K. Schumann, *A. klinkii* Lauterbach, and *A. schumanniana* Warburg.

d) Section *Bunya.*Wilde and Eames (1952). Leaves large, flat, spreading or slightly imbricate. Male cones axillary, female cones subsessile or on short (2 cm long) peduncles. Ovulate cone scales large, heavy with woody rings; dehiscent, large "seed" shed from the scale at maturity. The vascular supply of bract-scale unit double at source. Germination hypogeal. Cotyledons long-stalked at germination, retained within seed coats, the stalks fused into a hollow cylinder. Seedling fleshy with a long subterranean dormant period. One sp: *A. bidwillii* Hooker.

Morphology

The young evergreen trees are symmetrical, clothed with branches from base to apex, the old trees have the trunk clean of branches for the greater part of their height; branches are horizontal and in whorls, leaves persist for many years, and are extremely variable in morphology. They may be sharply pointed, needle-like and imbricate (*A. memorasa*; Fig. 15.5.1 A, B) or broad and imbricate (*A. angustifolia, A. bidwillii, A. klinkii;* Fig. 15.5.1 C, D).

Fig. 15.5.1 A-D. A, B *Araucaria memorosa*. **A** Juvenile foliage. **B** Mature foliage and young female cones (*fc*) at pollination. **C** *A. klinkii* and **D** *A. bidwillii*, mature foliage. (After Stockey 1982)

The leaves are spirally arranged, clasp the stem and overlap, or may occur in two or more ranks by means of a basal twist. They are leathery, lance-shaped and sharply pointed, or awl-shaped and four-angled, or triangular. The size and shape of the leaf varies on different parts of the same tree.

This taxon is dioecious, occasionally the male and female cones are borne on different branches of the same tree.

Anatomy

Stem.   The wood of *Araucaria* is mostly whitish or very light coloured. In *A. brasiliana*, irregular red streaks are present. Leaf traces occur and are often numerous.

The wood of extant araucarias can be distinguished readily from other conifers by the absence of resin canals or resin cells (Fig. 15.5.2 B). Resin occurs in the ray cells and axial tracheids as resin plugs (Jane 1970). Multiseriate circular bordered pits are present in two to five vertical rows, which are characteristically alternate and appear hexagonal due to crowding (Greguss 1955; Fig. 15.5.2 A). Pitting on the ray cells is cupressoid and rays are 1–20 cells high. The family is also peculiar in lacking crassulae.

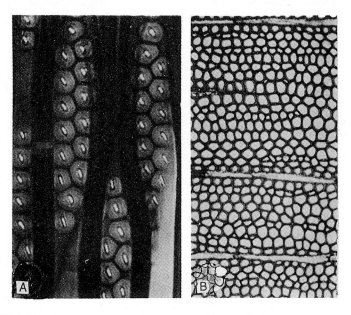

Fig. 15.5.2 A, B.   *Araucaria heterophylla*, wood anatomy. **A** Radial longisection shows characteristic alternate pitting. **B** Transection, note absence of resin canals. (After Stockey 1982)

Leaf.   The cuticle of *Araucaria* leaf is thick and the micromorphology of juvenile, adult, herbarium and preserved leaves has been studied with scanning electron microscopy (SEM). The micromorphological features reveal much more variability than was presumed previously (Stockey and Ko 1986). The external and internal characters of abaxial and adaxial cuticle have been characterized for the four sections of the genus.

The outer surface of *Araucaria* leaf is relatively smooth (Fig. 15.5.3 A). There are no stomatal papillae, "Florin rings" (Buchholz and Gray 1948), or other surface ornamentation. However, irregularly shaped platelets of unknown origin are scattered over the cuticular surface frequently on the small needle-like leaves of Section *Eutacta*. All the species have stomatal plugs (Fig. 15.5.3 B) sunken into the stoma on both adaxial and abaxial surfaces. These plugs (perhaps wax) are commonly composed of fused rods (Fig. 15.5.3 C), or are nearly solid (Fig. 15.5.3 D).

The stomata are sunken to the level of the hypodermis in discontinuous rows. They are arranged regularly on broad leaves regardless of the Section affinities. The stomata are also present in groups or clusters (*A. barnieri*). They occur on both adaxial and abaxial surfaces in widely spaced rows in species with needlelike imbricate leaves (*A. cunninghamii*; Stockey and Taylor 1978a, b).

The stomatal orientation varies. It is parallel (vertical) in the section *Buniya* (Fig. 15.5.3 E, F), *Columbea* and *Intermedia*. In the Section *Eutacta* it is oblique and perpendicular (horizontal; Fig. 15.5.3 G).

Subsidiary cells vary from three to seven (Stockey and Ko 1986). Four subsidiary cells (two polar and two lateral; Fig. 15.5.3 E, G) are common for the genus; five subsidiary cells are also common in smaller or needle-like leaves.

The cuticle on the subsidiary cells is slightly pitted. Depending on the species, regardless of sectional affinity, on the stomatal surface the cuticle may be oval, elliptical, central or polygonal. On the guard cell surface the cuticle is mostly reticulate over the entire or part of the surface, especially near the stoma. It is a diagnostic character for distinguishing the araucarian species.

The cuticular flange between the subsidiary cell and guard cells is usually serrated with a few species showing a smooth edge. Prominent polar extensions occur, more often in smaller leaves of the section *Eutacta*.

The epidermal cells are usually elongate and rectangular with pitted cuticular surface (Fig. 15.5.3 H).

According to Stockey and Ko (1986), study of the cuticular morphology indicates a need for further taxonomic work on *Araucaria*.

The leaves of *Araucaria* have a complete layer of sclerified hypodermal cells interrutped only by stomata (Vasiliyeva 1969). In *Agathis*, the sclerified hypodermal layer is incomplete. The mesophyll contains palisade and spongy parenchyma, sclerids and unusually enlarged cells. Bamber et al. (1978) examined these unusual cells (which are present only in *Araucaria*). The lack of cytoplasm and the presence of numerous partitions in the lumen characterize these cells. The partitions divide the cell lumen into

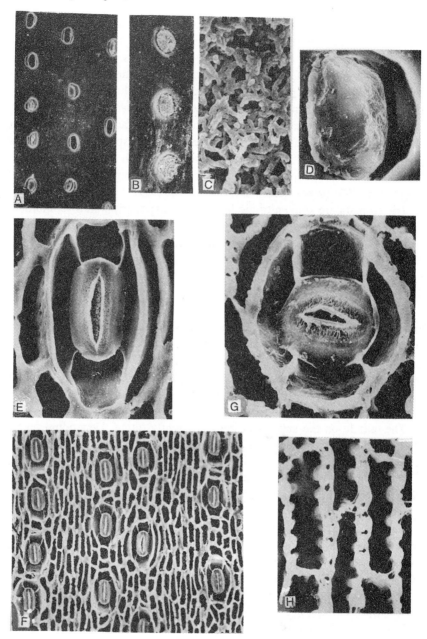

Fig. 15.5.3 A-H. Cuticle. **A, D, G, H** *Araucaria huntsteinii*. **A** Adaxial cuticle shows smooth surface and stomata. **B, C, E, F** *A. bidwillii*. **B** Stomatal plugs. **C** Stomatal plugs of fused rods. **D** Nearly solid stomatal plug. **E, F** Adaxial cuticle. **E** Stomatal apparatus with two polar and two lateral subsidiary cells. **F** Orientation of stomata parallel to the long axis of the leaf. **G** Horizontally oriented stomatal apparatus shows thick cuticle especially on subsidiary cell surfaces. **H** Adaxial cuticle, inner surface shows pitted epidermal flanges. (After Stockey and Ko 1986)

compartments. Bamber et al. (1978) suggest that these cells can be referred to as compartmented cells. Histochemical tests indicate that the partitions are pectinaceous.

The function of the compartmented cells has not been determined. The high proportion of the mesophyll zone (which they occupy) suggests some function in relation to the stroage of carbohydrates. This is supported by the strong positive PAS reaction for carbohydrates (by the compartmented cell partitions). There was no effect on cell contents after 7 days' complete shading from light. The abundance of these cells in the mesophyll zone of all *Araucaria* leaves examined indicates that they have (or have had in the past) an important physiological role. Compartmented cells appear to be of taxonomic significance as they are the only definite anatomical character by which *Araucaria* and *Agathis* can be distinguished.

Reproduction

A survey of the literature reveals that Araucariaceae, as compared to other families of the Coniferales, have received little attention in the study of reproductive biology. There are a few brief reports on the life cycle of *Agathis australis* (Eames 1913), *Araucaria brasiliensis* (= *A. angustifolia*— Burlingame 1913, 1914, 1915) and *A. araucana* (Favre-Duchartre 1960a, Hodcent 1971), and male cone development in *Agathis robusta* (Kaur et al. 1981). This is possibly due to the restricted distribution of the two taxa (Kaur et al. 1981).

Male cone

The male cones are large, ca. 20 cm in length (*Araucaria rulei*). There is a central axis with numerous microsporophylls arranged spirally (Fig. 15.5.4 A). Each microsporophyll bears 8–15 microsporangia on its abaxial surface (Fig. 15.5.4 B). The sporangia are elongated. The pollen output of each cone may be as high as 10 000 000 (see Sporne 1965). The pollen grains are wingless.

Microsporogenesis. Due to the amoeboid nature of only a few tapetal protoplasts in *Araucaria*, Hodcent (1965) has reported the formation of limited periplasmodium. Towards the end of meiosis in the microspore mother cell, the tapetal cells begin to degenerate and are consumed by the time pollen grains are formed.

Male Gametophyte. The microspore nucleus divides periclinally, and two prothallial cells are formed which divide repeatedly and produce 13–40 cells (cf. family Podocarpaceae, where three or four cells are formed). The walls between these cells break down, so that the free nuclei remain irregularly distributed in the cytoplasm. The antheridial cell divides into a tube cell and a generative cell. The latter divides anticlinally into a stalk cell and

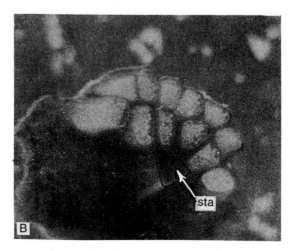

Fig. 15.5.4 A, B. *Araucaria columnaris.* **A** Shoot with microsporangiate cones.
B Transection microsporophyll with several microsporangia on abaxial surface. *sta*
stalk. (After Bierhorst 1971)

body (spermatogenous) cell (Fig. 15.5.5 A-E). The spermatogenous cell divides and produces two equal (Burlingame 1915, Ghose 1924) or unequal (Burlingame 1913, Eames 1913) male gametes.

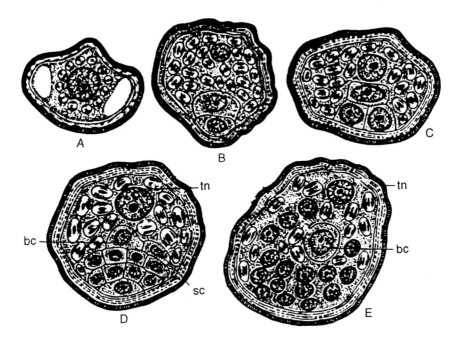

Fig. 15.5.5 A-E. *Araucaria cunninghamii,* male gametophyte. **A** Uninucleate pollen grain. **B** Pollen grain with two prothallial cells. **C** Lower prothallial cell divided anticlinally. **D** Further multiplication of prothallial cells; also note tube (*tn*), body (*bc*) and stalk (*sc*) cells. **E** The enlarged stalk cell, body cell, tube cell, and free nuclei of prothallial cells; the walls have collapsed. (After Chamberlain 1935)

Female Cone

The female cones are spherical (Fig. 15.5.6 A, B), ca. 30 cm in diameter, and have deciduous scales. The bract and the ovuliferous scale are only partially fused, the tip of the ovuliferous scale is free and constitutes the so-called ligule. There is normally only one or two (less common) reflexed ovule(s) on each cone scale. At a relatively early stage in development, the ovule fuses (to a large extent) to the subtending bract. A flap of cone scale tissue covers the ovule and its micropyle is exposed at a basal notch. This flap is free of the ovule and can be easily peeled off (Fig. 15.5.7 A-D). The seeds are deeply embedded in the scale tissue. The cones shed the scales at maturity; the entire cone shatters. In *A. bidwilli,* the scales are shed; however, the entire cone abscisses, scattering the scales on impact. The seeds can then be removed from the cone scale.

Megasporogenesis. A deep-seated megaspore mother cell differentiates

Fig. 15.5.6 A, B. **A** *Araucaria angustifolia*, immature ovulate cones and mature foliage. **B** *A. bidwillii*, female cone at pollination. (After Stockey 1982)

in *A. brasiliensis* (Burlingame 1914). It is assumed that the reduction divisions occur, tetrads are formed, and the chalazal megaspore functions (P. Maheshwari and H. Singh 1967).

Female Gametophyte.   Over 2000 free nuclei are produced (Burlingame 1914) before wall formation. According to P. Maheshwari and H. Singh (1967), alveolation in the Araucariaceae needs to be reinvestigated.

The number of archegonia varies from 3 to 25, remain confined to the upper half or one third of the gametophyte, and form a ring around the apex of the prothallus (P. Maheshwari and Sanwal 1963). Although apparently sunken, the archegonia probably arise from the superficial initials which later come to lie at a lower level, due to upward growth of the gametophytic cells. Not all the archegonia mature or are fertilized (P. Maheshwari and Sanwal 1963).

A single-layered archegonial jacket differentiates, and may rarely be absent between two archegonia. A binucleate condition of the jacket has been recorded by Burlingame (1914) in *A. brasiliensis*. Thick and thin areas

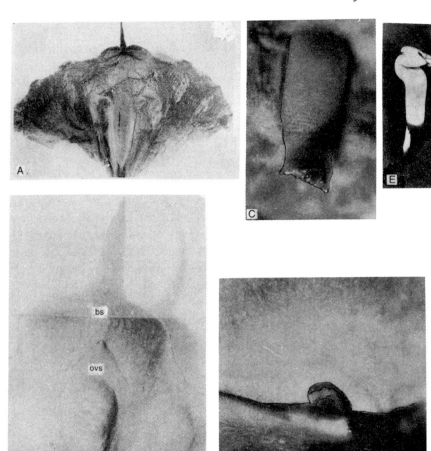

Fig. 15.5.7 A-E. A *Araucaria klinkii,* cone-scale complx. **B-D** *A. columnaris.* **B** Cone-scale complex, adaxial view, arrow indicates opening of the ovular pouch. *bs* bract scale. *ovs* ovuliferus scale. **C** Ovule (from **B**) after removing cone-scale flap. **D** opening of the ovular pouch (from **B**). **E** *A. bidwilli,* excised embryo shows a small embryo (at left), dark area corresponds to calyptro-periblem, the cotyledons are folded. (**A, E** After Stockey 1982, **B-D** after Bierhorst 1971)

occur in the inner wall of the jacket cells. In exceptional cases, the jacket may be double-layered at places (see Konar and Moitra 1980).

The neck is dome-shaped, comprise about 12 wedge-shaped cells in a single tier, and a narrow passage is left in the centre. Frequently, the entire neck is shed.

A ventral canal nucleus is present in the archegonium of *Araucaria* (P. Maheshwari and H. Singh 1967), but Burlingame (1914) failed to observe it.

Pollination

Probably there is no pollination drop mechanism. The pollen grains fall on the ovuliferous scale, germinate in situ, the pollen tubes grow (like fungal hyphae) towards the micropyle and are the longest (along with *Agathis, Saxegothea, Tsuga dumosa* and *T. heterophylla*) in gymnosperms (see H. Singh 1978). The micropyle is symmetrical and non-stigmatic and the nucellus has a beak which projects beyond the micropyle.

Embryogeny

The zygote divides and both the nuclei lie in an irregularly shaped mass of densely staining cytoplsm (Fig. 15.5.8 A), slightly below the centre of the archegonium (*A. cunninghamii*—Haines and Prakash 1980). The central location of the proembryo of *Araucaria* is a specialized feature in contrast to the basally situated proembryo in other conifers, cycads and *Ginkgo*. Four synchronous mitoses follow, and there is no evidence of any irregularity in

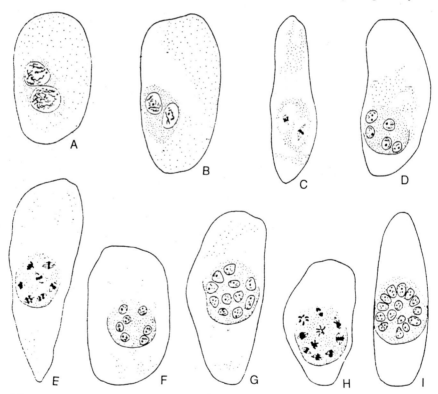

Fig. 15.5.8 A-I. *Araucaria cunninghamii*, proembryogeny. **A** Binucleate proembryo. **B** proembryo after second mitosis. **C** Metaphase of third division. **D** 16-nucleate (only 5 nuclei visible) proembryo. **E** Fifth mitosis, nuclei at metaphase. **F** 32-nucleate (only 7 nuclei are seen) proembryo, nuclei spherical and irregularly arranged. **G** Nuclei (only 11 are seen) angular and regularly arranged. **H** Sixth mitosis. **I** 64-nucleate (only 15 nuclei are seen) proembryo. (After Haines and Prakash 1980)

these divisions. The cytoplasm of the proembryo, containing the randomly scattered nuclei, is clearly demarcated from the degenerating archegonial cytoplasm, particularly at the lower end. The proembryo becomes spherical (Fig. 15.5.8 B-I), unlike other conifers. At the 32-nucleate stage (which lasts considerably longer than 2-, 4-, 8-, and 16-nucleate stages) the nuclei become angular and regularly arranged (Fig. 15.5.8 G). After the sixth mitosis (Fig 15.5.8 H), there are 64 free nuclei in the proembryo (Fig. 15.5.8 I) followed by wall formation in *A. cunninghamii, A. bidwilli, A. heterophylla* and *A. araucana*. Although the details are not known, the 64-nucleate proembryo is either free-nuclear or cellular.

The two-tiered proembryo consists of a primary upper tier (*pU*) and a primary embryonal group (p*E*) with a cap. The cells of the p*U* tier and the cap are in a single tier, each arranged in a hemi-spherical plane. They jointly enclose the embryonal group (Fig. 15.5.9 A, B; *A. bidwilli*). The cells of the p*U* tier remain open towards the distal end and subsequently elongate, as do also the cap cells, though not as markedly.

The cells of both the p*U* and p*E* tiers divide. In the former, the division leads to the formation of the upper tier (*U*) and the suspensor tier (*S*). In the p*E* tier, the cap cells divide anticlinally and the daughter cells become arranged in a single tier. The cells of the embryonal group divide irregularly and in *A. cunninghamii* the proembryo consists of a *U* tier of 31–28 cells, *S* of 31–38 cells and *E* of 18–26 cells + cap 32–44 cells. In *A. bidwilli* the *U* tier is 21–32 cells, *S* 21–32 cells and *E* 18–22 cells + cap 40–62 cells. In *A. heterophylla* the *U* tier comprises 32–35 cells, *S* 32–35 cells and *E* 18–24 cells + cap 36–44 cells.

In a cellular proembryo of *Araucaria*, the *U* and *S* tiers are curved (Fig. 15.5.9 A, B) while they are more or less horizontal in other conifers. Other features of interest in the proembryogeny are: (a) six free mitoses (the highest in conifers), (b) cell wall formation, (c) elongation of *S* cells independently of the final mitosis and (d) the formation of cap. According to Haines and Prakash (1980), *Araucaria* is the most primitive of the conifers, with affinity closest to the primitive podocarps.

The embryogeny of *Araucaria* resembles other conifers (Haines 1983). Closest to the axis of the proembryo, the laterally situated cap cells elongate (Fig. 15.5.9 A-D). The cell walls are not markedly thick, and intercellular spaces do not occur (*A. bidwilli*; Fig. 15.5.9 F, G). During this phase, the suspensor cells also elongate, begin to separate at their distal end, and accumulate abundant starch grains (Fig. 15.5.9 C). The elongation is subsequently evident in two regions of the suspensor:
a) The separated distal part which curls and exhibits thickening of cell walls to form an anchorage (Fig. 15.5.9 C).
b) Elongation in the unseparated region pushes the proembryo out of the archegoinum through the nutritive tissue of the corrosion region. The latter is formed below the archegonia in the central region of the gametophyte concurrently with the development of proembryo.

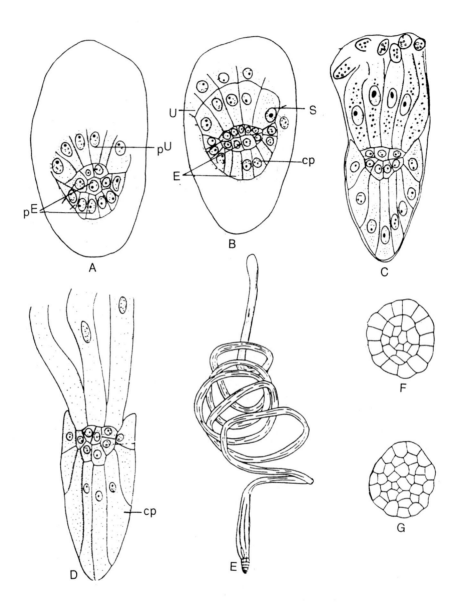

Fig. 15.5.9 A-G. *Araucaria bidwillii*, embryogeny. **A-D** Longisection proembryo.
A Two-tiered proembryo shows primary upper tier (*pU*) and primary embryonal tier
(*pE*). **B** Three-tiered proembryo; upper tier (*U*), suspensor (*S*) and embryonal (*E*) tier;
the latter comprises embryonal cells and cap cells (*cp*). **C, D** Proembryo with elongated
cap and suspensor cells. **C** Suspensor anchorage. **E** Coiled suspensor with embryonal
mass at tip. **F, G** Transection embryonal cells surrounded by cap cells. (After Haines.
and Prakash 1980)

The suspensor cells do not divide. In a fully elongated suspensor, the cells remain attached to one another as a column, although there are spaces between them. The region next to the tip of the proembryo and the comparatively longer region towards the archegonium remain more or less linear, while the intervening area is markedly coiled (Fig. 15.5.9 E). The individual suspensor cells increase in diameter from the archegonium towards the tip, ultimately reach four to five times their original diameter. Only four or five suspensor cells remain in contact with the tip of the proembryo. The remaining cells cease to elongate during the development of the suspensor.

As the proembryo grows downward to the gametophytic tissue, the corrosion region is digested and a cavity is formed. Although there is some degeneration of cell contents, ahead of the embryo the major activity occurs in the vicinity of the suspensor. The corrosion region and corrosion cavity develop in both pollinated and unpollinated ovules. The cap cell contents often show signs of degeneration during later stages of suspensor elongation.

Simple polyembryony is a regular feature; cleavage polyembryony has not been reported in the family Araucariaceae (Haines and Prakash 1980). In this respect, according to Doyle and Brennan (1971, 1972), the Araucariaceae must be considered as the most primitive conifer family.

A mature excised embryo of *A. bidwillii* (Fig. 15.5.7 E) shows a small embryo, a dark calyptro-periblem and folded cotyledons (Stockey 1982). In the species of *Araucaria*, in the mature embryo, there are variations in size, number, length of the cotyledons, hypocotyl and root cap, differentiation of stomates, and the extent and arrangement of the vascular and secretory tissues. According to Haines (1983), all the above features lend support to the generally recognized division of the taxon into four sections.

Chromosome Number

The karyotypes are similar in *A. angustifolia*, *A. bidwillii* and *A. columnaris*, studied from the endosperm and root tip squashes (Mehra 1988). In *A. bidwillii* and *A. angustifolia* the endosperm squashes (n = 13) showed four chromosomes heterobrachial and nine isobrachial (Fig. 15.5.10 A). The root tip squashes in *A. columnaris* (2n = 26) showed 8 heterobrachials and 18 isobrachials (Fig. 15.5.10 B).

15.6   CEPHALOTAXACEAE

The meagre fossil record of the family Cephalotaxaceae extends from the Jurassic. It is speculated that this family may have evolved from the *Ernestiodendron* line of evolution (through *Palissaya*) because the sterile part of each fertile dwarf shoot is highly reduced. According to Miller (1977), the present fossil record of the family does not confirm this hypothesis.

Cephalotaxus

The Cephatotaxaceae is represented by a single living genus, *Cephalotaxus*,

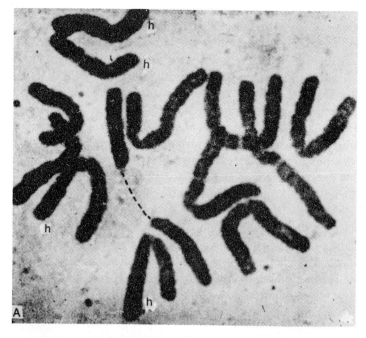

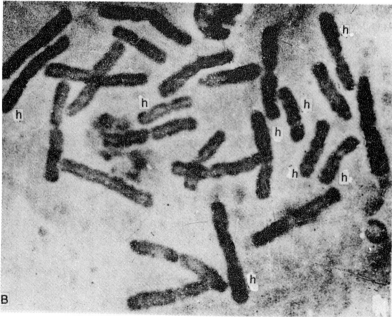

Fig. 15.5.10 A, B. A *Araucaria angustifolia*, squash preparation of endosperm cells shows n = 13; four chromosomes heterobrachial (*h*), nine isobrachial. **B** *A. columnaris*, squash of root-tip cell, 2n = 26; 8 chromosomes heterobrachial, 18 isobrachial. (After Mehra 1988)

with six species (Pilger and Melchior 1954).

Morphology

The plant is s shrub or small tree up to 15 m high; all species are dioecious. The branches may be opposite or in whorls. The young branchlets are green, prominently grooved, and show stomata as minute white dots. The buds have numerous overlapping scales. The leaves are arranged in a decussate and distichous fashion (Fig. 15.6.1 A). They remain functional for at least

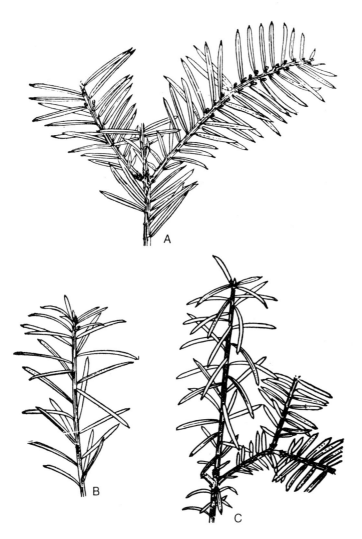

Fig. 15.6.1 A-C. *Cephalotaxus drupacea.* **A** Young twig, three branches from previous year's shoot, leaf arrangment decussate and distichous. **B** Alternate arrangement of leaves on shoots from near-soil level. **C** Shoots with branches after 2 or 3 year's growth. (After H. Singh 1961)

3 years. The leaves are linear and pointed at the apex (Fig. 15.6.1 B). The dorsal surface is dark shiny green with a conspicuous mid-rib. The lower surface shows two broad silvery bands of stomatic lines. On vertical shoots the leaves are spreading, while on lateral shoots they are arranged in two opposite rows. In *Cephalotaxus drupacea*, some branches arise from the main stem near the ground level, grow vigorously and bear alternate leaves (Fig. 15.6.1 B). Lateral buds on these branches lack scales and unfold in the same year in which the shoot is produced. After 2–3 years, the branches bear decussate and two-ranked leaves (Fig. 15.6.1 C). Winter buds are covered by numerous scales.

Anatomy

The anatomy of root and stem has not been studied.

Leaf. The leaves are distinctly bifacial (Kausik and Bhattacharya 1977). A vertical section of the leaf of *C. harringtonia* shows: (a) Epidermis covered by thick stratified cuticle, the cells are thick-walled with a prominent lumen. (b) Sunken stomata in longitudinal rows on the abaxial surface. (c) Mesophyll of uniseriate palisade layer, spongy parenchyma with irregular cells and large intercellular spaces. In *C. fortunei* transversely elongated, loosely arranged parenchyma cells occur between the palisade and spongy tissue (Fig. 15.6.2 B). Several cells of the mesophyll contain tannin. A single median resin duct is present (Figs. 15.6.2A; 15.6.3 A-C). The duct is surrounded by two or three (Fig. 15.6.2 C) layers of compactly arranged cells; some of the cells of the outer layer contain tannin. (d) The single vascular bundle does not have a bundle sheath. Both protoxylem and protophloem become crushed. The metaxylem and metaphloem are almost equal in amount. The transfusion tissue is associated with the vascular bundle throughout its length. The transfusion tissue is especially prominent in the terminal region (Fig. 15.6.3 A), and forms an arc adjacent to the xylem. In the middle of the leaf, this tissue extends in a wing-like arrangement from each side of the xylem (Figs. 15.6.2 A; 15.6.3 B). At the base, only a few transfusion cells are present close to the xylem (Fig. 15.6.3 C). Richly cytoplasmic albuminous cells are present adjacent to the phloem (Fig. 15.6.2 D). The lignified cells of the transfusion tissue are designated tracheids because of their wall structure (Hu and Yao 1981). They are variable in size and shape (Fig. 15.6.3 D-G) but shorter and wider than the tracheids of the xylem. Transfusion tracheids next to the xylem are generally longer and narrower than those further from it. The secondary walls have bordered pits and reticulate thickening (Fig. 15.6.3 E-G). A similar lignification and type of wall is also common in the metaxylem tracheids of the vein. The transfusion tracheids of mature leaves lack protoplasts (Griffith 1971).

According to Hu and Yao (1981), the transfusion tissue in the family

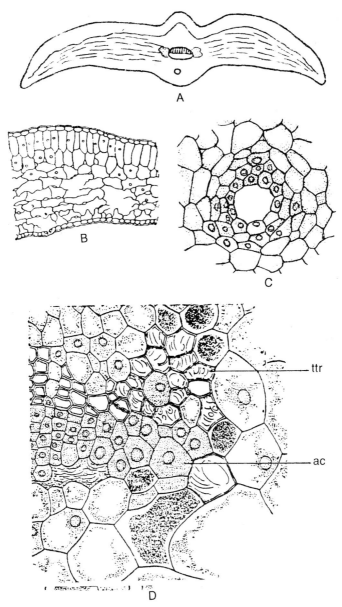

Fig. 15.6.2 A-D. *Cephalotaxus fortunei.* **A, B** Vertical section of leaf. **A** Diagrammatic. **B** Transversely elongated loosely arranged parenchyma cells between palisade and spongy tissue. **C** Resin duct. **D** Vein and transfustion tissue (*ttr*). *ac* albuminous cell. (After Kausik and Bhattacharya 1977)

Cephalotaxaceae is similar to that in the Taxaceae. It is of the Taxus type, i.e. lateral in groups, wing-like and mostly with bordered pits and spiral thickenings.

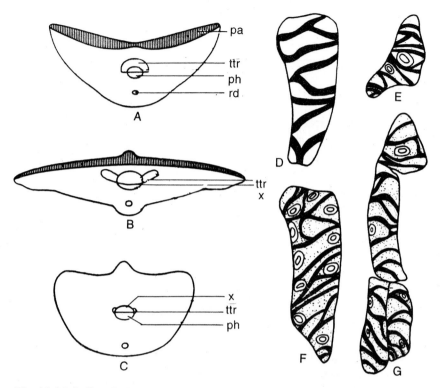

Fig. 15.6.3 A-G. *Cephalotaxus harringtonia.* **A-C** Diagrammatic vertical sections of leaf at tip (**A**), middle (**B**) and base (**C**). *pa* palisade. *ph* phloem. *rd* resin canal. *ttr* transfusion tissue. *x* xylem. **D-G** Transfusion tracheids. **D** Tracheid shows first-order frame work. **E-G** Mature tracheids. (After Griffith 1971)

Reproduction

Male Cone

The short, unbranched lateral shoots, covered with scales, mostly arise in the axils of the newly emerged shoot of the male tree. In the axils of the upper scales of these shoots, six to eight compactly arranged cones apear (Fig. 15.6.4 A-C). A male cone consists of a short sessile axis with 15–20 microsporophylls arranged spirally.

The microsporophylls vary in shape (Fig. 15.6.4 D-G). A sporophyll consists of a stalk bearing three or four (rarely two) abaxial sporangia pointing toward the cone axis, and a sterile flattened region which points outward and upward (Fig. 15.6.4 D, G).

Microsporangium. A plate of hypodermal cells differentiates on the abaxial surface near the base of each microsporophyll (Fig. 15.6.5 A-C). Later, due to partial sterilization, two to four groups of archesporial cells become delimited (Fig. 15.6.5 D); each is the site of a microsporangium. Periclinal

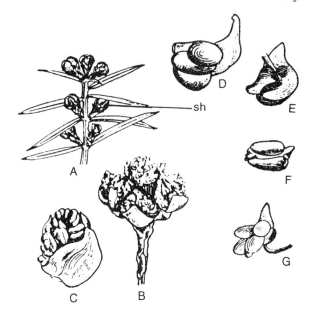

Fig. 15.6.4 A-G. *Cephalotaxus drupacea.* **A** Male cones covered by scales borne on short lateral shoot (*sh*). **B** Short lateral shoot, male cones exposed. **C** Male cone with subtending bract. **D-G** Microsporophylls. (After H. Singh 1961)

divisions in archesporial cells give rise to the primary parietal layer, which organizes a four-layered wall, and the primary sporogenous layer. The latter divide repeatedly so that a mature microsporangium has a mass of sporogenous cells surrounded by the wall layers and the epidermis (Fig. 15.6.5 E). The innermost wall layer forms the tapetum (Fig. 15.6.5 K), its cells becoming binucleate at the microspore mother cell stage (Fig. 15.6.5 L). The tapetum is of the glandular type and collapses during the reduction divisions in the mother cells (Fig. 15.6.5 M). At maturity, the epidermis develops fibrous thickenings, while the remaining wall layers become crushed (Fig. 15.6.5 N). The sporangia dehisce along a line facing the stalk of the microsporophyll. The epidermal cells do not develop thickening in this region (Fig. 15.6.5 O-Q). The upper portion of the sporophyll forms the sterile flattened region. Its cells take a uniformly dense stain, and occasionally show a resin duct (Fig. 15.6.5 F, G). The microsporangia grow toward the axis of the cone, remain united at the base, but become free of each other at the tip (Fig. 15.6.5 G-J).

Microsporogenesis. The microspore mother cells (mimc) contain abundant starch (Fig. 15.6.6 A). During meiosis, the cytoplasm contracts from the wall and a special callose wall is secreted around it. Twelve bivalents can be counted at diakinesis (Fig. 15.6.6 B). Cytokinesis occurs after meiosis II (Fig. 15.6.6 C-E), and tetrahedral and isobilateral microspore tetrads are formed. Several abnormalities during microsporogenesis have been reported (see H. Singh 1961).

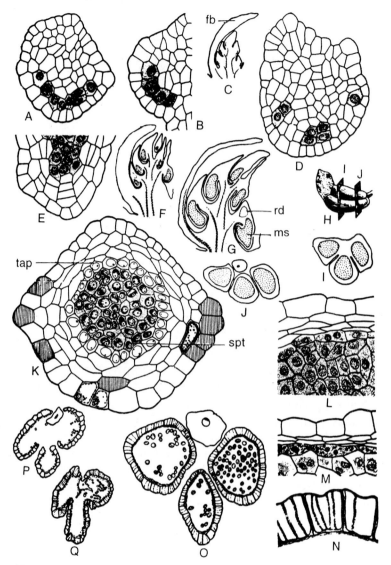

Fig. 15.6.5 A-Q. *Cephalotaxus drupacea.* **A, B** Cross- and longisection of young sporophylls show hypodermal archesporium. **C, F, G** Longisection male cones. *fb* fertile bract. *ms* microsporangia. *rd* resin duct. **C** Young male cone. **D** Transection sporophyll, initial stage of three sporangia. **E** Differentiation of wall layers. **F, G** Male cones at early (**F**) and late (**G**) stage. **H-J** Microsporophyll, sections at *I, J* marked in **H.** **K** Microsporangium (cross section) shows sporogenous tissue (*spt*) and tapetum (*tap*). **L** Sporangium at resting stage; some of the tapetal cells are binucleate. **M, N** Wall layers at reduction division of microspore mother cell (**M**), and exodermis with fibrous thickenings at shedding stage of sporangium (**N**). **O-Q** Cross sections of mature (**O**) and dehisced sporophyll (**P, Q**). (After H. Singh 1961)

Male Gametophyte.   The microspores are released from tetrads after the digestion of the callose wall and break down of the original wall of the

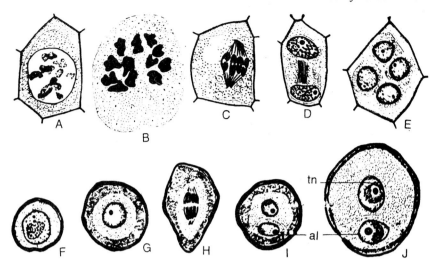

Fig. 15.6.6 A-J. *Cephalotaxus drupacea*, microsporogenesis and male gametophyte.
A-D Meiosis I of microspore mother cell. **E** Cytokinesis. **F, G** Uninucleate microspores.
H Division of microspore nucleus. **I, J** Two-nucleate pollen grains. *al* antheridial cell.
tn tube nucleus. (After H. Singh 1961)

mother cell. The microspores are spherical and small with a centrally situated
nucleus (Fig. 15.6.6 F), and a few starch grains in the cytoplasm. The intine
is thick while the exine remains thin (Fig. 15.6.6 G). According to Gullvag
(1966), the exine consists of two layers. The inner layer is lamellated and
the outer one comprises orbicules and/or a sculptured granular layer. The
nucleus divides to form a lenticular antheridial cell and a large tube cell
(Fig. 15.6.6 H, I). Prothallial cells are absent. The pollen grain enlarges to
about twice its original size, the exine ruptures, and the intine appears
thinner due to expansion (Fig. 15.6.6 J). Starch grains disappear and the
pollen grains are shed at the two-nucleate stage.

Female Cone

The female cones usually develop close to the shoot apex, since the resting
buds mostly arise in that position (Fig. 15.6.7 A, B). Occasionally, the
cones develop in the middle or at the base of the branch. It is somewhat
difficult to determine the point of origin of the female cones because the
bud scales from which they arise and the "leaves" which subtend them look
alike. These lateral organs are not true leaves but are termed so as they are
produced by the shoot apex during the period of foliage leaf formation
(H. Singh 1961). A study of the formation of the lateral organs, in different
seasons in a year, shows that the cones are borne in the axil of the lowermost
two to four leaves on some newly formed shoots.

The female cones are small with a short stalk and consist of five to
seven pairs of opposite and decussate bracts (Fig. 15.6.7 C). The fleshy
nature of the cone axis conceals the exact arrangement of bracts, and is

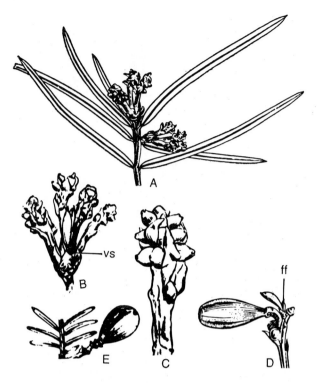

Fig. 15.6.7 A-E. *Cephalotaxus drupacea*, female cones. **A** Cones near the apex and middle of branch. **B** Four cones and vegetative shoot (*vs*), from a resting bud. **C** Cone. **D** Two ovules of a seed scale complex replaced by foliage leaves (*ff*). **E** Mature ovule on a cone. (After H. Singh 1961)

revealed only by a study of the vascular anatomy. Two ovules are borne in the axil of each bract (Fig. 15.6.7 C), except the lowest pair. There is a small ridge-like outgrowth of the cone axis between the two ovules of each bract. Florin (1938–1945) termed the two ovules and the outgrowth between them the seed-scale complex. The ovules that enlarge in the following spring are green, and become red and fleshy on ripening. Generally, one or two ovules mature in a cone (Fig. 15.6.7 D, E) although as many as five seeds may sometimes mature.

Normally, the apex of the female cone becomes inactive after producing the bracts and the ovules. However, in a few cones, the apex remains surrounded by numerous bud scales. According to Favre-Duchartre (1957), the proliferation of such a bud may give rise to a female cone in the following year.

Ovule.   An outgrowth develops in the axil of every fertile bract of a young female cone (Fig. 15.6.8 A, B). Its apex can be demarcated into four cytohistological zones typical of a vegetative shoot apex (Fig. 15.6.8 C). Thus, the outgrowth apparently represents a scondary axis. The two ovules

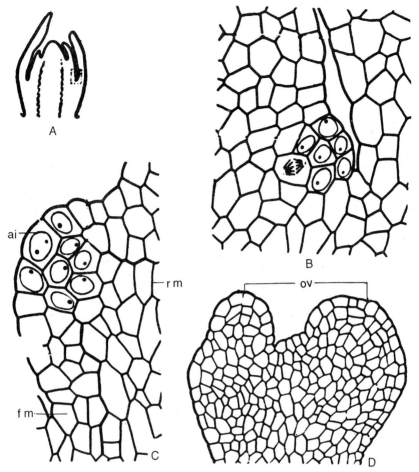

Fig. 15.6.8 A-D. *Cephalotaxus drupacea*, megasporangium. **A** Longisection young female cone. **B** Inset from **A** Initiation of secondary axis. **C** Later stage, secondary axis; tip shows typical organization of a shoot apex. *ai* apical initials. *fm* flank meristem. *rm* rib meristem. **D** Initiation of two ovules (*ov*), from secondary axis. (After H. Singh 1961)

originate as lateral protuberances from this secondary axis (Fig. 15.6.8 D). In younger stages, the growth occurs mainly by periclinal divisions which characterize a leaf primordium.

A young ovule, with a conspicuous integument, shows three to six hypodermal archesporial cells (Fig. 15.6.9 A). The latter divide periclinally to form the sporogenous layer and the primary parietal layer (Fig. 15.6.9 B). A massive nucellus is formed (Fig. 15.6.9 C, D) by repeated periclinal divisions of the primary parietal layer. The nucellar cells in the vicinity of the sporogenous layer, at the chalazal end, also undergo periclinal divisions to form a distinct tussue, which in later stages contains large compound starch grains (Fig. 15.6.9 E, F). This tissue (designated pavement tissue by H. Singh 1961) becomes crushed during the enlargement of the free-nuclear female gametophyte.

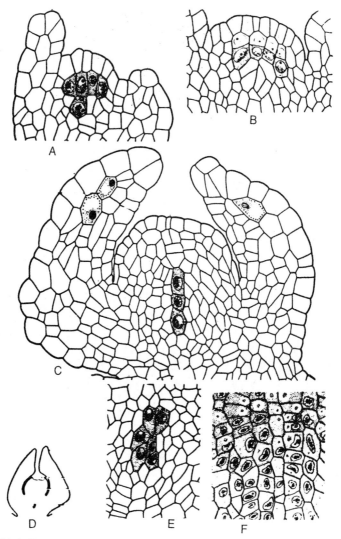

Fig. 15.6.9 A-F. *Cephalotaxus drupacea*, megasporangium. **A-C** Longisection, ovules.
A Hypodermal archesporium. **B** Periclinal division in archesporium forms primary parietal
and sporogenous layer. **C** Ovule in the first (winter) resting stage, some integumentary
cells contain tannin. **D** Ovule at the time of pollination. **E** Sporogenous tissue. **F** Pavement
tissue. (After H. Singh 1961)

A few days before pollination, at the micropylar region, two or three
layers of nucellar cells degenerate and form a rudimentary pollen chamber
(Fig. 15.6.10 A-C). Simultaneously, some of the inner epidermal cells of
the integument lining the micropyle become active (Fig. 15.6.10 B). After
pollination, these cells grow inward, undergo a few transverse divisions
close the micropylar opening (Fig. 15.6.10 D-F), and effectively seal the
freshly deposited pollen. These cells eventually acquire a very thick wall.

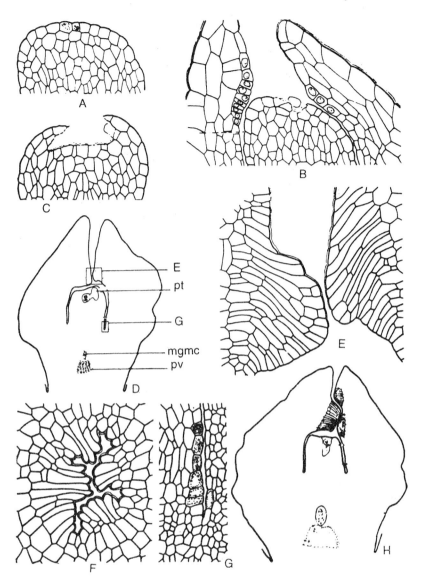

Fig. 15.6.10 A-H. *Cephalotaxus drupacea*, megasporangium. **A-C** Longisections upper part of nucellus show formation of pollen chamber, and active inner epidermal cells of integument (**B**). **D** Outline diagram of ovule. *mgmc* megaspore mother cell. *pt* pollen tube *pv* pavement tissue. **E** Inset marked *E* in **D** to show closure of micropyle. **F** Transection ovule shows closed micropyle. **G** Inset marked *G* in **D** shows tannin-containing epidermal cells of nucellus and integument. **H** Longisection ovule in the second winter. (After H. Singh 1961)

Some epidermal cells at the junction of the nucellus and the integument accumulate tannin (Fig. 15.6.10 G). Figure 15.6.10 H represents the condition of the ovule during the second winter.

Megasporogenesis.   One of the sporogenous cells enlarges, especially along its longer axis, the nucleus also enlarges, cytoplasm becomes "frothy", and it functions as the megaspore mother cell (Fig. 15.6.11 A). The non-functional sporogenous cells persist in the vicinity of the mother cell. Meiosis I and II result in a linear tetrad, and the chalazal megaspore functions (Fig. 15.6.11 B, C). The non-functional megaspores degenerate without any definite sequence (Fig. 15.6.11 B, C).

Female Gametophyte.   The functional chalazal megaspore becomes vacuolate, its nucleus divides repeatedly, and an elongated free-nuclear female gametophyte is formed (Fig. 15.6.12 A). During this process, the circumjacent nucellar cells are crushed. The free nuclei become arranged in a single layer along the periphery, leaving a central vacuole. At the 32-nucleate stage, the gametophyte has a two-nucleate (rarely four-nucleate) finger-like projection at the micropylar end (Fig. 15.6.12 B, C). This projection remains conspicuous until about 128 nuclei have been formed (Fig. 15.6.12 D). Later, it is no longer prominent (Fig. 15.6.12 E). A similar micropylar projection of the free-nuclear gametophyte has also been reported in other taxa. In *Taxus* it is enucleate (Dupler 1917), in *Torreya* (Coulter and Land 1905), *Austrotaxus* (Saxton 1934) and *Saxegothaea* (Looby and Doyle 1939) it contains only one or two nuclei.

During later stages, the oblong gametophyte is limited by a thin membrane (Fig. 15.6.12 E, F). Wall formation is initiated after several hundred nuclei have been formed (4096 nuclei according to Favre-Duchartre 1957). Cellularization occurs by centripetally advancing walls (Fig. 15.6.12 G) and the cells show a large central vacuole with the nuclei distributed either along the periphery of the gametophyte or the lower end (Fig. 15.6.12 H). The earlier formed cells divide periclinally even before the central region becomes cellularized. As the gametophyte matures, the nuclei gradually become reduced in size. Thin-walled cells have scanty cytoplasm and are arranged in files which converge towards the centre (Fig. 15.6.12 G, H). Binucleate cells are common. As the embryo matures, the cells of the gametophyte become densely cytoplasmic and accumulate starch grains and oil globules.

After the gametophyte has become cellular, two to five cells at the micropylar end enlarge, become densely cytoplasmic, the nuclei enlarge and function as the archegonial initials (Fig. 15.6.13 A). The adjoining cells undergo several periclinal divisions and differentiate as the jacket layer. At first, these cells are isodiametric, but later elongate. The jacket cells may contain two nuclei.

The archegonial initials develop into the archegonium (Fig. 15.6.13 B-F). The neck of the archegonium comprises two cells (occasionally three or four) mostly arranged in one tier; their outer tangential walls are always thick. During the "foam" stage of the archegonium, the cells of the

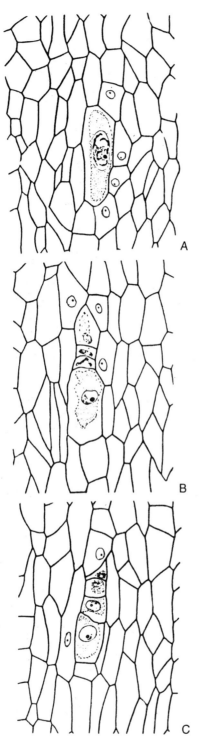

Fig. 15.6.11 A-C. *Cephalotaxus drupacea*, megasporogenesis. **A** Megaspore mother cell. **B, C** Linear tetrad; degenerated non-functioning megaspores. (After H. Singh 1961)

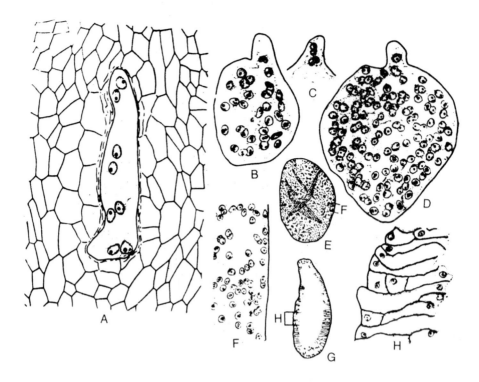

Fig. 15.6.12 A-H.   *Cephalotaxus drupacea*, female gametophyte. **A** Longisection nucellus shows eight-nucleate gametophyte. **B, D, E** Whole mount of free-nuclear gametophytes. **B, C, D** Micropylar projection with nuclei. **F** Inset marked *F* in **E. G** Part of gametophyte (longisection), to show wall formation. **H** Inset marked *H* in **G**. (After H. Singh 1961)

gametophyte at the micropylar end elongate upward and divide periclinally. As the archegonium matures, these cells become richly cytoplasmic and frequently show two nuclei. They are especially prominent during the development of the embryo. A similar large cone of cells is also present in Podocarpaceae (see H. Singh 1961). The neck cells appear sunken, so that each archegonium has its own archegonial chamber (Fig. 15.6.13 G, H). The central cell elongates considerably, has a pointed chalazal end (Fig. 15.6.13 H), and during the foam stage it contains many irregular and darkly staining bodies. Its nucleus is densely chromatic, has a prominent nucleolus and lies just below the neck cells. In a mature central cell, the cytoplasm is dense and contains round protein granules. The nucleus divides to form the egg and an ephemeral ventral canal nucleus of equal size; they are not separated by a wall (Fig. 15.6.13 D, E). The egg nucleus migrates downward and enlarges considerably (Fig. 15.6.13 F). Minute darkly staining bodies appear along the inner surface of the nuclear membrane, the nucleolus and the chromatin become indistinct. A mature archegonium is long, narrow

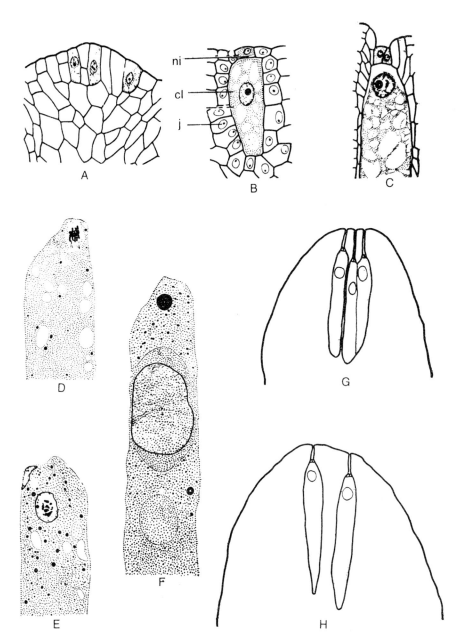

Fig. 15.6.13 A-H. *Cephalotaxus drupacea*, archegonium. **A-E** Development of archegonium. *cl* central cell. *j* jacket. *ni* neck initial. **F** Mature archegonium (upper part). **G, H** Micropylar region of female gametophyte shows archegonial complex (**G**) and mature archegonia (**H**). (After H. Singh 1961)

and pointed at the chalazal end. It contains dense cytoplasm and numerous protein globules (Fig. 15.6.13 E, F). In a mature egg nucleus, stained with

Schiff's reagent, the chromatin appears as a small dot in the nucleus (H. Singh 1961). At the time of fertilization, the large nucleus is full of nucleoplasm and several darkly staining round bodies which are unlike nucleoli. The archegonia are situated slightly below the tip of the gametophyte (Fig. 15.6.13 G, H), and separated from each other by four to eight layers of cells. Occasionally, the archegonia lie in a somewhat lateral position. Rarely, two or three adjacent archegonia have a common archegonial cavity (Fig. 15.6.13 G).

Pollination

The pollen grains land on the pollination drop exuded from the micropyle, imbibe water and other substances from the fluid, become heavy and sink down in the micropyle, settling on the degenerated tip of the nucellus (a weakly developed pollen chamber). It is likely that the ruptured exine is cast off when the pollen grain is caught in the pollination drop. The intine enlarges at the (tube cell end) proximal end to form a short but broad pollen tube which grows through the nucellus (Fig. 15.6.14 A, C). During its growth, the pollen tube crushes the nucellar cells and reaches the female gametophyte when the archegonia are almost mature. Rarely, a pollen tube may reach the archegonia in the foam stage. Such an archegonium matures earlier than the other archegonia in the same gametophyte. This behaviour indicates a correlation between the adjacence of the pollen tube and maturation of the archegonium (H. Singh 1961).

The tube nucleus migrates into the pollen tube. The antheridial cell enlarges considerably, and divides at right angles to the long axis of the pollen tube to form the stalk and body cells (Fig. 15.6.14 B). The stalk cell lacks a clear wall, but is surrounded by dense cytoplasm. Both the cells move down and lie free in the cytoplasm of the tube (Fig. 15.6.14 C). When they are close to the tube nucleus, the cytoplasm around the stalk nucleus becomes indistinguishable. The pollen tube contains a body cell, and stalk nucleus and tube nucleus—the latter two being similar to each other (Fig. 15.6.14 C). During the growth of the pollen tube, the stalk and tube nuclei usually precede the body cell, which remains closely associated. The contents of the pollen tube lie at its tip (H. Singh 1961), or may lie far behind (Favre-Duchartre 1957). The pollen tube persists in this stage for nearly a year.

Post-Pollination Development of Male Gametophyte.   In the following year, the pollen tube grows rapidly and reaches the tip of the female gametophyte when the archegonia are almost mature (Fig. 15.6.14 D). The pollen tube enters the archegonial cavity, comes in contact with the neck cells of the archegonium, and swells at the tip (Fig. 15.6.14 E). By this time, the body cell has enlarged considerably. Its prominent nucleus divides, but no wall is laid down between the two nuclei (Fig. 15.6.14 F, G). These

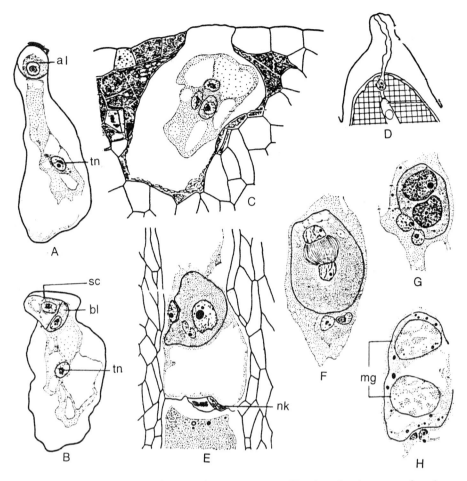

Fig. 15.6.14 A-H. *Cephalotaxus drupacea*, post-pollination development of male gametophyte. **A** Two-celled pollen tube. *al* antheridial cell. *tn* tube nucleus. **B** Antheridial cell has divided. *bl* body cell. *sc* stalk cell. **C** Pollen tube in the nucellus; stalk and body cells are lying close to the tube nucleus. **D** Longisection upper part of ovule. **E** Pollen tube has grown up to the tip of archegonium. *nk* neck cells. **F** Division of body cell. **G, H** Mature male gamete (*mg*). (After H. Singh 1961)

are small, and have dense chromatin and a prominent nucleolus. At maturity, the male gametes enlarge, the nucleolus is no longer distinguishable and the chromatin also becomes inconspicuous (Fig. 15.6.14 H). They may be equal (Lawson 1907; Favre-Duchartre 1957, H. Singh 1961), or unequal (Coker 1907, Sugihara 1947, Kaur 1958). The cytoplasm of the body cell and the pollen tube contains small, round, densely staining bodies (Fig. 15.6.14 G, H).

Fertilization

The neck cells begin to collapse as the pollen tube comes in contact with them, and a passage is formed for the tube to enter the egg cell. The male gametes separate from each other due to the dissolution of the wall of the body cell; the stalk and tube nuclei degenerate. The tip of the pollen tube comes to lie inside the archegonium, where it bursts, and one or both the male gametes, together with some cytoplasm of the body cell, enter the egg cell (Fig. 15.6.15 A). As the male gamete approaches the egg nucleus, the cytoplasm in the upper part of the archegonium becomes vacuolate (Fig. 15.6.15 B-D). When the male gamete and egg come in contact with each other, the male nucleus becomes lenticular and both become enveloped in the male cytoplasm (Fig. 15.6.15 B). The nuclear membrane "dissolves" at the point of contact, and the chromatin of the two nuclei come to lie close to each other (Fig. 15.6.15 C, D). Numerous fibrils appear in the fusion nucleus. The two chromatin groups integrate and the fibrils become arranged in a bipolar spindle. The metaphase plate is slightly oblique to the long axis of the archegonium (Fig. 15.6.15 E).

Embryogeny

The division of the zygote nucleus is intranuclear and occurs in situ. The resulting two nuclei, enclosed in the male cytoplasm, migrate to the lower part of the proembryo. Numerous vacuoles in the cytoplasm mark the path of descent of these nuclei (Fig. 15.6.16 A-C). Both the proembryonal nuclei divide simultaneously (Fig. 15.6.16 A) and the resultant four nuclei separate from each other as the sheath of the male cytoplasm disappears. The upper part of the proembryo becomes much vacuolated while the lower portion has dense cytoplasm with numerous protein glubules. The two regions are separated by a large vacuole. These four nuclei divide twice simultaneously (Fig. 15.6.16 B, C) and wall formation is initiated at the 16-nucleate stage. the nuclei become progressively smaller as the proembryo advances in age.

A variable number of large nuclei, which degenerate later, have been observed in the upper region of the proembryo. They could represent: (a) the second male gamete and its derivatives, (b) supernumerary sperms, (c) persistent ventral canal nucleus or (d) relict and lagging nuclei (see H. Singh 1961).

Wall formation in the proembryo results in a lower embryonal group of 10–13 small cells (pE) and an upper tier of 3–6 cells open towards the archegonium (pU). The lowermost one or two cells of the embryonal group are large, have dense cytoplasm, a prominent nucleus, and constitute the cap cells (Fig. 15.6.16 D). One more division in the embryonal group increases the number of cells to 20-26. The upper tier divides transversely to form the lower suspensor tier (S) and an open tier (U) which later degenerates (Fig. 15.6.16 E, F). By this time, the cytoplasm of the archegonium forms a dense mass or plug (Fig. 15.6.16 F). The suspensor tier elongates and pushes the embryonal tier into a cavity in the centre of the female gametophyte (Fig. 15.6.16 G, H). This cavity is formed by the

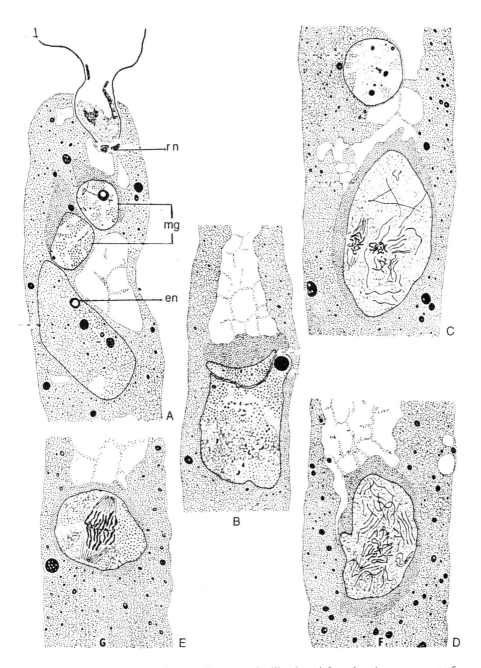

Fig. 15.6.15 A-E. *Cephalotaxus drupacea*, fertilization. **A** Longisection upper part of archegonium to show pollen tube and two discharged male gametes. *en* egg nucleus. *mg* male gamete. *rn* relict nuclei. **B** Middle portion of archegonium shows egg nucleus and lenticular male gamete. **C, D** Integration of male and female chromatin. **E** Zygote nucleus in metaphase. (After H. Singh 1961)

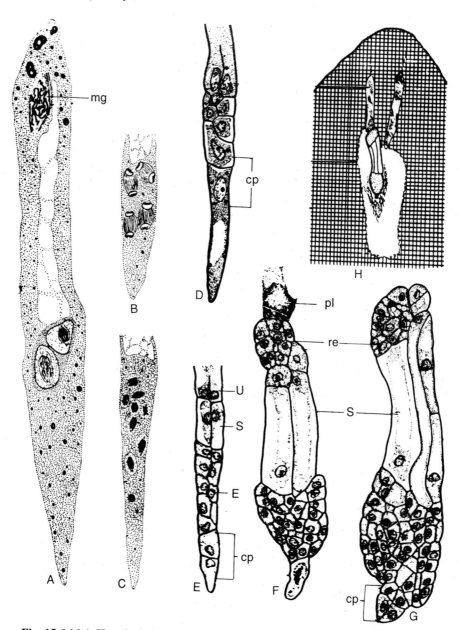

Fig. 15.6.16 A-H. *Cephalotaxus drupacea.* **A-C** Development of proembryo. **A** Two proembryonal nuclei in division; the second male gamete (*mg*) is also in division in the upper part of proembryo. **B, C** Four-and eight-nucleate proembryo, nuclei in division. **D-H** Embryogeny. **D** Proembryo after wall formation. *cp* cap. **E** Proembryo shows upper tier (*U*), suspensor tier (*S*) and embryonal tier (*E*) with cap cells. **F** Degenerated remains of upper tier and cytoplasm of archegonium (*pl*); the cap cells have degenerated and rosette embryos (*re*) initiated. **G** Later stage, elongated suspensor. **H** Longisection micropylar region of female gametophyte shows embryos in the corrosion cavity. (After H. Singh 1961)

degeneration of cells and progresses toward the chalazal end. Some of the suspensor cells divide transversely and give rise to numerous small rosette embryos, which eventually degenerate. All the cells of the young embryo contain starch.

By the time the suspensor has elongated, the cap cells degenerate and are replaced by three or four adjacent embryonal cells which enlarge. Rarely, a cap cell is binucleate. Ultimately, these cells also degenerate.

The basal cells of the embryonal mass elongate to form embryonal suspensor, which pushes the embryonal mass further down into the female gametophyte. The suspensor cells collapse by this time. Occasionally, the cells of the embryonal suspensor proliferate to form a small embryo. As the embryonal mass grows, the suspensors become massive. Of the one or two embryonal masses (derived from as many fertilized archegonia), one takes the lead and develops into a mature embryo. Cleavage of the embryonal mass is rare.

The leading embryo remains undifferentiated. Its outermost layer undergoes periclinal divisions, and the dermatogen, periblem and root initials become distinct. The two cotyledonary primordia also differenttate. In a mature embryo, the root tip initials are very distinct while the stem tip is still undifferentiated. The various histological regions of the stem tip can be discerned when the seed germinates.

Seed

At the time of pollination, the integument comprises several layers of thin-walled cells and some cells contain tannin. As the megaspre mother cell differentiates in the ovule, resin ducts appear in the chalazal region. They are narrow and lined by an inner layer of epithelial and an outer layer of tannin cells. The lower part of the integument becomes meristematic, and several layers thick. These cells contribute to a major part of the seed coat. At the free-nuclear stage of the gametophyte, the lower portion of the integument enlarges considerably. The innermost layer of the integument comprises narrow and vertically elongated cells which becqme crushed by the enlarging gametophyte. There is a prominent band of tanniniferous cells in the central region of the integument. Stomata are present in the epidermis (Fig. 15.6.17 D). At a later stage of development, some of the hypodermal cells acquire thick pitted walls, and seven or eight layers of cells inner to the tanniniferous band form the sclerenchymatous stony layer (Fig. 15.6.17 A, B, E, F). Outer to the stony layer is the red fleshy layer, which contains scattered thick-walled and tannin cells, numerous resin ducts, and is traversed by two vascular bundles. The epithelial lining of the resin duct degenerates. The cells inner to the stony layer form the papery layer, which has a mosaic surface due to the presence of patches of tannin cells (Fig. 15.6.17 G, H).

A mature seed is large, consists of a thick coat, and the embryo extends

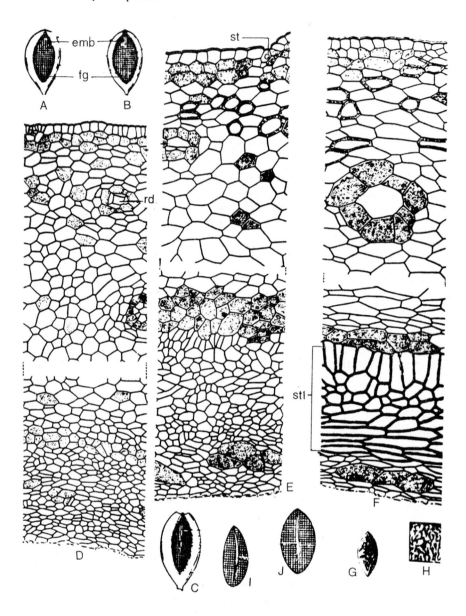

Fig. 15.6.17 A-J.   *Cephalotaxus drupacea,* seed and seed-coat. **A-C** Longisection of seeds containing embryos at different stages of development. *emb* embryo. *fg* female gametophyte. **D-F** Portion of integument in transection. *rd* resin duct. *st* stoma. *stl* stony layer. **D** Free-nuclear gametophyte. **E, F** At the stage of ovule shown in **A, B. G** Seed after removal of stony and fleshy layers. **H** Mosaic surface of papery layer. **I, J** Longisections of female gametophyte, from seeds shed from the plant, show elongation of embryo. (After H. Singh 1961)

to a short length of the massive female gametophyte (Fig. 15.6.17 C). On shedding, the outer fleshy layer disintegrates and the seed is covered by the stony layer, which is sharply pointed at the two ends. The embryo grows and extends to almost the entire length of the gametophyte (Fig. 15.6.17 I, J). The seeds undergo a long period of dormancy lasting 2 to 3 years.

Germination. During germination, the stony layer cracks laterally and longitudinally and the radicle emerges. As in other conifers, the germination is epigeal. The hypocotyl and cotyledons are fleshy, and the spirally arranged juvenile leaves are longer but narrower than the mature leaves. Cotyledonary leaves persist for more than 2 years after germination.

Chromosome Number

The haploid chromosome number in *C. griffithii, C. mannii, C. harringtonia* var. *drupacea* and *C. fortunei* is n = 12, as determined from endosperm squashes, and their karyotypes are identical (Mehra 1988). Eleven chromosomes are isobrachial. The single heterobrachial chromosome bears a terminal satellite on the short arm. One isobrachial chromosome possibly bears a satellite at/ near the terminal end but it is not distinct (Fig. 15.6.18 A). Meiosis has been studied in *C. mannii*. Eleven chromosome pairs have terminal chiasmata at both ends and form O-shaped configuration. One pair has a single chiasma at one end (Fig. 15.6.18 B).

Fig. 15.6.18 A, B. Cytology. **A** *Cephalotaxus grifithii,* endosperm nuclear mitosis shows n = 12. *h* heterobrachial chromosome, arrow shows constriction. **B** *C. manii,* microspore mother cell shows bivalents. (After Mehra 1988)

Temporal Considerations

Based on cultivated plants in Dehra Dun (India), *Cephalotaxus drupacea* has a 2 year reproductive cycle.

The male cones initiate during August, and the microsporangia over winter in the microspore mother cell stage. Development is resumed in the middle of March. Dehiscence and shedding of pollen occurs during spring, pollen grains land on the ovule, and produce short pollen tubes. The antheridial cell divides, and stalk and body cells migrate to the lower portion of the pollen tube by April. No further development takes place for about a year. Then, development is resumed, and the male gametes are formed and liberated into the archegonium by mid-April.

The female cones initiate during late autumn, and the ovules overwinter in the sporogenous cell stage. Pollination occurs during spring, followed by ovule enlargement, megasporogenesis and formation of the free-nuclear gametophyte. The ovule undergoes the second winter rest. Development is resumed during spring, and archegonia are formed. Fertilization takes place by mid-April. The embryo develops quickly and the seeds mature by August. The ovule needs 1 year and 9 months from appearance to maturation into a seed (H. Singh 1961).

15.7 PHYLOGENETIC CONSIDERATIONS OF CONIFERALES

The knowledge of primitive conifer groups has expanded as a result of investigations by a large number of researchers since the pioneering work of Rudolf Florin (see Florin 1939, 1944, 1945, 1950a, 1950b, 1951, 1954). Florin compared the fertile dwarf shoot of *Cordaianthus* (Cordaitales) with the ovuliferous scale of modern conifers and documented several intermediate stages within the Voltziales. Steps in the evolution (of an ovulate conifer cone from Cordaitean strobilus) are as follows: (a) a reduction in internode length, (b) a flattening of the short shoot, (c) fusion of the sterile parts of the short shoot, (d) fusion of the ovule axis to the scale and eventual elemination of this axis, (e) total reduction in the number of fertile parts, and (f) reduction in the numbers of ovules (and a change in ovular position in some genera).

Since the work of Florin, much has been learnt about the evolution and reproductive biology of Coniferales. Systematic investigations of fossil cone vasculature and resin canal distribution, leaf cuticles, seed integuments and embryo structure have increased our knowledge of conifer evolution (Stockey 1982). The living conifer families appear to have originated earlier than was formerly presumed. All the families of modern conifers (with the exception of Cephalotaxaceae, and perhaps Pinaceae) appear to have an Upper Triassic to Lower Jurassic appearance (Miller 1977). Several present-day taxa are evident in the Mesozoic and can be recognized by the Late Triassic or Early Jurassic. Many genera described by Florin are being reinvestigated. Miller (1977) pointed out that many of the intermediate Voltzian conifers are present in the Triassic and overlap the earliest of the modern conifer families. Thus, conifers have an extensive geological record, with many of the modern groups representing very ancient lineages. According to Miller (1977), it is doubtful that modern families evolved from the Voltziaceae, as they are mostly Late Triassic to Jurassic in age (see Chap. 14). Precursors should be sought in Palaeozoic to Early Triassic sediments, (early records of the families based on foliage alone are disputed). New families have been erected for many of the newly discovered fossils. The Cheirolepidaceae have seed cones with deciduous scales and large persistent bracts and are unlike any other known conifer. With the discovery of pollen cones (*Tamaxellia, Classostrobus*) containing *Classopollis* pollen and their associated foliage types (*Frenelopsis, Pseudofrenelopsis*), the family is now better understood. The Cheirolepidaceae apears to have been widespread during the Mesozoic, and its genera may have occupied a number of diverse habitats (see Stockey 1982). Further work is necessary, especally of the plants within the Cordaitales and Progymnospermopsida to understand their diversity and specialization. This approach will give a clearer picture of possible conifer ancestors. The studies of fossil conifers has added to our understanding of the extinct forms. With the discovery of additional fossil conifers in Late Palaeozoic to Early Mesozoic sediments, we are coming to a closer understanding of their origin and subsequent radiation. Additional well-preserved Triassic conifers will further expand our knowledge of the diversity of coniferophytes (Corditales, Voltziales, Coniferales), a group that probably was at its zenith during the Mesozoic.

16. Taxales

Taxales is a small group of plants comprising 5 living taxa and 20 species, included in a single family, Taxaceae (Dallimore and Jackson 1966): *Amentotaxus* (4 spp.), *Austrotaxus*, *Pseudotaxus* (= *Nothotaxus*), both monotypic, *Taxus* (8 spp.) and *Torreya* (6 spp.).

Taxaceae

The Taxaceae appeared during the Late Triassic or Early Jurassic (Miller 1977). The earliest remains attributed to the family are leafy twigs and attached arillate fruits of *Palaeotaxus* (Florin 1948) from the Lowermost Jurassic. *Amentotaxus* is present by the Late Cretaceous. Thus, the fossil record extends from the onset of the Jurassic, and the early forms show ovulate reproductive structures. However, the fossils are so like those of living forms that they provide no clue about the ancestry of the family.

The plants are evergreen shrubs or small trees, profusely branched, with small spirally arranged linear leaves. The wood is pycnoxylic and the tracheids have tertiary spirals; resin canals are absent. The plants are mostly dioecious, only sometimes monoecious; male cones are small and microsporangiophores have two to eight peltate pollen sacs. Solitary, arillate ovules terminate a dwarf shoot with decussate bracts. The embryo is dicotyledonous.

Taxus

Taxus, commonly known as yew, the leaves, shoots and seeds have poisonous properties. The active principle is taxine although other alkaloids are also present. Both fresh and partly dried shoots contain alkaloids; the withered/dried shoots are considered more toxic in action than the fresh foliage. The poison content may vary in male and female trees, or in different trees. The scarlet aril around the seed is, however, harmless. The plants have to be fenced from cattle. From the inner bark (*T. brevifolia*) taxol is obtained which has therapeutic qualities (see Chap. 23).

Morphology

Taxus is one of the common hardy evergreen shrubs or trees. *T. baccata* can reach a height of 20 m, with a massive trunk 7 m or more in girth, while *T. wallichiana*, exceptionally, may reach 30 m in height and 5 m in girth. The dorsiventral leaves are linear, 2–3 cm long with recurved margin, tapering apex, dark green above and yellowish green beneath (Fig. 16.1). They are spirally arranged but appear distichous due to a twist in the short leaf base (*T. wallichiana*). The mid-rib is prominent, with two bands of stomata on either side on the under surface. The scale-like leaves on the fertile shoots are opposite and decussate.

Fig. 16.1. *Taxus wallichiana,* twig to show arrangement of leaves and arillate ovules. (Photograph, courtesy Professor B.D. Sharma, Jodhpur)

Anatomy

Stem. A distinct tunica is absent. The zonation of the apical meristem is similar to Taxodiaceae and Pinaceae, except that the frequency of periclinal divisions (in all regions) varies considerably as the growing season advances.

The secondary wood is formed from a single persistent cambium. The wood of *Taxus* shows specific xylotomical features (Greguss 1955). The wood parenchyma is generally absent, but may be present in *T. baccata* and *T. brevifolia* (and in the root of *T. cuspidata*). Distinct growth rings develop due to the difference between the flattened and thick-walled late wood tracheids and the thin-walled early tracheids of longer lumen (*T. brevifolia*). The transition from early to late wood is usually gradual.

The most characteristic feature is the presence (in longitudinal tracheids) of spiral thickenings, which originate from the tertiary layer and lie side by side with bordered pits. The horizontal ray cells have an uneven thickness with pit-like depressions. Tangential ray-cell walls are always smooth and identures are fairly frequent. Pits (cupressoid) are relatively small and three, four or six pits are present in the cross-field. The pit apertures are slit-like, oblique/vertical but never horizontal.

Leaf. A vertical section of leaf shows mesophyll differentiated into palisade and spongy tissues; Resin ducts absent (which distinguishes it from the conifers; Fig. 16.2 A). The distinct bundle sheath is a ring of regularly arranged prominent cells (Fig. 16.2 B). The vascular bundle is large and dorsiventrally flattened, parenchyma develops between the rows of xylem and phloem; transfusion tissue occurs around the vascular bundle, and tracheids and some included parenchyma are confined to wing-like extensions lateral to xylem. However, most of the parenchyma also extends adaxially and abaxially to the bundle. The albuminous cells are especially prominent in this region, those lying within the bundle sheath being very large and also vesicular (Fig. 16.2 B).

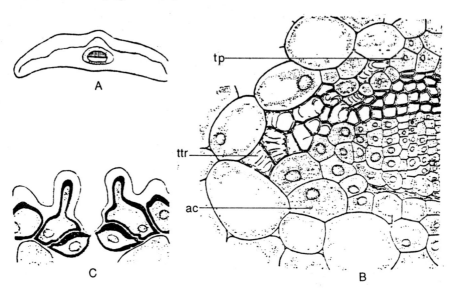

Fig. 16.2 A-C. *Taxus baccata*, leaf, vertical sections. **A** Outline diagram. **B** Transfusion tissue and vascular bundle. *ac* albuminous cells. *tp* transfusion parenchyma. *ttr* transfusion tracheid. **C** Stomata. (After Kausik and Bhattacharya 1977)

The stomata are haplocheilic; subsidiary cells develop papillate extensions (Fig. 16.2 C) which form a deep epistomal chamber so that the guard cells appear to lie deep in it.

Reproduction

Male Cone

The male cones are stalked globose heads in the axil of leaves. Each male cone axis bears ca. 10 decussate sterile scales and 6–14 symmetrical and perisporangiate microsporophylls, each bearing 6–8 reflexed sporangia. They are terminally arranged in small cones on short shoots (Fig. 16.3 A).

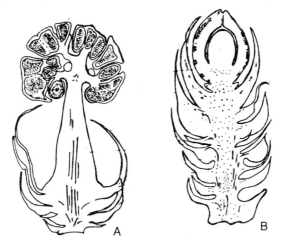

Fig. 16.3 A, B. *Taxus* sp. **A** Longisection male cone. **B** Longisection, fertile shoot with terminal ovule. (After Bracegirdle and Miles 1973)

Microsporangium. An ultrastructural and cytochemical investigation of the development of the microsporangium of *T. baccata* reveals many distinctive features (Pennell and Bell 1985). The early development of the archesporial cells is not synchronized (cf. other gymnosperms investigated). Groups of degenerated cells are present in the archesporium which disappear, presumably by resorption. The tapetum differentiates from the outer layer of the archesporium. The sporogenous cells can be recognized about 3 weeks before meiosis. There is a tendency towards synchrony and diad formation in a sporangium, almost simultaneously.

Microsporogenesis. After meiosis I is completed, vesicles accumulate in the equatorial plane of the spindle, coalesce and give rise to a sinuous wall. This wall fluoresces brilliantly when stained with Calcofluor White and with aniline blue, while the wall surrounding the diad fluoresced only with aniline blue. This wall (and not the dividing wall) contains a fibrillar layer. After meiosis II is completed, the four microspores lie in one plane, so that the symmetry of the tetrad is regularly isobilateral (Pennell and Bell 1986). The partitioning walls formed at telophase of meiosis II fluoresce brilliantly with aniline blue. In the electron microscope these walls appear similar to those giving rise to the diads, but there is a thin layer of fibrillar material evident at the site of the middle lamellae of all the partitioning walls. The successive wall formation in the mimc of *Taxus* distinguishes it from many conifers. The cytoplasm of the four spores is often different, although no differences are detectable in the parent cells, those derived from a diad being less dense than those derived from the other. A small proportion (ca. 5%) of the tetrads in a sporangium degenerate. Although there is inequality in the frequency of ribosomes between the spores of a tetrad, partial degeneration within a tetrad has not been observed (Pennell and Bell 1986).

The microspores are released into the loculus by rapid dissolution of the callosed wall of the pollen mother cell; the formation of the sporoderm begins. Osmiophilic droplets emerge from the spore protoplast and enter the wall. The droplets coalesce to form an outer layer on which up to six sporopollenin lamellae, probably of tapetal origin, are deposited (the fibrillar layer can no longer be recognized). The accretion of a single layer of sporopollenin droplets, in no recognizable pattern, gives rise to the outer verrucose part of the exine. The sporoderm becomes resistant to acetolysis and fluoresces when stained with Auramine O.

The cytoplasm of the mature spores is vacuolate. Plastids dedifferentiate, nucleoli reappear but synaptonemal-like complexes and nuclear vacuoles are no longer present. Large vacuoles persist in the cytoplasm and many others develop from the dictyosomes. A single cisterna of endoplasmic reticulum often encircles the nucleus. The microspores are shed in this condition (Pennell and Bell 1986).

During meiosis, the tapetal cells elongate tangentially, later they become round. Many cells are binucleate and later become trinucleate. The sinuous radial walls contain a single fibrillar layer. Elongated amyloplasts, dictyosomes and isolated cisternae of endoplasmic reticulum are frequent. Large vesicles and vacuoles give the tapetum a foamy appearance (under the light microscope). About 3 weeks before anthesis, plastids are no longer visible in the tapetum. Orbicules appear simultaneously in the walls of the tapetal cells and in the loculi of the sporangium. With the rupture of the inner membrane, the tapetum becomes partly amoeboid. A peritapetal membrane (often overlain with orbicules) can be identified at this time. Cytochemical tests indicate that, from the beginning of meiosis, the tapetum is rich in acid phosphatase (Pennell and Bell 1986).

A mature microsporangium splits, curls and liberates pollen grains. The pollen grains are wingless and at the uninucleate stage.

Ovule

The ovules are solitary, arillate and terminate a modified dwarf shoot with decussate bracts (Fig. 16.3 B).

The initiation of the ovule (in *Taxus*) is by a transformation of the shoot meristem (Loze 1965). The apical initials give rise to the nucellus, the integument arises from the flank meristem and the pith meristem. The subapical initials and the flank meristem contribute to the chalazal portion of the ovule (Fig. 16.4 A, B). The epidermal cells of the dome-shaped nucellus undergo periclinal divisions to form a nucellar cap. A discreet epidermis is not distinguishable in the young nucellus. The cells of the outermost exposed layer divide periclinally to form the inner archesporial cells and the outer layer (Fagerlind 1961). The cells of the outer layer again divide periclinally and the derivatives divide repeatedly to form the nucellar tissue. The dual origin (parietal tissue and nucellar cap) of the nucellus is well established in *Taxus* (Dupler 1917, Sterling 1948a, Pankow 1962).

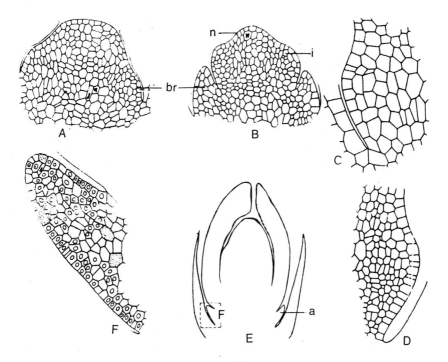

Fig. 16.4 A-F. A-C, F *Taxus baccata*, D, E Taxus sp. A, B Longisection, tip of fertile shoot. A After initiation of last pair of bracts (*br*). B Nucellus (*n*) and integument (*i*). C, D Development of aril. E Longisection ovule with a short aril (*a*) at the base. F Inset F marked in E. (A, B, E, F After Loze 1965, C, D after Pankow 1962)

The integument is fused to the nucellus, except at the apex, and becomes stony. Usually, two feebly developed vascular bundles (represented by phloem strands) traverse upward inner to the stony layer. The xylem of these strands extends up to the chalaza. They alternate with the uppermost pair of scale leaves, and correspond to the two primordia from which the integument arises. The aril also has a stub of vasculature at the base. There is a plate meristem across the axis, at the base of the ovule. The meristem grows upward at the periphery and around the ovule, and forms the aril (Fig. 16.4 C-F). The latter persists in a primordial state until after fertilization, when it develops and ultimately extends beyond the seed. The solitary erect seed borne in a fleshy cup-like aril ripens in one season. The aril becomes scarlet and is attractive to birds (Fig. 16.1).

Megasporogenesis. At the time of differentiation of the megaspore mother cell (mgmc), the cell lineages of nucellus (from above) converge towards the centre. The cells in the upper half of the nucellus are larger, isodiametric, and have large vacuoles and thick walls. In the lower half the meristematic cells are smaller, stain densely, and divide transversely. Approximately at the junction of these two zones, there is a small central area with ca. 20 cells (as seen in longisection). These cells have dense cytoplasm, large nuclei, take deeper stain than the circumjacent cells and differentiate as sporogenous tissue. Simultaneously, the adjoining cells divide parallel to the periphery and produce rows of cells which do not enlarge and appear to radiate from the central area (Fig. 16.5 A-E).

The megaspore mother cell lies almost in the centre of the sporogenous tissue. In *T. canadensis*, usually one cell, occasionally two, functions (Dupler 1917). In *T. baccata* the mgmc can be distinguished from the nucellar cells by their larger size (up to 50 μm in diameter) and the composite nature of their boundary walls. The combined wall is fibrillar, 300–500 nm wide, and the middle lamella cannot be readily distinguished. Adjacent to the protoplast of the mother cell, there is an additional layer (up to 120 nm thick) of denser amorphous material. Plasmodesmata have not been observed in the wall of the mother cell, although they are frequent in the nucellar cells. Within the protoplast of the mother cell, the mitochondria and plastids lie close to the chalazal pole and form a distinct layer beneath the nucleus. The envelopes of these organelles are associated with lipid droplets (Pennell and Bell 1987).

The development of ovules on the same tree is synchronized (*T. baccata*). The prophase in the mgmc lasts ca. 1 week and corresponds with the germination of the pollen on the nucellus (Pennell and Bell 1986). Dividing nuclei have rarely been seen. The cytoplasmic organelles in the diads are distributed unequally, the mitochondria and plastids are confined to the chalazal cell. Subsequent division in the micropylar and chalazal dyad leads to a T-shaped tetrad (*T. baccata*—Pennell and Bell 1987). Most of the mitochondria and plastids are included in the chalazal megaspore, which alone develops further, while lipid droplets are present in all the megaspores.

In *T. canadensis* after meiosis a linear tetrad is formed and any one or all the four megaspores may develop (Fig. 16.5 A-E) so that supernumerary gametophytes are common. Sometimes, the upper two megaspores enlarge and divide while the lower two abort. Normally, the larger chalazal megaspore functions (Sterling 1948a). Although as many as three megaspores may germinate (Fig. 16.5 C), only one forms the gametophyte (*T. cuspidata*); the adjoining sporogenous tissue breaks down. The localized growth of the

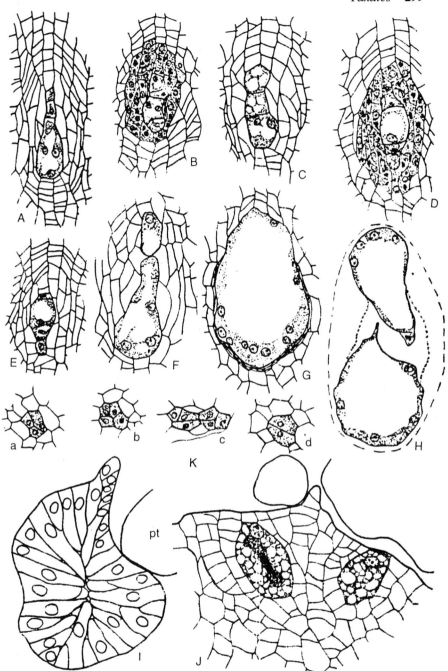

Fig. 16.5 A-K. *Taxus cuspidata*, longisection ovule, megaspore and female gametophyte. **A-E** Linear megaspores, the position of functional megaspore varies. **F-H** Free-nuclear female gametophyte. **H** Two gametophytes with opposite tentpole. **I** Cellular gametophyte, indentation on the right is due to pressure of pollen tube (*pt*). **J** Upper portion of mature gametophyte with tent pole, archegonia and pollen tube. **K a-d** Neck cells in surface view. (After Sterling 1948a)

gametophyte gives it a flask-shaped appearance with a tent-pole-like tip. The micropylar tip of the gametophyte does not enlarge and the adjoining sporogenous tissue persists as a deeply staining cap. When twin gametophytes develop, the persistent sporogenous tissue is crushed by the second gametophyte. Such gametophytes are usually superposed, the upper one enlarges upward so that its tent pole faces the chalaza and the base the micropyle (Fig. 16.5 A).

The megaspore membrane of *Taxus* resembles that of gymnosperms generally (Pettitt 1966). This membrane is formed later and is not a direct continuation of the wall of the megaspore mother cell. It may be formed from the interactions of proteins characteristic of the two phases. It probably continues to isolate the gametophyte from potentially informational molecules from the sporophyte (Pennell and Bell 1987).

The tapetum is poorly developed or absent. Following the absorption of the sporogenous tissue, the developing gametophyte comes into contact with the nucellar cells.

A central vacuole appears before the division of the megaspore nucleus (Fig. 16.5 D,E). The division continues until (*T. baccata*) 250 after ca. 8 mitoses (Pennell and bell 1987), 256 (Jäger 1899) or more than 280 (*T. cuspidata*, Sterling 1948a) nuclei are formed. The nuclei form a single layer in the peripheral cytoplasm (Fig. 16.5 F-H). During the free-nuclear stage, the gametophyte is enveloped in a thin, tenuous, pliable megaspore membrane, often intimately associated with the circumjacent nucellar tissue. In the living material, it can be dissected with much difficulty.

At the time of cellularization the gametophyte is pear-shaped/spherical with an apical tent pole. Frequently, during development, the gametophyte becomes distorted due to the growth of the pollen tubes (Fig. 16.5 I). The gametophyte may be at various stages of development, four-nucleate to mature archegonia. The pressure by the pollen tubes indents the gametophyte and its outline changes; even the tent pole may become indented. However, there is no "erosive" effect of the tube on the gametophyte (Sterling 1948a). Frequently, while the tent pole is present at the younger stage, it may not remain distinguishable in a mature gametophyte.

Initiation of walls follows alveoli formation (Sokolowa 1890, Sterling 1948a). The alveolar cells form a honeycomb-like structure containing nucleus in each hexagon (Fig. 16.5 I). According to Pennell and Bell (1987), in *T. baccata*, wall formation begins shortly after completion of nuclear divisions and the gametophyte becomes cellular in about 1 month. New walls arise between the nuclei in positions marked out by the accumulation of cisternae of the endoplasmic reticulum. It is not clear how the cisternae, which coalesce to give rise to the walls, come to lie symmetrically between the nuclei. Secondary spindles, or any other form of microtubular participation in the siting of these walls, have not been detected. Jäger (1899) suggested that the nuclei and associated cytoplasm of the gametophyte of *Taxus* become enclosed directly by growing walls. According to Pennell and Bell (1987), in the cellular development of the female gametophyte, *Taxus* therefore resembles *Gnetum* (Sanwal 1962) and *Welwitschia* (Martens 1963), where the nuclei and the associated cytoplasm of the gametophyte become enclosed directly by growing walls.

The first formed cells of the gametophyte usually have large vacuoles and lightly staining cytoplasm. Subsequently, divisions occur at the micropylar and, later, at the chalazal end. The latter cells are small and stain deeply. The micropylar zone, with archegonial initials, is generally three to five cells deep. The enlargement of the gametophyte crushes the basal and lateral nucellar tissue. At the same time, the enlargement of the archegonia results in the widening of the apical part of the gametophyte. At the time of fertilization, the gametophyte is often cordate.

The superficial cells (in the micropylar region), which function as archegonial initials, have a basal vacuole. All the initials develop simultaneously. A periclinal division cuts off a neck initial and a central cell. The neck initial divides to form a single tier of two to four neck cells (Fig. 16.5 Ka-d). The central cell has a conspicuous basal vacuole. As it elongates, the nucleus situated near the neck in dense cytoplasm moves to the centre of the archegonium; the cytoplasm becomes frothy—the foam stage (Fig. 16.5 J). With the maturation of the archegonium, the dense granular cytoplasm forms a sheath around the egg nucleus. Numerous archegonia develop simultaneously and some of them are in contact with each other. There are 4–8 archegonia in *T. canadensis*, 5–11 in *T. baccata* and 6–25 (usually 8-14) in *T. cuspidata*. In *T. cuspidata*, the archegonia may form a sheath around the invaginated gametophyte. Occasionally, when there are two or more gametophytes in the same ovule, the archegonia may develop towards either the chalazal or the micropylar end (Dupler 1917). In the lower gametophyte, the archegonia abut on the pollen tube, which grows between two gametophytes. In *T. cuspidata* some archegonia become superposed and are presumed to be derived from periclinal division of a single initial; the internal archegonia lack neck cells.

In *T. canadensis*, the nucleus of the central cell has not been observed to divide, and probably functions directly as the egg (Dupler 1917). In *T. cuspidata* the ventral canal nucleus is rarely formed. The archegonial jacket is single-layered around an archegonium, or a group of archegonia (*T. cuspidata* — Sterling 1948a). The jacket cells are smaller and denser than the adjoining cells.

According to Pennell and Bell (1987), earlier stages in the development of the archegonium in *Taxus* is as follows. Two or four archegonial initials appear. The cells are large (ca. twice the diameter of the nucellar cells), occupy peripheral position, and are in contact with the megaspore membrane. Each cell divides to give rise to a large cell. It is 100 μm in diameter, surmounted by a single tier of neck cells, and surrounded by tangentially elongated jacket cells. The primary cell divides and gives rise to a well-defined neck canal cell (this is an unusual observation by the authors since neck canal cell is absent in gymnosperm. It is likely to be ventral canal cell) of 140 by 25 μm, and a central cell with highly organized cytoplasm. The transverse wall which separates the two cells in the archegonial canal is generally uniformly thick. Further nuclear divisions do not occur. A mature archegonium is round, the egg cell is ca. 250 by 175 μm. According to Pennell and Bell (1987), it remains to be resolved whether the cell equivalent to the central cell becomes the egg cell directly (P. Maheshwari and H. Singh 1967), or the nucleus divides and the resultant upper nucleus degenerates quickly (Konar and Moitra 1980).

The cytoplasm of the egg cell is zoned. The outer cytoplasm has prominent granular bodies, 2–5 μm in diameter (cf. "grandes inclusions" in the egg cells of conifers, see Chap. 2). They are derived from hypertrophied plastids which have encapsulated part of the cytoplasm (there is a complete absence of recognizable plastids in the mature cell). A conspicuous feature is the presence of bundles of microtubules which radiate from the nuclear envelope and possibly stabilize the zonation of the cytoplasm.

Pollination

Pollination occurs around the time the megaspore mother cell is undergoing meiosis. The collections made soon after pollination show pollen tubes on the irregular surface of the nucellus, the megaspore mother cell undergoes meiosis, or a four nucleate gametophyte (*T. cuspidata*).

A pollination drop is secreted by the apical cells of the nucellus, and the secretion continues even when the micropyle is amputated (Ziegler 1959). The chemical composition of the pollination drop is comparable to that of the extract of nucellar cells. The high concentration of sucrose in the pollination drop is attributed to the occurrence of acid phosphatases in the nucellus. The enzymes release sucrose from sucrose phosphate. Several amino acids, peptides, malic and citric acids are also present in the pollination drop.

The wingless pollen grains caught in the pollination drop imbibe the fluid, shift from the micropyle and come to lie against the lowermost surface of the drop. The fluid (along with the pollen) is eventually reabsorbed by the ovule and pollen reaches the nucellus.

Post-Pollination Development of Male Gametophyte.

The development of male gametophyte following pollination to fertilization was investigated by Dupler (1917) and Sterling (1948a). The pollen is shed at the uninucleate stage, a prothallial cell is not cut off, and the microspore nucleus functions directly as the antheridial initial. The nucleus divides ca. 2 months before fertilization, when the pollen grain has germinated and the tubes have grown in the nucellus. The antheridial and tube cell are formed, and the antheridial cell divides to form the stalk and body cell. In *T. cuspidata*, the stalk nucleus remains closely adpressed to the body cell on its entry into the pollen tube. Eventually, the stalk nucleus moves close to the tube nucleus located ahead of the enlarging body cell, and just behind the tip of the tube. The two sterile nuclei, embedded in abundant cytoplasm, are identical in size and structure. In the beginning, the body cell is more or less conical with its enlarged nucleus situated at the base. It enlarges, becomes spherical, and has dense cytoplasm with a distinct membrane. The body cell nucleus is eccentric, located diametrically opposite to the two sterile nuclei. Before marked expansion of the pollen tube, the sterile nuclei usually lie below the body cell, but their relative position changes during later stages. Thus, the entire complex may revolve as much as 180° along its horizontal axis so that the sterile nuclei come to lie above the body cell. The much-enlarged body cell (60 μm in diameter) forms two unequal male cells. According to Favre-Duchartre (1960b), a wall is absent between the two gametes which should be considered as nuclei. In *T. baccata*, under in vitro conditions, Rohr (1973a) observed two equal male cells unlike those produced in vivo. The pollen tubes expand in a characteristic way and the sac-like structure may cover the entire apex of the female prothallus. The dense cytoplasm at the tip of the tube becomes vacuolate. The sperm cells mature and the sterile nuclei can still be distinguished.

As the pollen tube grows through the nucellus, it becomes wider, occasionally twists and branches. Just before it reaches the gametophyte, the tube often branches once, and one branch grows on each side of the tent pole. The tip of each branch expands a good deal, either on the shoulders of the gametophyte, or over its apex, and often becomes as broad as the prothallus itself. When several tubes grow simultaneously, they "pile up" on each other so that the sac-like expansions of successive tubes form a heap above the prothallial apex. Occasionally, the tubes grow up to the gametophyte and meet below the prothallus (Saxton 1936, Sterling 1948a). The wall of the pollen tube is quite thick.

Fertilization

According to the variability in the development of the gametophyte (*T. cuspidata*), fertilization may take place in 4 weeks or more. At fertilization, the prothallus becomes relatively smaller as compared to its size at maturity. After fertilization in *Taxus* sp., conspicuous growth occurs in the gametophyte, which is in contrast to the condition in the conifers (particularly Pinaceae), where the gametophyte is fully grown, or nearly so, at fertilization.

The male gamete contains dense granular material and stains more deeply than the much longer egg nucleus. The functional male gamete comes to lie close to the egg nucleus in the centre of the egg. The cytoplasm above the mating nuclei is highly vacuolar and looks frothy. The male gamete "sinks" into the egg nucleus, and they move to the base of the archegonium.

The mating gametes, as well as the early proembryonic nuclei, are embedded in a highly granular, deeply staining cytoplasm at the base of the egg. The dense cytoplasm appears different in consistency and staining capacity from the cytoplasm around the unfertilized egg nucleus, and may even include some cytoplasm of the male gamete. Whether the male and female nuclear material integrates completely or not requires further study. In two preparations, Sterling (1948a) observed only a single resting nucleus at the base of the egg. All other preparations of this stage showed either the two mating nuclei or the telophase of the first mitotic division.

Embryogeny

During division of the zygote, the spindle lies transversely to the long axis of the archegonium. The resultant two free nuclei lie close together in the dense cytoplasm at the base of the egg cell. Occasionally, the division may take place in the centre of the archegonium, and the two nuclei migrate to the base. The next division is also in the transverse plane, but at right angles to the earlier division, and produces a tier of four nuclei perpendicular to the long axis of the archegonium. Following further division, the nuclei become progressively smaller and irregularly distributed in the proembryonic cytoplasm.

The walls are laid down at the 16- or 32-nucleate stage on spindle fibers which extend between all the nuclei. With wall formation, the nuclei become evenly distributed in the dense cytoplasm, and the cells lie approximately in two tiers: a primary upper tier (p^U) and a basal primary embryonal tier (p^E). The p^U tier divides transversely and produces an upper tier (U) of "open" cells, and a lower tier of closed cells which develop into the suspensor (S). Simultaneously, or just before the suspensor elongates, the cells of the basal group divide irregularly. Thus, from the base upward, a proembryo (*T. cuspidata*) comprises a group of (6–14) deeply staining embryonic cells (E), followed by another group of (9–13) suspensor cells (S), topped by a tier of (9–3) cells (U) "open" to the archegonial cytoplasm (Fig. 16.6 A). The proembryo occupies the lower half of the archegonium (Fig. 16.6 A, B). Towards the neck of the archegonium, several supernumerary nuclei (from the pollen tube) may persist in the archegonial cytoplasm; eventually they disintegrate.

Soon after the organization of the proembryo, the suspensor cells elongate and thrust the embryonal cells through the base of the archegonium (Fig. 16.6 B). There is dissolution of the cells in the central region of the gametophyte to form a corrosion cavity (cf. conifers). Meanwhile, at its chalazal end, the gametophyte enlarges considerably. In the lower region, the cells divide repeatedly and the numerous cells appear small. With further development, the gametophyte becomes spherical and later pear-shaped and remains so until seed ripening.

The densely cytoplasmic embryonic cells at the tip of the suspensor

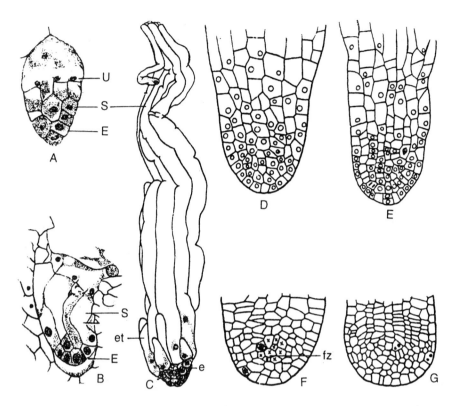

Fig. 16.6 A-G. *Taxus cuspidata*, embryogeny. **A** Longisection three-tiered proembryo shows embryonal (*E*), suspensor (*S*) and upper tier (*U*). **B** Suspensor. **C** Early embryo shows embryonal mass (*e*) at the tip of elongated suspensor, and embryonal tubes (*et*). **D-G** Development of embryo, longisections. **D, E** Embryonal mass shows anticlinal and periclinal divisions. **F** Cells marked *X* represent the focal zone (*fz*). **G** Cell walls oriented around the focal zone. (After Sterling 1948b, 1949a)

remain quiescent during their passage into the endosperm. Divisions in the embryonic group result in the formation of successive secondary suspensors, the embryonal tubes (*et*; Fig. 16.6 C) which elongate autonomously. The latter may elongate, while the original suspensor cells remain relatively short. Most of the suspensor cells elongate synchronously and present a tiered configuration. The suspensor cells enlarge continuously and their diameter increases. The nucleus and most of the cytoplasm is located at the tip. After elongation, a variable number of suspensor cells may separate from one another with or without embryonic cells attached to their tip. These embryonic cells form a meristematic group simulating the original embryonic group. This cleavage or splitting results in the production of two to four embryos, each consisting of embryonic cells borne on several suspensor cells. More often, the embryo develops without such cleavage. Separated single suspensor cells may also show some proliferation.

The open cells generally degenerate, but some of the completely walled cells may undergo a few divisions.

The embryonal primordium gradually becomes massive by periclinal and anticlinal divisions at the free tip; the rib-meristem activity (behind the apex) contributes to its mass as well as to that of the suspensor (Sterling 1948b).

Sterling (1949a) has studied the meristematic development, and tissue differentiation of the embryonic group of cells in *T. cuspidata.*

After elongation of the suspensor tier, the embryonic cells become active and, along the surface of the young embryo, a group of densely cytoplasmic cells divides periclinally and anticlinally (Fig. 16.6 D, E). As the tip enlarges, oblique divisions also occur in the superficial cells. With continued enlargement, anticlinal divisions predominate on the forward flanks of the apex, while periclinal divisions continue at the summit. Growth activity is limited to a small group of initials along the surface of the massive embryonal apex. From these initials, all the other cells of the embryo (appear to) diverge in periclinal rows. The anticlinal cells appear as concentric arcs radiating from the initial region, indicating successive growth increments in the apical development. After apical growth has occurred in the young embryo for some time, a focal area of lighter-staining cells differentiates just behind the apex (Fig. 16.6 F). The focal area enlarges and adjacent cells divide with walls concentric to this area (Fig. 16.6 G). The concentric cells divide along the lateral limits of the focal area, are arranged in a cup-shaped group, somewhat flared below the apex of the embryo. These cells constitute the procambial cylinder; pith is absent. There is a group of distinct cells with relatively large spherical nuclei at the base of the procambial cylinder—the root generative meristem.

As the focal group becomes less active, the cells produced by the lateral concentric divisions build up buttresses of tissue on the flanks of the free apex of the embryo (Fig. 16.7 A, B). Deeply staining cells in these buttresses have a histological continuity with the cells of the elongating procambial core of the hypocotyl. Cotyledonary primordia develop from the buttresses by predominantly subepidermal activity; later growth of the cotyledon occurs by the activity of the superficial initials. The procambial strand in each cotyledon, continuous with the procambial core, develops acropetally behind the apex of the cotyledon (Fig. 16.7 C-F).

The root generative meristem produces a tissue simulating the "column" of the pinaceous root cap, only after marked elongation of the hypocotyl and enlargement of the cotyledon. This tissue is very indistinct, there is no juncture zone on the outer surface of the embryo, and the epidermal and several hypodermal layers continue from the cotyledons into the suspensor (Fig. 16.7 C). There are usually two, occasionally three, short cotyledons.

The fleshy aril (which covers the seed) comprises thin-walled cells rich in cell sap. The epidermis has numerous stomata. The cells situated below

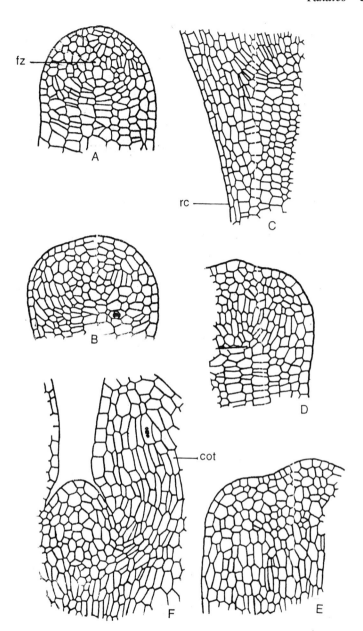

Fig. 16.7A-F. *Taxus cuspidata*, longisections of early and late embryos. **A** Early embryo shows concentric divisions around focal zone (*fz*). **B** Root, generative meristem. **C** Part of hypocotyl and root cap (*rc*) region, the epidermal and two hypodermal layers are continuous from the hypocotyl to the root cap. **D-F** Initiation and development of cotyledons (*cot*). (After Sterling 1949a)

the outer epidermis are isodiametric, those next to the inner epidermis are flattened. The cells in the middle zone are large and vacuolate. The base has a stub of vasculature. During earlier stages the cells are chlorophyllous, later several of them contain tannin.

On the basis of morphological and ontogenetical evidence, Andre (1956) and Loze (1965) interpret the aril as the second integument, which originates as a foliar organ. This interpretation is erroneous, since an integument cannot have a foliar origin.

Seed
Both chalaza and integument contribute to the seed coat. There is an increase in the number of cell layers, the sarcotesta is absent (in Taxales), while sclerotesta and endotesta differentiate. A plate of thick-walled cells develops at the chalaza, is one to several layers of palisade-like simple pitted cells, and is contiguous with the sclerotesta. Several traces enter the chalaza, pass through the basal plate, and traverse the length of seed coat through endotesta (see Schnarf 1937).

Germination.   The germination of seed is hypogeal and the first two linear cotyledonary leaves pierce the soil. The primary root persists.

Chromosome Number
In *Taxus wallichiana*, the haploid chromosome number (as determined from endosperm squashes) is n=12 (Fig. 16.8). Mehra (1988) states that the karyotype could not be analyzed precisely but there are definitely three isobrachial chromosomes (Fig. 16.8 I, II, VI), three hetero-brachial chromosomes (Fig. 16.8 III, V, VII), and the rest apparently have

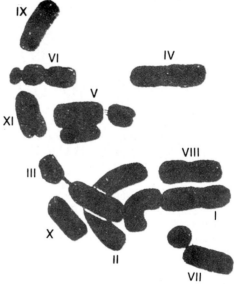

Fig. 16.8 *Taxus wallichiana*, endosperm cell with 12 chromosomes (squash preparation). (After Mehra 1988)

terminal or near-terminal primary constriction (Fig. 16.8 IV, VIII, IX, X, XI, XII).

Temporal Considerations
Taxus shows a 1-year life cycle. According to Sterling (1948a), the entire development from megaspore formation to seed maturity takes place within the space of a single season.

Phylogenetic Considerations
The position of the Taxaceae in the Coniferales has long been disputed. Sahni (1920) first suggested that *Taxus, Torreya* and *Cephalotaxus* should be separated and placed in a phylum of their own. Florin (1948, 1951) agreed, but considered *Cephalotaxus* a true conifer. He (Florin 1954, 1955, 1958) considered the taxads to be sufficiently distinct to form a separate class, the Taxopsida, and segregated the Taxaceae from the other conifer families. The main basis for such a treatment is the distinctive type of reproductive organs in the taxads. The female reproductive structures are not organized into cones, in either living or fossil genera. The taxads appear to deviate from the conifers in the single terminal seed (enclosed in an aril) borne on fertile shoots, and the perisporangiate microsporophylls. In the Coniferopsida, the ovule is a lateral organ and its position is consistent with the general interpretation of a conifer cone as a conduplicate strobilus (see Miller 1977). Similar axial and terminal ovules are associated with the extinct taxa *Paleotaxus* (Triassic) and *Taxus jurassica* (Jurassic), a feature of great antiquity. However, the epidermis of *Paleotaxus* is quite different from that of *Taxus*. According to Stewart (1983), this is surprising, because several characters are in common with *Taxus*.

Because of lack of clear fossil evidence, the origin of the perisporangiate microsporophyll, notably in *Taxus* and later in *Pseudotaxus*, has long been a subject of speculation. Although the taxads have been recognized as advanced in several morphological characters, the perisporangiate microsporophyll is considered primitive. Wilde (1975) interpreted the male cones in *Taxus* and *Pseudotaxus* as evolved from compound structures (similar to those of *Cephalotaxus*), presenting evidence to show that the axis of these vestigial cones is a reduced fertile branch on which each fertile branch unit (cone) is reduced to a single terminal perisporangiate microsporophyll, attached directly on the fertile branch. Such sporophylls resulted from the phylogenetic fusion of two or more apical, hyposporangiate microsporophylls. The male structures in *Taxus* are, therefore, interpreted as extremely reduced and not primitive (Wilde 1975).

According to H. Singh (1978), the only novel feature in taxads may be their ovules, and this character alone cannot be a sufficient justification to treat them as a separate order or class. Embryologically, the taxads are like the conifers in several respects: (*a*) unwinged pollen and absence of prothallial cells in the male gametophyte (as in Cupressaceae, Taxodiaceae and Cephalotaxaceae), (*b*) archegonia occur singly (as in Pinaceae, Podocarpaceae), and (*c*) a general pattern of embryogeny, as in conifers.

While various studies (Keng 1969, Harris 1976 and others) have not solved the problem, they indicate a disagreement with the present separation of the taxads and conifers, and emphasize a reconsideration of the relationships. The subject therefore deserves prime attention in future research (Miller 1977).

17. Gnetopsida

The extant taxa *Ephedra*, *Gnetum* and *Welwitschia* were earlier placed in a single order, Gnetales, and a single family, Gnetaceae. These three taxa (considered to be the highest evolved among the gymnosperms) lack fossil record except for the remains of pollen of *Ephedra* from the Eocene (54 my B.P.), and pollen-like remains of *Ephedra* and *Welwitschia* from the Permian (280 my B.P.) (Delevoryas 1962). Certain common features of these taxa are comparable to angiosperms: (a) vessels are present in the wood, (b) the microsporangia and ovules are borne on fertile shoots which form compound strobili, (c) the ovules are enclosed within one or two envelopes, in addition to the integument and (d) the upper part of the inner integument extends into a long tubular structure, the micropylar tube.

In spite of these common features, the heterogeneous nature of the order has been apparent, and resulted in its constantly changing taxonomic treatment. By the end of the 19th century, the order Genetales comprised three families with one genus each. Further knowledge, especially about reproduction, revealed significant differences between the three genera. Florin (1931, 1933b) and Eames (1952) supported the establishment of three independent orders: Ephedrales, Welwitschiales and Gnetales, each comprising a monogeneric family. The phylogenetic consideration of these taxa are discussed in Chap. 21.

18.   Ephedrales

The plants are herbs, woody shrubs or lianas, leaves free, scale-like, and the stem jointed and green. Secondary wood has vessels. The plants are dioecious with compound male and female strobili. The ovule is surrounded by two envelopes, the inner projects as a long tube and archegonia are present. The embryo is dicotyledonous.

Ephedraceae

Ephedra

Morphology

The plants are highly branched herbs, shrubs or lianas. *Ephedra campylopoda* is a cultivated ornamental with pendulous branches, and is commonly grown in hanging baskets. It grows naturally between the crevices of the Wailing Wall in Jerusalem (Fig 18.1 A). *E. triandra* grows into a small tree, and *E. gerardiana* is a perennial herb. *E. foliata* is a scrambling shrub that may reach a height of about 6m, may climb up the trees, or walls, or, in the absence of a support, may spread along the ground.

The stem is long, jointed, fluted and green. It has distinct nodes and internodes (Fig 18.1 B). The internodes in most species are longitudinally ridged and the ridges of the successive internodes stand on alternate radii. The shoot comprises indeterminate as well as determinate branches, although the distinction between the two is not as well marked, as in *Pinus*. On each node, the indeterminate shoot bears three, occasionally four, leaves in a whorl. The axil of some of the leaves bears determinate shoots with leaves in opposite and decussate pairs (*E. foliata*, Tiagi 1966). On a node, the leaves stand on a radii of the ridges of the internode below. Since the ridges of the successive internodes alternate, so do the leaves of the successive nodes. Branching is axillary (Fig. 18.1 D), and additional accessory buds arise below and at the base of axillary buds. An intercalary meristem occurs above each node (Fig. 18.1 E, im, ima), and it produces most of the internodal tissue. Later, it forms either abscission layers, or may mature as transverse bands of sclerified parenchyma.

Fig. 18.1 A–E. A *Ephedra campylopoda,* rooted plants between the stones of the Wailing Wall in Jerusalem. **B–D** *E. gerardiana.* **B** Vegetative twig. **C, D** Scale-like leaves fused at the base to form a sheath. **E** *E. antisyphilitica,* longisection of node with axillary bud (*ab*), intercalary meristem (*im*), and intercalary meristem of axillary shoot (*ima*). (After Bierhorst 1971)

The scaly leaves are deciduous. They may be opposite or whorled (Fig. 18.1 C, D). Each leaf receives two traces which have distinct origin (see below). The leaves are ultimately shed after the formation of an abscission layer.

Anatomy

Root. The roots of *E. foliata* have a long cap. The apical meristem can be distinguished into: (a) a discrete layer of stelar initials, (b) the columella and its initials, and (c) a common region for the initiation of cortex and peripheral region of the cap (Pillai 1966).

In cross section, the root shows epiblema, cortex, endodermis, pericycle and a diarch vascular region (Fig. 18.2 A).

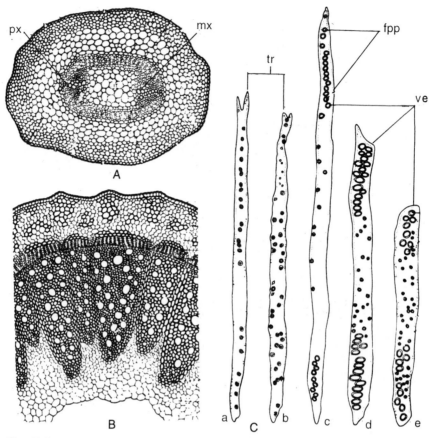

Fig. 18.2 A–C. Root and stem anatomy. **A, B** *Ephedra* sp. **A** Transection young root. *mx* metaxylem. *px* protoxylem. **B** Transection of old stem. **C.** *E. californica*, tracheary elements from secondary xylem; **a, b** tracheids (*tr*) with bordered pits, **c–e** vessel members (*ve*) with foraminate plates (*fpp*) on end walls. (**C** after Esau 1953)

Stem. In longisection the shoot apex in *E. altissima* (Strasburger 1872, Gifford 1943) and *E. fragilis* var. *campylopoda* (Seeliger 1954) shows a well-defined tunica and a corpus (Fig. 18.3). A tunica layer is also reported in *Gnetum, Araucaria brasiliana*, (Strasburger 1872), and permanent shoots of *Taxodium distichum* (Cross 1939). In *Cryptomeria japonica* (Cross 1941), *Agathis lanceolata* (Sterling 1958) and a few *Araucaria* spp. the tunica layer is rather restricted (Griffith 1952). The presence of tunica and corpus is a regular feature in angiosperms. The stem of *Ephedra* is also capable of elongation by an intercalary meristem present at the base of each node (Fig. 18.1 E).

Transection of the stem exhibits ridges and furrows (Fig. 18.2 B). A single-layered epidermis is covered with a thick cuticle. Sunken stomata occur in rows in the furrows only. The stomata are of the haplocheilic type, i.e the two guard cells originate from a common mother cell, and the

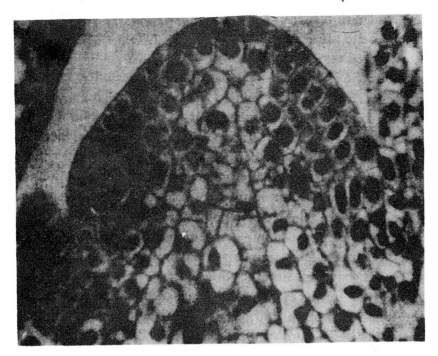

Fig. 18.3. *Ephedra altissima.* Longisection of shoot apex, the tunica layer is well defined. (After Foster and Gifford 1959)

neighbouring epidermal cells become modified as subsidiary cells. Below the epidermis, in the region of the ridge, patches of sclerenchymatous tissue are very conspicuous. The cortex comprises palisade-like chlorenchyma, which actively photosynthesizes and compensates for the scale-like leaves. Besides the photosynthetic zone, chloroplasts also occur in thin-walled parenchymatous cells of the cortex. The endodermis is distinct, while the pericycle is ill-defined.

In *Ephedra,* the leaves are opposite and decussate, or in whorls of three or four. Accordingly, there is variation in the number of vascular bundles from species to species (see Sporne 1965). In *E. foliata* (Tiagi 1966), the vascular anatomy of the indeterminate shoots which bear three or four leaves in whorls is fundamentally alike. A transection through the middle region of an internode of a stem, which bears leaves in whorls of three, shows three ridges and three alternating furrows (Fig. 18.4 A). There are 15 vascular bundles arranged in three groups of five each, below the three ridges (Fig. 18.4 A). In the subnodal region all the five bundles (under a ridge) become laterally united to form a single vascular arc (Fig. 18.4 B, C). Slightly higher up, each of these three arcs splits radially into two bundles (Fig. 18.4 D). Two traces each, one from either side of the gap thus formed, are first given off to the leaf, subsequently another pair to its axillary branch (Fig. 18.4 E, F). The two branch traces quickly divide twice in a

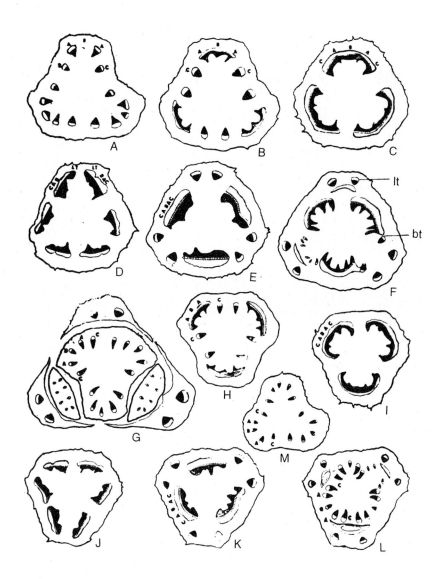

Fig. 18.4 A–M. *Ephedra foliata*, serial transections through successive nodes and two internodes of an indeterminate shoot with leaves in whorls of three. *bt* branch trace. *lt* leaf trace. (After Tiagi 1966)

radial manner to form the stelar ring of eight bundles of the axillary shoot (Fig. 18.4 F, G). Simultaneously, two vascular strands between any of the three gaps become laterally united to form a single arc (Fig. 18.4 H, I). As the leaf sheath and the axillary shoots separate from the stem at the node, each of the arcs radially splits into a group of five separate vascular strands below each of the three ridges of the next higher internode (Fig. 18.4 J–M).

This process of fusion, branching and regrouping is repeated below every node. Consequently, the three sets of five bundles below each ridge in an internode alternate in position with their counterparts in the internode above but, again, become superposed in the alternate internodes. The longitudinal course of the vascular bundles in the stem through three successive internodes and two nodes in shown in Fig. 18.5.

In the determinate shoots with leaves in opposite and decussate pairs, the vascular anatomy is similar to that of the indeterminate shoots except that under each ridge there are only four bundles (the middle bundle is absent).

According to Tiagi (1966), the primary vascular system in *Ephedra* (based on his work on *E. foliata*) is best interpreted as a perforated, ectophloic type of dictyostele where the breaks in the vascular cylinder internal to the furrows and the ridges are leaf-cum-branch gaps and perforations, respectively. Marsden and Steeves (1955) made a detailed study of the primary vascular system in a number of species of *Ephedra* bearing two or three leaves at a node. They interpreted the primary vascular system as an eustele in which the internal bundles are continuous with the leaf traces.

The xylem consists of tracheids, vessels and xylem parenchyma. The vessels originate from the pitted tracheids. Transitional forms between typical tracheids and foraminate vessel members, characteristic of *Ephedra*, have been observed (Fig. 18.2 Ca–e). The plates show a uniseriate arrangement of pores in early-formed elements, while in the later-formed plates most pores are two-ranked, or irregularly grouped. There is a persistent primary cambium but not much secondary wood formation occurs, so that the vascular cylinder does not become thick.

A transection of stem of *Ephedra* with secondary growth appears more like an angiosperm than a gymnosperm, because of the deep origin of the periderm, slight increment of phloem, multiseriate and dilated phloem ray and ring-porous xylem (Fig. 18.2 B).

A peculiar anatomical feature is the presence of a diaphragm-like plate of cells at the base of each internode. This makes the stem easily separable at the nodes.

The phloem in *Ephedra* combines characteristics of both angiosperms and gymnosperms, and has some unique features of its own.

Alosi and Alfieri (1972) investigated the structure of phloem in *E. californica* and *E. viridis*. The conducting phloem lies outside the cambial zone. It consists of parenchymatous rays, sieve cells, and axial parenchyma which includes albuminous cells. In the non-conducting phloem the sieve elements cease to function. This region is comparatively small due to the formation of annual rings of periderm which originate deep within the previous season's phloem (Fig. 18.6 A). The annual periderm arises in spring at the same time as the vascular cambium is reactivated. Fusiform prenchyma cells which lie internal to the periderm develop into fibre sclereids.

Fig. 18.5. *Ephedra foliata*, vascular skeleton of three successive nodes and two internodes of an indeterminate shoot with leaves in whorls of three. The skeleton has been cut longitudinally on one side and spread open. Approximate levels of transections **A–L** are shown in Fig. 18.4. *bt* branch trace. *lt* leaf trace. (After Tiagi 1966)

All sclerified cells in both the axial and ray systems generally contain crystal sand in the region of the middle lamella.

The immediate phloem derivatives resemble the cambial initials. They are long, have primary pit fields in their primary cell walls, small radial

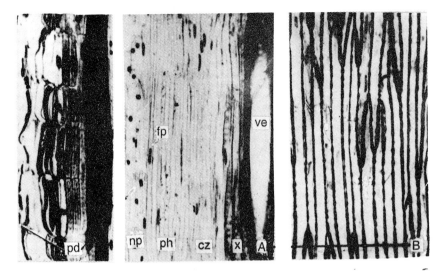

Fig. 18.6 A, B. *Ephedra californica.* **A** Radial sections of bark to show (left to right) old periderm (*pd*), collapsed phloem, non-conducting phloem (*np*), conducting phloem (*ph*) with fusiform parenchyma (*fp*), cambial zone (*cz*), xylem (*x*) and a vessel (*ve*). **B** Tangential section to show differentiated phloem with oblique anticlinal divisions, sieve areas and sieve elements. (After Alosi and Alfieri 1972)

diameter, and oblique end walls. The cellular contents take a light stain, the nucleus is granular with several nucleoli. An ovoid body appears near the nucleus in the differentiated cell. It resembles the largest nucleolus in shape and staining. Because of its positive staining reaction with Ponceau S, this body has been referred to as the slime body (as in the angiospermous sieve elements). However, ultrastructural studies by Behnke and Paliwal (1973) failed to confirm this.

The nucleus flattens and in some cells appears to curl around the perphery of the cell in a bracelet-like fashion. It may be irregular in outline, but maintains its granular character and chromaticity. Degeneration of the nuclear membrane has not been observed. Necrotic nuclei (as in Pinaceae) are frequent in the mature sieve cells.

Mature sieve elements vary in shape from extremely tapered (Fig. 18.6 B) to blunt-ended. They remain thin-walled with no apparent secondary thickening. The secondary sieve elements are of two categories based on their length: ca. 400 and 220 μm long. The former are direct derivatives of the fusiform initials, whereas the latter arise after a transverse division in their precursors. This shortening of sieve elements has been considered an advanced character (see Paliwal 1992).

In a radial view, the end walls comprise a row of closely spaced sieve areas (Fig. 18.7 A) similar in size and shape to lateral sieve areas

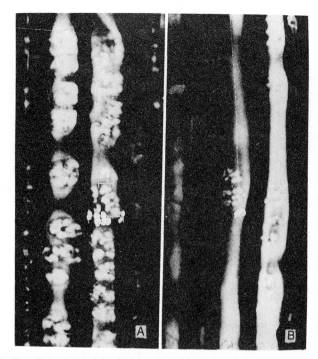

Fig. 18.7 A, B. *Ephedra californica*. **A, B** Radial sections of phloem. **A** Sieve areas with sieve fields on overlapping radial end walls of sieve elements. **B** Lateral sieve areas on radial walls. (After Alosi and Alfieri 1972)

(Fig. 18.7 B). Within the sieve areas, the pores are generally aggregated into groups (Fig. 18.7 A), termed sieve fields. Each pore measures ca. 0.8 μm. Sieve areas do not occur on transverse walls between sister cells (two sieve cells or a sieve cell and albuminous cell).

The differentiation of sieve areas begins very early, and progresses very rapidly in developing sieve elements. Presumably, full perforation of the sieve-area pores occurs just before the element becomes functional.

Numerous conspicuous albuminous cells are present in the axial system of *Ephedra*. They may occur in radial files, or scattered in the conducting phloem. The albuminous cells have dense protoplasm, rich in elongated mitochondria, plastids, granular ER, free ribosomes and a prominent nucleus. They are connected to the sieve cells by branched plasmodesmata on their side, fusing with a sieve pore on the sieve cell side (see Paliwal 1992). The two daughter cells from a transverse division may not differentiate into a similar cell type; one may develop into a sieve cell, the other into an albuminous cell or phloem parenchyma cell.

There is a remarkable intergradation between the elements of phloem: sieve cells → albuminous cells → parenchyma. It is difficult to distinguish between certain sieve cells and albuminous cells, and also between some albuminous cells and other phloem parenchyma cells. Although most sieve elements and some parenchyma cells may occur as long fusiform cells, any of the three cell types may arise from a daughter cell following transverse division of a fusiform phloem mother cell. That an albuminous cell of *Ephedra* forms directly from a single division of a phloem mother cell distinguishes it from albuminous cells of most other gymnosperms, and is comparable to the specialized parenchyma cells, including companion cells, of higher plants.

Reproduction

Terms like strobili, inflorescence, flower, bract (commonly used for describing angiosperms) are often used for *Ephedra*, *Gnetum* and *Welwitschia* for comparison only; there are no phylogenetic implications.

Ephedra is typically dioecious, although bisporangiate cones (Pearson 1929) have been recorded on the same plant.

Male Strobilus

The male strobili are in clusters at the nodes of branches (Fig. 18.8 A), each strobilus arising in the axil of a scale leaf so that their number depends on the number of scale leaves. The microsporangiate cone is regarded as a compound structure. It consists of a number of pairs of decussately arranged, broad, cupped bracts, the lowest pair of which is sterile (Fig. 18.8 B). Each remaining pair of bracts develops a solitary microsporangiate shoot in its axil. This shoot continues into a short axis. The microsporangiophore (antherophore) bears a pair of bracteoles or perianth fused at the base. In their upper free region, the posterior bracteole overlaps the anterior one, and completely encloses the antherophore. The latter outgrows the sheath and protrudes at maturity, bearing a variable number of terminal anthers (Fig. 18.8 C).

Microsporangium. Both *E. foliata* and *E. gerardiana* show a similar development of the antherophore and microsporangium (H. Singh and K. Maheshwari 1962). In the axil of each bract of a male cone, an antherophore arises as a protuberance; two perianth primordia differentiate laterally (Fig. 18.9 A, B). The protuberance has a tunica-corpus organization which is disturbed due to periclinal divisions in the tunica soon after the differentiation of perianth (Fig. 18.9 B). A group of hypodermal archesporial cells appear in the protuberance after the demarcation of the perianth (Fig. 18.9 C, D). After some time, the protuberance becomes lobed. Each lobe is the primordium of a sporangium (Fig. 18.9 E, F), and contains a plate of hypodermal archesporial cells. Later, in each sporangium, a band of cells becomes sterile, forming two chambers (Fig. 18.9 G). The outermost cells of the archesporial tissue divide periclinally and give rise to an inner sporogenous layer and the outer wall layer. The latter divides again to form a middle layer and the tapetum (Fig. 18.9 H, I). Anticlinal divisions occur

Fig. 18.8 A–C. Microsporangiate strobilus. **A** *Ephedra chilensis*, twig with microsporangiate strobili. **B, C** *Ephedra* sp. **B** Male strobilus. C Mature male flower (from B) after removal of bracts, the antherophore bears sporangia at its apex and protrudes by displacing the bracteoles. (**A** After Foster and Gifford 1959, **B, C** after Bierhorst 1971)

in the wall layers to keep pace with the growth of the anther. With further growth of the sporangium, the middle layer becomes flattened and crushed. The tapetal cells enlarge, become densely cytoplasmic, and show two to four nuclei when the microspore mother cells enter prophase I (Fig. 18.9 J). The tapetum begins to degenerate soon after the reduction divisions are over. Only the epidermis persists in the mature microsporangium; its walls

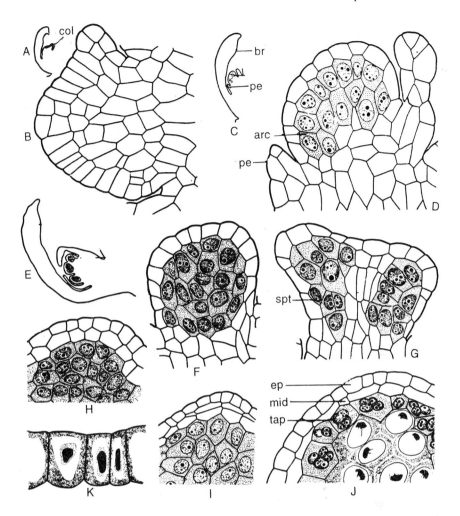

Fig. 18.9 A–K. *Ephedra gerardian*a, development of microsporangium. **A, C, E** Longisection of male flowers at progressive stages of development. *br* bract. *col* column. *pe* perianth. **B** Column from **A. D** Later stage to show perianth and a plate of hypodermal archesporial cells (*arc*). **F** Sporangium from **E. G** Two groups of archesporial cells formed by sterilization of sporogenous (*spt*) tissue. **H–J** Transection of sporangium. *ep* epidermis. *mid* middle layer. *tap* tapetum. **K** Epidermis from mature sporangium. (After H. Singh and K. Maheshwari 1962)

thicken and show a wavy outline (Fig. 18.9 K). The sporogenous tissue increases due to repeated mitotic divisions.

In a young microsporangium of *Ephedra foliata* (see Moitra and Bhatnagar 1982), there is a difference in the pattern of staining of the walls of sporogenous and tapetal cells, and the other wall layers. The cell walls of wall layers are thick, and take a brighter magenta stain with periodic Schiff's (PAS) reaction than the walls of tapetum and sporogenous cells (Fig. 18.10 C).

The cell walls of the epidermis and middle layer are mostly of cellulose, and of tapetal and sporogenous cells mostly pectinaceous. There is no starch in the microsporangium until the microspore mother cell stage (Fig 18.10 A); it then appears in the tapetum and wall layers when the microspores are formed. In the wall layers, starch is consumed during and after meiosis. The polysaccharides become depleted in the wall layers but, simultaneously, increase in the locular fluid.

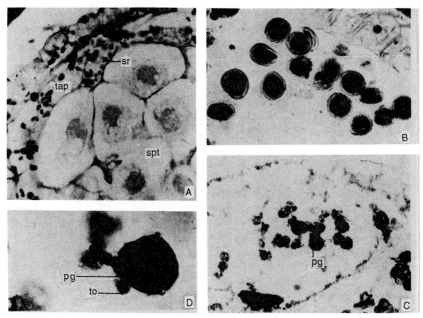

Fig. 18.10 A–D. *Ephedra foliata*, microsporogenesis. **A** Longisection of microsporangium to show starch (*sr*) in wall layers and tapetum (*tap*). *spt* sporogenous tissue. **B** Mature pollen grains. **C** Longisection of microsporangium stained with spirit blue, tapetal orbicules and pollen grains (*pg*) take up similar stain. **D** Association of pollen grains with tapetal orbicules (*to*). (After Moitra and Bhatnagar 1982)

In *E. foliata*, an increase in Feulgen stainability occurs at three stages: (a) just before meiosis, (b) at the interphase between the mother cell meiosis and microspore mitosis, and (c) in the antheridial cell.

Throughout their development, the sporongenous cells are rich in RNA. The maximal level in mother cells reaches prior to meiosis. A loss of cytoplasmic basophilia has been observed, especially during meiosis II. The young microspore is poor in RNA but the level rises again at the time of antheridial cell formation.

The archesporial, later sporogenous, cells are rich in total proteins. The level is high in the microspore mother cells, but low in microspores. The protein content in the microspores increases sharply before microspore mitosis (Fig. 18.10 B). The antheridial nucleus in the pollen grain has less protein than the tube nucleus

Microsporogenesis.   Before meiosis, the microspore mother cells undergo a gradual stretching, thinning and loss of PAS stainability. Their walls dissolve and a thin layer of callose develops between the cell wall and the plasma membrane. With the advance of meiosis, callose increases in thickness. The wall formation is simultaneous (as in most gymnosperms), the callose grows centripetally, and separates the four nuclei into a tetrad (Fig. 18.11 A-C). Numerous vesicles of Golgi origin are active at anaphase II and telophase II. They become arranged in the region of the future wall. Later, a cellulosic wall separates the resultant cells. The four microspores are usually arranged in a tetrahedral or decussate fashion. Meiosis is asynchronous, so that meiotic stages range from prophase I to telephase II, in various microspore mother cells in a microsporangium.

A peritapetal membrane completely covers the tapetum and developing pollen grains, in *Ephedra foliata*. It is acetolysis-resistant and can be separated as a bag containing pollen grains.

Alongwith the development of tapetal membrane, tapetal orbicules are formed in the tapetal cytoplasm which completely mask the developing pollen grains (Fig. 18.10 C, D). The staining intensity of these orbicules increases, accompanied by an increase in the intensity of the pollen exine. Young microspores have bright autofluorescent nuclei, which suggests that the haploid microspore may have the capacity to synthesize sporpollenin (see Moitra and Bhatnagar 1982).

Changes in DNA, RNA and protein level in the tapetal cells have been studied. The tapetal nuclei show a sharp increase in chromatin stainability before meiosis in mirospore mother cells. The nuclei in a number of tapetal cells undergo regular mitotic divisions followed by inhibited cytokinesis. This results in the formation of binucleate, rarely multinucleate cells. (It is somewhat difficult to distinguish the changes in tapetal cells with respect to mother cells because of asynchronous meiosis). The tapetal nuclei show a higher Feulgen stainability in younger sporangia than in older sporangia.

The role of tapetal cells as a source of nucleic acid is not fully understood. In *E. foliata*, Feulgen-positive material extruding from tapetal cells has not been observed. There is, however, a reduction in staining intensity of nuclei at later stages of meiosis and thereafter.

Mehra (1949) studied the effect of sulphanilamide on the division mechanism of the body nucleus in pollen grains of *Ephedra foliata* ($n = 7$), *E. sinica* ($n = 14$), *E. saxatilis* ($n = 14$) and *E. intermedia* ($n = 14$). The grains were germinated in artificial culture in the natural secretion which oozes out during pollination. In cultures containing germinating grains at metaphase, anaphase or telophase, the addition of 0.5% sulphanilamide caused immediate collapse of the spindle mechanism followed by the splitting and scattering of chromosomes. With 0.2% sulphanilamide, there was no effect on the mitotic phenomena; while 0.4% was at the critical level and had a paralyzing effect on the spindle mechanism. The effects of the chemical

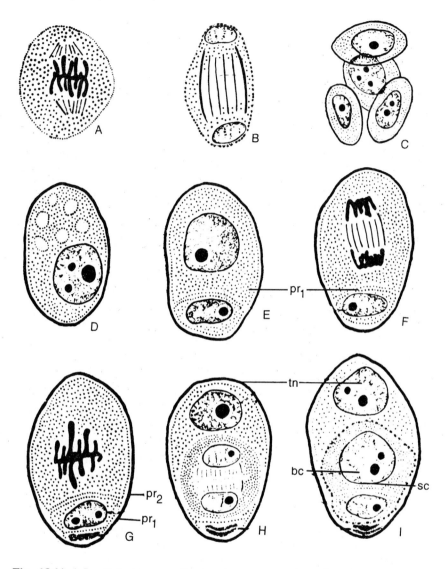

Fig. 18.11 A-I. *Ephedra gerardiana*, microsporogenesis and male gametophyte.
A-C Formation of microspore tetrad. **D** Uninucleate microspore. **E-H** Development of
male gametophyte. *pr*1 first prothallial cell. *pr*2 second prothallial cell. *tn* tube nucleus.
I Mature pollen grain at shedding stage. *bc* body cell. *sc* stalk cell. (After H. Singh and
K. Maheshwari 1962)

on chromosomes are: the kinetochore region becomes prominent; occasionally
a high degree of contraction occurs, and the chromosomes become straightened
due to the absence of a spindle mechanism. Sulphanilamide (in effective
dose) produces effects similar to colchicine which is, however, required in
relatively diluted strength (Mehra 1949).

The pollen grain wall shows an outer exine and an inner intine. In

Ephedra monosperma (Gullvag 1966) the exine comprises two layers, the outer of orbicules and/or a sculptured granular layer, and a lamellated inner layer. In *E. foliata* the exine is ornamented with 16 meridional ridges with crests which show prominent undulations (Tiagi 1966). The intine is the last layer to be laid down and consists of cellulose only (*E. foliata*). *Ephedra* pollen is polycolpate (see Khoshoo and Ahuja 1963).

Male Gametophyte. After separation from the tetrad, the microspores become ovoid in *E. gerardiana* (H. Singh and K. Maheshwari 1962). The nucleus moves to the lower end and divides to form the first lenticular prothallial cell (Fig. 18.11 D, E). The next division cuts off the second prothallial cell followed by cleavage of the cytoplasm (Fig. 18.11 F, G). Sometimes, the second prothallial cell may lack the upper wall; both prothallial cells are ephemeral. The central large nucleus divides and produces the tube cell and antheridial nucleus. The latter has a distinct cytoplasmic sheath and gives rise to the stalk and body cell (Fig. 18.11 H, I). The mature pollen grain is five-celled at shedding (Fig. 18.11 I); it is spindle-shaped and yellow.

Dehiscence of microsporangia is poricidal, the pores are located at the tip of sporangia.

Female Strobilus

The female strobili are borne in the axil of leaves at the nodes (Fig. 18.12 A). Rarely, there may be a single strobilus which terminates the main axis. The number of strobili varies with species: *Ephedra distachya* L. usually produces three strobili from each axil of foliage leaves, while *E. equisetina* bears only one (Takaso 1984). Each strobilus (cone) consists of three or four pairs of decussately arranged bracts (Fig. 18.12 A) followed by one or two (or three) ovules (Fig. 18.13 A-I) per strobilus. The number of ovules varies from species to species. Each ovule has an outer (*oe*) and an inner envelope (*ie*) which enclose the nucellus (Fig. 18.13 C-G). At maturity the *ie* extends to form a long micropylar tube, the longest known is gymnosperm (Fig. 18.13 I).

Ovule. The ovule is initiated by a transformation of the apical meristem (Fagerlind 1971). This is marked by periclinal divisions in the outermost layer of the lateral shoot apical meristem. Takaso (1984) studied the histological changes in the apex of the female strobilus, and the initiation of the ovule in *E. distachya* and *E. equisetina*. There is a uniseriate dermal layer in the shoot apex, and residual shoot apex. Two relatively large ovular primordia, semi globose in outline, originate from the axil of the uppermost bract (Fig. 18.13 A, B). In contrast to the vegetative shoot apex, periclinal divisions occur frequently in the dermal layer of apex of the female strobilus and of the ovule. These divisions begin at the apex of the female strobilus

Fig. 18.12 A, B.   A *Ephedra chilensis*, twig bearing megasporangiate strobili (*str*).
B *Ephedra* sp., note pollination drop at the tip of ovule. (**A** After Foster and Gifford
1959, **B** after Bierhorst 1971)

before the initiation of the ovule, or simultaneously at the apex as well as
in the peripheral region. The shoot apex increases slightly in width towards
the uppermost pair of bracts and the two ovular primordia and the
residual shoot apex (between the primordia) become clearly discernible
(Fig. 18.13 A, B).

The early ontogeny of the two ovular envelopes (*ie* and *oe*) was studied
by Takaso (1985). The *oe* arises from all round the base of the ovular

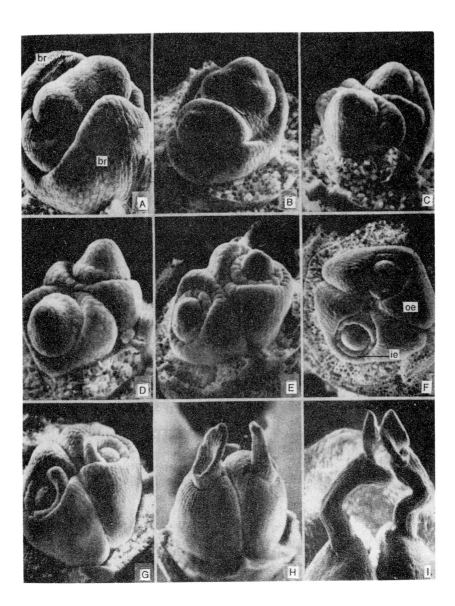

Fig. 18.13 A-I. Scanning electron micrographs of ovules at various stages of development. **A-G** *Ephedra distachya*. **A, B** Apical and lateral views of (swollen) ovular primordia. *br* barct. **C-G** Initiation of outer (*oe*) and inner (*ie*) envelopes. **H, I** *E. equisetina.* **H** Later stage of development of ovule, the inner envelope has one-sided growth. **I** Nearly mature ovule. The inner envelope forms a long micropylar tube with an elliptic opening which extends beyond the outer envelope. (After Takaso 1985)

primordium, except at the dorsal side, so that it is horseshoe-shaped (Fig. 18.13 E, F). Later, it becomes continuous but grows more rapidly laterally so that *oe* appears as two opposite projections (Fig. 18.13 F). The mature envelope is a stout structure; most of its cells are tanniferous. It is laterally vascularized.

The *ie* appears as a small annular swelling after the initiation of *oe*. It grows faster on the ventral side than on the dorsal side and become asymmetric (Fig. 18.13 G). At maturity, the *ie* forms a long, asymmetric micropylar tube which grows beyond the nucellus and has an elliptic orifice (Fig. 18.13 H, I). The greater part of *ie* is uniformly thin, except the basal part which is thick and massive.

In a young ovule the two envelopes are inserted at nearly the same levels, while in later stages the *ie* is set at a much higher level.

The morphology of the *oe* has long been disputed and, at one time, it was regarded as the outer integument. According to Mehra (1950) and Eames (1952), the *oe* is the result of fusion of two perianth leaves or bracteoles. Eames regards the inner envelope alone as the true integument. According to Takaso (1985), histologically, the *oe* in *Ephedra distachya* and *E. equisetina* is formed by derivatives of both the dermal and subdermal cells of the ovular primordium, i.e. it is of dual origin. The inner envelope is formed exclusively by derivatives of the dermal cells (except at the basal region), i.e. it is of dermal origin. Thus, morphology and anatomy do not support the homology between the two envelopes. Rather, the outer envelope resembles vegetative leaves more than the inner one.

The young nucellus shows one to several hypodermal archesporial cells which divide periclinally, and form the inner primary sporogenous cells and the outer primary parietal layer. The latter undergoes further divisions and a massive parietal tissue is formed. The epidermal cells of the nucellus also divide perclinally and form a nucellar cap (H. Singh and K. Maheshwari 1962) which adds to the already massive nucellus. In *Ephedra americana* (Pankow 1962) and *E. gerardiana* (H. Singh and K. Maheshwari 1962), the formation of parietal tissue (with some exceptions) has been confirmed. In *E. foliata* the hypodermal archesporial cell divides to produce a primary parietal cell and a megaspore mother cell (P. Maheshwari 1935), while in *E. helvetica* and *E. intermedia* (Mehra 1950) the archesporial cell functions directly as megaspore mother cell. Occasionally, two hypodermal cells and two tetrads have been observed in *E. foliata*. The megaspore mother cell (Fig. 18.14 B) undergoes reduction division and gives rise to linear (Fig. 18.14 C) or T-shaped tetrad.

A well-defined thin-walled hypostase is present deep in the chalaza, where vasculature for central portion of the ovule ends (Fig. 18.14 A, E).

Female Gametophyte

The chalazal megaspore of the tetrad usually functions. Following the first

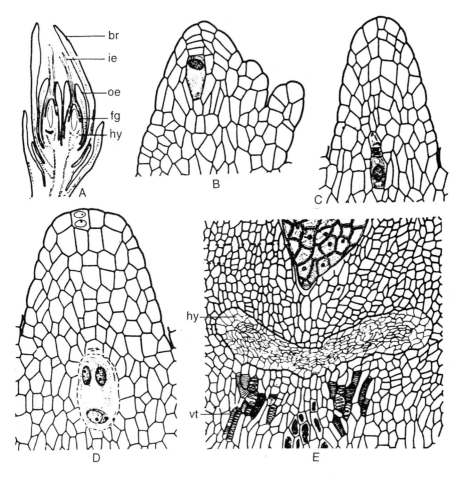

Fig. 18.14 A-E. *Ephedra gerardiana,* megasporogenesis. **A** Longisection female cone. *br* bract. *fg* female gametophyte. *hy* hypostase. *ie* inner envelope. *oe* outer envelope. **B-D** Longisection nucellus to show sporogenous cell, linear megaspore tetrad, and four-nucleate gametophyte. **E** Longisection lower portion of ovule to show hypostase and vascular trace (*vt*). (After H. Singh and K. Maheshwari 1962)

division, the two nuclei move towards opposite poles and a vacuole is formed between them. Further nuclear divisions result in a number of free nuclei which lie on the periphery of a large central vacuole (Figs 18.14 D; 18.15 A-D). The free-nuclear divisions are not simultaneous. Fig. 18.15 D shows a gametophyte with most of its nuclei at metaphase, the remaining nuclei at the micropylar end being at anaphase or early telophase. Fig. 18.15 E is a wholemount of a top-shaped free-nuclear gametophyte of *E. foliata.* The maximum number of free nuclei in the gametophyte varies with species: 256 in *E. trifurca* (Land 1904), 500 in *E. foliata* (P. Maheshwari 1935), and nearly 1000 in *E. distachya* (Berridge and Sandy 1907).

Ephedra lacks a spongy tissue. The mechanics of enlargement of megaspore

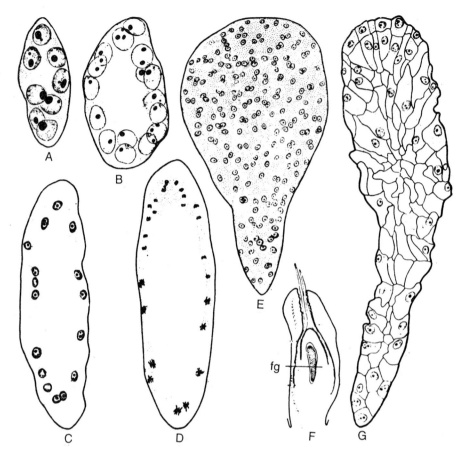

Fig. 18.15 A-G.   Female gametophyte. **A-D, F, G** *Ephedra gerardiana.* **A-D** Longisection of free-nuclear female gametophyte at progressive stages of development. **E** *E. foliata,* wholmount of free-nuclear female gametophyte. **F** Longisection of ovule, outline diagramme for **G**. *fg.* female gametophyte. **G** Cellular female gametophyte from **F**. (After H. Singh and K. Maheshwari 1962)

membrane during the coenocytic phase of female gametophyte has been described by B. and C. Moussel (1973), in *E. distachya*. During the mitotic cycle, the Golgi apparatus gives out numerous vescicles (of variable morphology) (Fig. 18.16) which store polysaccharides. These vescicles may fuse and form concentration vescicles with fibrillar contents, and an eccentrically placed blob of electron-dense material. The simple or concentration vescicles pour their contents outside the plasma membrane and increase the surface area of the megaspore membrane (Fig. 18.16).

Cell wall formation in *Ephedra* seems to follow the usual plan of alveolation (P. Maheshwari and H. Singh 1967). The gametophyte differentiates into two distinct regions: (a) the micropylar part composed of radially-elongated thin-walled hyaline cells with scanty cytoplasm, and (b)

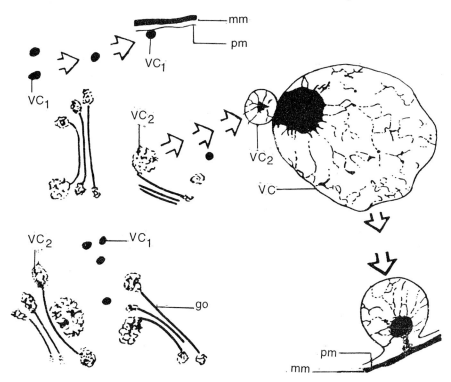

Fig. 18.16. *Ephedra distachya*, formation of megaspore membrane. Diagrammatic representation of golgi (*go*) activity in female gametophyte during mitotic cycle. Production of two types of vescicles (*vc1* and *vc2*), formation of concentration vescicles (*cv*), deposition of contents of *vc1* or concentration vescicles on the megaspore membrane (*mm*). *pm* plasma membrane. (After B. Moussel and C. Moussel 1973)

the chalazal part of compact polygonal cells of smaller size but with dense cytoplasm (Fig. 18.15 F, G). By the time archegonia are mature, the lower half of gametophyte differentiates further into an upper region of actively dividing polygonal cells, and a lower region of relatively larger cells with dense cytoplasm and large nuclei. The lower part of the gametophyte is narrow and tubular and grows deep into the chalazal region. It consumes the protoplasmic contents of the circumjacent cells, which ultimately collapse. Thus, the mature gametophyte can be differentiated into three zones: (a) the upper fertile, (b) the middle storage, and (c) the lower haustorial zone.

In *E. trifurca* (Land 1907) and *E. distachya* (Lehman-Baerts 1967), during embryo formation the apical cells of the gametophyte divide actively and form a plug, which closes the pollen chamber. In *E. foliata* (P. Maheshwari 1935), this plug is formed even before the archegonia are mature, and simulates the tent pole.

Archegonial initials appear (Fig. 18.17 A) at the micropylar end of the cellular gametophyte. One to six archegonia develop singly. P. Maheshwari (1935) reported two archegonia within a common jacket in *E. foliata*.

A mature archegonium has a long and massive neck (longest in any

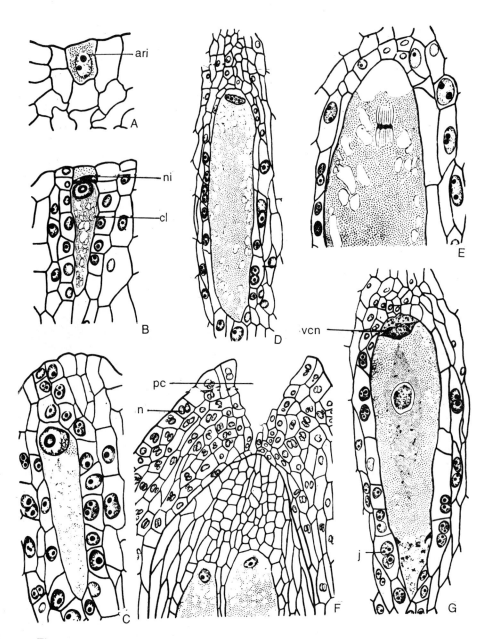

Fig. 18.17 A-G. *Ephedra gerardiana*, development of archegonium. **A, B** Longisection of micropylar part of female gametophyte. **A** Archegonial initial (*ari*). **B** Archegonial initial divides and forms a central cell (*cl*) and a neck initial (*ni*). **C-E** Development of archegonium. **E** Central cell nucleus at metaphase. **F** Portion of longisection of nucellus (*n*) and female gametophyte; note the pollen chamber (*pc*). **G** Longisection of mature archegonium, note ventral canal nucleus (*vcn*) and jacket (*j*). (After H. Singh and K. Maheshwari 1962)

gymnosperm) of 30–40 cells. They merge into the adjoining cells of the gametophyte and it is not possible to make an exact count (Fig. 18.17 D). The newly formed neck cells have a thin wall and show vacuolate cytoplasm rich in ribosomes and mitochondria (B. Moussel 1972). During the maturation of the neck, Golgi vescicles deposit a thick wall across the plasma membrane. Eventually, the cytoplasm and its organelles degenerate and the cell wall becomes prominent. The neck cells facilitate the passage of the pollen tube. Each archegonium is surrounded by a two-layered jacket of thin-walled elongated cells (Fig. 18.17 G). The jacket cells originate by repeated transverse divisions of the cells adjacent to the archegonial initials. The cells are mostly multinucleate. Frequently, the nuclei of the neck and jacket cells migrate into the archegonium, where they divide and produce many micronuclei, or may enlarge and simulate embryonal nuclei. H. Singh and K. Maheshwari (1962) observed (in one preparation of *E. gerardiana*) one of the jacket nuclei so close to the egg that it could easily be mistaken for a male nucleus.

The central cell enlarges (Fig. 18.17 B), its cytoplasm becomes thinly distributed, and is highly vacuolate in initial stages (Fig. 18.17 B-F). This is the foam stage. The vacuolated archegonium gradually matures. Its large nucleus at the tip (Fig. 18.17 C, F) divides mitotically to give rise to the egg and ventral canal nuclei (Fig. 18.17 G). A wall is not formed between the nuclei. The egg nucleus moves to the center of the archegonium. It is surrounded by a dense sheath of cytoplasm which sometimes extends downward to a considerable extent in *E. gerardiana* (Fig. 18.17 G).

Pollination·

Ephedra is believed to be wind-pollinated, but there are also reports of insect pollination (see Bino et al. 1984). The stickiness of pollen also suggests entomophily. In *E. aphylla* and *E. campylopoda*, pollination is partially entomophilous. The adaptation to insect pollination is shown by nectarial secretion which occurs in both male and female flowers. In *E. aphylla,* the male flowers produce one to ten nectar droplets, before anthesis, mainly on the outer cover (perianth) in the functional reproductive unit (Bino et al. 1984). The droplets persist for 2–4 days. In the female flowers, the single droplet of nectar is secreted on the outer cover before anthesis and continues till anthesis. Amino acid is present in the exudate in appreciable quantity and remains constant, irrespective of the stage of the anthers. In addition to nectar, there is a single pollination droplet, at anthesis, at the tip of the micropylar tube (Fig. 18.12 B). It persists for 2–4 days when accessible to insects or 10–16 days when there are no insect visitors. Usually, the amino acid is not detectable (Bino et al. 1984). There is no appreciable difference between the volume and sugar concentration of nectar drops of male and female flowers and pollination drop. The sugar content is independent of temperature, time and location on the plant, but there is relation with relative humidity (RH). During the day, in the area studied, RH is usually below 60%, the sugar content is over 80%, while during the night, the RH is occasionally 90% or higher and the sugar content is lower. A chemical analysis of the pollination drop also shows the presence of peptides, malic and citric acid (Ziegler 1959), inorganic phosphates and sugars (especially high—25%).

Among the frequently encountered insect visitors are *Lucilia caeson* (Calliphoridae), *Metasyrphus corollae*, *Episyrphus balteatus* and *Scaeva albomaculata* (Syrphidae). Only nectar is consumed while pollen consumption is negligible (as shown by an analysis of gut content).

The pollen grains are caught in the glistening pollination drop, and are sucked with the fluid column (present in the micropylar tube) which becomes thicker due to evaporation (Strasburger 1871).

During archesporial development, the nucellar tissue, directly above the female gametophyte, disorganizes and becomes obliterated. Thus , a deep and conspicuous pollen chamber is formed. The upper portion of the gametophyte is exposed at the base of this chamber (Fig. 18.18 A), a unique feature in *Ephedra*. The pollen grains come into direct contact with the female gemetophyte.

The pollen swells when caught in the pollination drop. The exine ruptures irregularly and is cast off before the pollen germinates. The intine is thin.

Pollination in *Ephedra* occurs when the ovule has fully formed archegonia. After pollination, the outer envelope of the ovule shows small papillate outgrowths directed towards the inner envelope. As the ovule matures, these papillae elongate, become thick-walled (Fig. 18.18 B, C), and help to close the space between the two envelopes. It may also seal off the micropyle by pressing inward on the micropylar tube (H. Singh and K. Maheshwari 1962).

Fertilization

The pollen grains germinate within a few hours of entering the pollen chamber. The body cell divides and forms two sperm nuclei. The latter may be of the same size (*E. foliata, E. sinica*), or markedly unequal (*E. altissima, E. saxatilis*). The pollen tube is narrow and short, and pushes its way between the neck cells of the archegonium. The tip ruptures and discharges four nuclei (tube, stalk cell nucleus, and two male gametes) into the archegonium. One of the gametes fuses with the egg and the other may fuse with the ventral canal nucleus near the upper part of archegonium (Fig. 18.19 A, B_1, B_2). According to Khan (1943); "... The type of double fertilization seen in *Ephedra* may have no phylogenetic significance at all and may simply be the natural outcome of a tendency towards fusion between any two nuclei of opposite sexual potencies that happen to lie in a common chamber". Moussel (1978), Friedman (1990a, b, 1991) have also reported

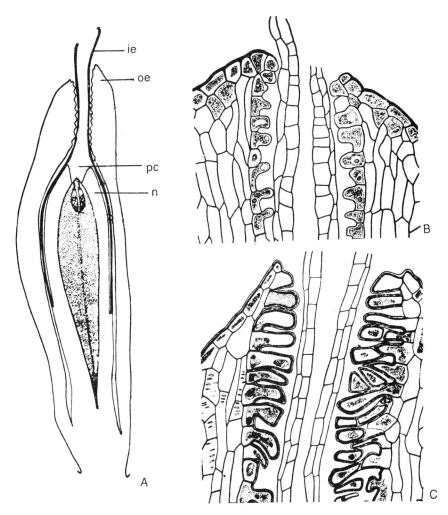

Fig. 18.18 A-C. A *Ephedra foliata,* longisection of ovule at pollination. *ie* inner envelope. *n* nucellus. *oe* outer envelope. *pc* pollen chamber. **B, C** *E. gerardiana*, longisection of micropylar canal at two stages of development of papillate projections from outer envelope. (**A** After P. Maheshwari 1935, **B, C** after H. Singh and K. Maheshwari 1962)

double fertilization in *E. distachya, E. nevadensis* and *E. trifurca*. Double fertilization in *Ephedra* (and in angiosperms) is a significant occurrence in reproduction (for details see Friedman 1992a, b, 1994).

According to Land (1907), the interval between pollination and fertilization in *E. trifurca* may be as short as 10 h.

Embryogeny

The embryogeny of only a few species of *Ephedra* has been studied. The

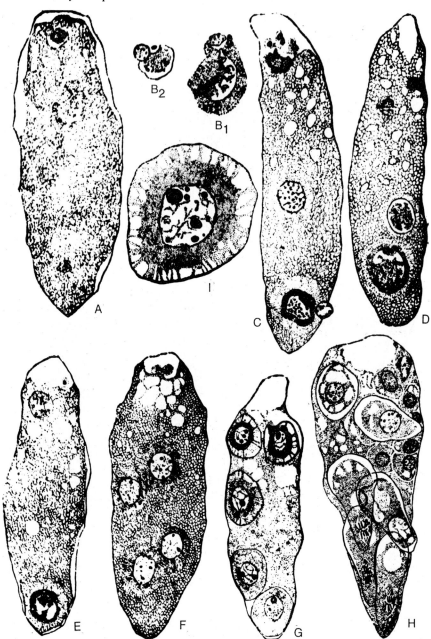

Fig. 18.19. A-I. *Ephedra foliata,* fertilization and proembryogeny. **A** Longisection of ovule at double fertilization. **B₁, B₂** Nuclei from **A. B₁** Male nucleus in contact with egg nucleus. **B₂** Second male nucleus in contact with ventral canal nucleus. **C.** Longisection of two-nucleate proembryo, note radiating cytoplasm. At the micropylar end persists a large ventral canal nucleus, and a few small nuclei of the neck or adjacent cells, at the lower end a jacket nucleus prior to entry into proembryo. **D-F** Development of eight-nucleate proembryo. **G, H** Longisection egg cell shows five (**G**) and six (**H**) proembryos. **I** Mature proembryo. (After Khan 1943)

zygote nucleus divides in situ; the two nuclei move apart (Fig. 18.19 C), occasionally to the two poles of the zygote. Two more mitoses follow which result in eight nuclei (Fig. 18.19 D). Each nucleus is surrounded by a cytoplasmic sheath which radiates strands (Fig. 18.19 C-F) and takes a dense stain. At the eight-nucleate stage, a wall develops around each nucleus (Fig. 18.19 G, H) followed by cleavage, which results in eight units. According to Foster and Gifford (1959), this is precocious type of cleavage polyembryony. The proembryo is spherical. The cytoplasm next to the nucleus takes a denser stain and does not show vacuoles; the outer part is lighter and highly vacuolated. The nucleus contains numerous nucleoli and granular, irregularly scattered chromatin (Fig. 18.19 I).

Each cell of the proembryo puts out a tubular outgrowth (Fig. 18.19 H). The nucleus may divide before the outgrowth is formed, or it may move into the tube and divide. A transverse wall is laid down, which gives rise to an embryonal cell and a suspensor cell (morphologically embryonal suspensor), which elongates (Fig. 18.20 A). A periclinal division in the embryonal cell is followed by a longitudinal division in the lower of the two cells to form three cells (Fig. 18.20 B). Further divisions give rise to the embryo proper; a multicellular secondary suspensor differentiates adjoining the suspensor cell (Fig. 18.20 C). The lower end of the embryo produces two cotyledons and the shoot apex. A root cap of column (horizontally oriented cells in the central region) and pericolumn (peripheral cells which are steeply inclined or almost vertical) differentiates (Deshpande and Bhatnagar 1961).

Several embryos begin to develop within a single ovule, but only one reaches maturity (Fig. 18.20 D, H). The large number of embryos is due to a combination of simple and cleavage polyembryony. A well-defined tunica layer originates in the embryo even before the development of cotyledons.

Seed

Following fertilization, the ovule undergoes marked changes, in the zygote (development of embryo), female gametophyte (development of endosperm) and integument (development of the seed coat).

During post-fertilization stages, in restricted areas the female gametophyte goes through meristematic activity. The micropylar part grows upward into the deep pollen chamber and may plug it partially. Starch accumulates in the cells of the female gametophyte in the micropylar region before fertilization, and in the axial cells after fertilization (B. Moussel 1974). Many of the gametophytic cells become binucleate. During the development of embryo, storage products accumulate only in the lower half of the endosperm. In a mature seed the endosperm is derived mainly from the lower part of the gametophyte. It can be distinguished into four zones from periphery inwards:

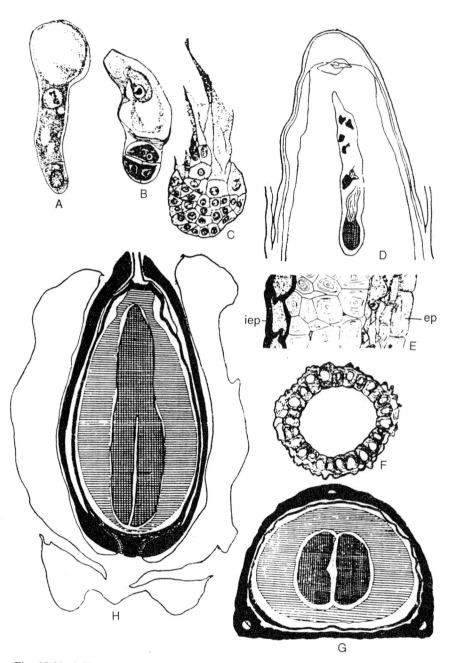

Fig. 18.20. A-H. *Ephedra foliata*, development of seed. **A, B** Two- and three-celled embryonal unit with elongated suspensor. **C** Later stage in embryo development. **D** Longisection of upper part of ovule shows six embryos. **E** Transection (part of) outer envelope. *ep* outer epidermis. *iep* inner epidermis. **F** Transection inner envelope passing through the micropylar region. **G, H** Trans- and longisection of mature seed. There are three vascular bundles in the outer envelope in **G**. (After Khan 1943)

a) **The haustorial zone** is a single layer of especially differentiated outer tangential cells which are richly cytoplasmic. The outer walls have a thick and uneven deposition which projects like stalacites into the cell cavity. The cytoplasm has several small vacuoles, mitochondria and different profiles of ER, but has only a few plastids. The nucellar cells in the vicinity of haustorial zone degenerate and become compressed.

b) **The proteolipid and starch zone** differentiates as the apical embryonal cell divides to initiate the formation of embryonal mass.

c) **The starchy zone**

d) **The central zone** is very narrow and devoid of storage products.

The accumulated food reserves in the endosperm are utilized at the time of seed germination.

The nucellar cells, in the micropylar region, become multinucleate (Khan 1943). Those around the pollen chamber persist in the seed and some of them become thick-walled. The remaining nucellus becomes crushed and appears as a band lying near the endosperm.

The inner envelope (integument) has two layers of cells. The cells of the outer layer are slightly thick while those of the inner layer are thick-walled (Fig. 18.20 F). The cells of the outer envelope (*oe*) are thick-walled (Fig. 18.20 E) and form the main protective layer in the seed.

The bracts surrounding the seeds become thick and fleshy. The lower pair is small while the upper pair of bracts is large and covers the seed (Fig. 18.20 H). In *Ephedra,* only the outer envelope is vascularized (Fig. 18.20 G). Three bundles enter the ovule and remain unbranched; only a few tracheids differentiate at the base of the nucellus (H. Singh and K. Maheshwari 1962).

Chromosome Number

The haploid chromosome number of 7 and 14 is characteristic of *Ephedra: E. foliata* n = 7; *E. sinica, E. saxitilis and E. intermedia* n = 14 (Mehra 1949). The basikaryo type is composed of five median-submedian and two subterminal chromosomes (see Khoshoo and Ahuja 1963).

E. saxatilis is an allo-polyploid (Mehra 1947). Besides normal haploid pollen grains, a small percentage of diploid pollen grains is also formed. The latter condition arises due to incomplete cytokinesis of the microspore mother cell after meiosis II. The tetrad may contain two diploid microspores or one diploid and two haploid microspores.

Temporal Considerations

Ephedra shows a non-synchronous reproductive behaviour. In a plot of *E. foliata* in the Delhi University Botanical Garden, initiation of cones, anthesis, pollination and ripening of seeds occurred (in different cones) during February-May (see H. Singh 1978). Thus, at the same time, young ovules and mature seeds can be collected from any one plant. In *E. gerardiana*, the cones appear in April and the seeds ripen towards the end of September. The bracts of the female cone become fleshy and red on ripening, which makes the plant very attractive (H. Singh and K. Maheshwari 1962). In *E. campylopoda*, the bracts, perianth leaves and other parts of female flowers are bright yellow and later turn fiery red.

19. Welwitschiales

Welwitschiaceae

Welwitschia

Morphology

The adult sporophyte of *Welwitschia* has no parallel in the plant kingdom. It resembles a gigantic turnip, and reaches a diameter of more than 1 m. The stem consists of a tough axis shaped like an inverted cone, and has a deep apical depression. Most of the stem is buried in the sandy soil, its lower part tapering abruptly to an extremely long carrot-like tap root which penetrates the soil between 1.0 and 1.5 m deep before it splits into numerous thin roots. How far downward the roots extend is still uncertain. The exposed portion of the stem consists of a massive, woody, concave disc which bears two opposite and decussate strap-shaped leaves (Fig. 19.1 A, B). These leaves grow continuously throughout the life of the plant but very slowly, ca. 10–15 cm in a year (Von Willart 1985), from a basal meristem, and the tip dies back continuously. A mature leaf is ca. 3 m long and 1–1.5 mm thick. The distal end of the leaves splits and becomes frayed and extends in a twisted or contorted manner along the surface of ground. Old plants may be more than 1000 years (Bierhorst 1971). A specimen ca. 2000 years old is shown in Fig. 19.1 A.

The plant produces only three pairs of foliar organs (Martens 1977a, Martens and Waterkeyn 1963a, b, 1964a, b):

a) The first is a pair of cotyledons which have a limited growth, and are not shed.

b) The second is the pair of large, persistent leaves which appear at right angles to the cotyledons.

c) Lastly, two more leaf primordia, the scaly bodies, appear which has a long life, and increase in height, breadth and thickness, due to an intercalary basal meristem comparable to that of the persistent leaves. The initial shoot apex also produces two short epicotyledonary internodes which increase in diameter in a manner similar to the hypocotyl. The apical meristem then aborts (cause unknown); its loss does not lead to the development of axillary

Fig. 19.1 A, B.   *Welwitschia mirabilis.* **A** Ca. 2 000-year-old plant, note opposite strap-shaped leaves. **B** Central part shows woody concave disc. (**A** After Von Willart 1985, **B** after Rodin 1953a)

buds and branches (as in most plants) in *Welwitschia*. Rather, an intercalary and permanent meristem, which widens indefinitely, arises at the leaf base. The vegetative activity is thus restricted to the basal leaf meristem, and extends to neighbouring tissues so that the hypocotyl increases in diameter due to intercalary growth. After a gap of 12–50 years, buds are formed on the centrifugal and axillary crests of the crown. They produce branches with scale-like leaves (nodal bracts according to Rodin 1963), internodes and axillary buds. These leaves are different from the two persistent leaves. The buds and leafy branches are vegetative in the beginning. Later, cones are formed due to hormonal determination of the branches and further vegetative activity is completely lost (Martens 1977a). The adult and fertile plant, therefore, produces an indefinite number of leaves, in addition to the three whorls on the primary axis. The succession of these two phases (primary axial growth and development of fertile branches) is separated by a long time-interval. Thus, *Welwitschia* can produce new roots, buds, branches (vegetative production) even after 100 or 1000 years (Martens 1977a).

The leaves grow slowly and the tip dies back continuously. Consequently, the leaves have a continuous developmental gradient spanning many years, which makes it possible to study the behaviour of every tissue age, unlike other plants (Von Willart 1985). The length of the leaf is determined by environmental factors. It may dry completely and resume growth when conditions improve. During the period of drought, antelopes and zebras feed on the leaves, and occasionally pull the ribbons out of the groove. The leaf will still grow if the meristem is not damaged.

Welwitschia is classified as a hydrostable plant because even a prolonged drought does not affect its water content significantly (see Von Willart 1985). There is not much diurnal change in the water content of the leaf. The stomata remain open during the day, and there is tremendous loss of water. However, the water conduit present functions, and water loss by transpiration is balanced without much delay. The sponge-like roots may act as a water store, balancing temporary water deficits, but the demand for water is undoubtedly supplied from a water source to which it must be connected. The accompanying flora cannot exploit this source and suffers much more from prolonged drought. It is not conclusively proved yet whether *Welwitschia* is able to make use of dew and fog to balance the daily water loss. Wherever the water conduit is disturbed, water loss appears to be regulated by reduction of leaf area.

The photosynthetic capacity and the carbon balance, over a period of 24 h, show that leaves of all ages are able to take up CO_2 during (parts of) the day. The CO_2 uptake decreases with increase in leaf age. Only the first half of the leaf has a carbon gain while the older parts show carbon loss. Nevertheless, the carbon balance of the entire leaf is positive, otherwise the leaf length is reduced. With improved water supply, even old parts contribute to a carbon gain. A plant with such a slow growth rate must keep as much of the leaf blade alive as possible.

The *Welwitschia* leaf has a high reflectivity. Only 55% of the incident global radiation is absorbed, nearly 40% is reflected. The temperature of the soil shaded by the leaf is cooler than the temperature of the lower surface, which results in a net loss of long wave radiation. Transpiration also plays an important part in energy dissipation. High transpiration is necessary to prevent lethal temperature of the leaf. When it cannot be maintained (due to shortage of water) the tissue at the tip dies (probably due to lethal temperature). Besides carbon balance, lethal temperature at the leaf tip may regulate leaf length.

According to Von Willart (1985), the key to understanding the biology of this peculiar plant lies in the fact that *Welwitschia* has a connection to water sources in the soil (which cannot be exploited by other plants) and, therefore, does not normally suffer from water shortage. With a guaranteed supply of sufficient water, a high transpiration rate is possible, which prevents lethal leaf temperature, and a large leaf can be maintained.

The surface of the stem is covered with a ridged, corky layer nearly 2 cm thick in the upper region. The portion of stem above the point of insertion of leaves is called the crown, while the lower portion is called the stock. The crown is marked by a series of ridges, each of which extends to nearly half the circumference of the stem. They bear scars of old, fallen inflorescences. The ridges on the stock are less prominent.

The taproot is long. It forms lateral roots, at various depths, depending on the soil.

Anatomy

Root. Primary Structure. A transection of root of a young seedling shows (Fig. 19.2 A) a piliferous layer which bears root hairs followed by one- or two-layered parenchymatous cortex, and an endodermis with Casparian bands. The pericycle remains single-layered above the protoxylem (px) points, while in the remaining area it undergoes a precocious periclinal division. The cells of the outer layer are large, their inner cell wall is thin and weakly suberized. The inner layer consists of smaller cells with cell contents. The stele is diarch. The two xylem groups join in the middle or remain separate. The metaxylem (mx) consists of large and the px of narrow tracheids. The two phloem groups are separated from the xylem by one or two small cells with dense contents which later form the cambium. Thick cellulosic fibres are present between the phloem and the pericycle (Fig. 19.2 A, B).

Two groups of secondary xylem tracheids alternate with large primary mx. The internal layer of the two-layered pericycle becomes the initial of the phellogen. This protective tissue is weak and is compensated for by several layers of cellulosic fibres which completely encircle the stele cylinder (except above the px points). An extrafascicular cambium is absent, which results in reduced radial growth above the two vascular plates. This double depression of the radicle is manifested externally in its bilateral symmetry (Martens 1977b).

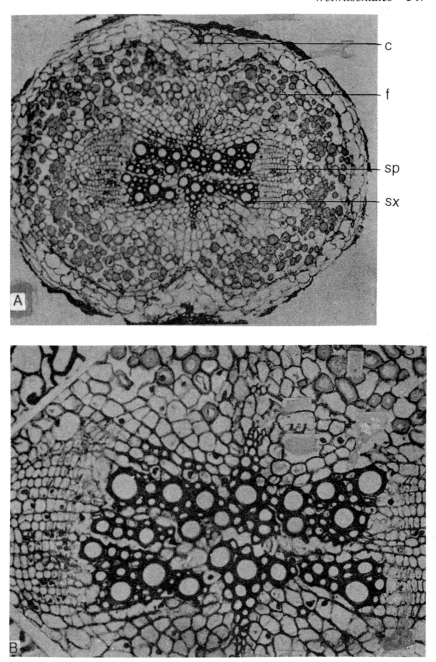

Fig. 19.2 A, B. *Welwitschia mirabilis.* **A** Transection of main root from 5-month-old seedling. *c* cortex. *f* fiber. *sp* secondary phloem. *sx* secondary xylem. **B** Vascular region from **A.** (After Martens 1977b)

Stem. In a young plant the main axis, below the crown, has a ring of collateral vascular bundles (Fig. 19.3 A). In the old plant there is a

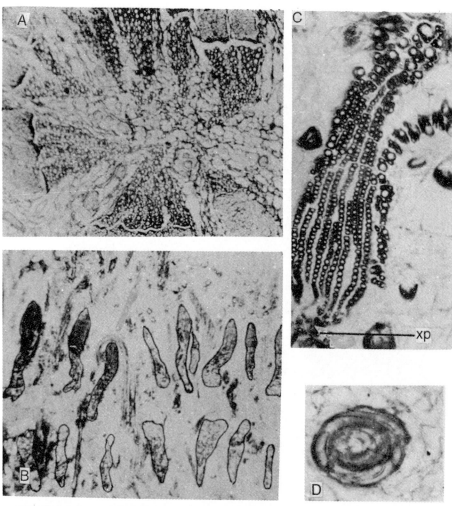

Fig. 19.3 A-D. *Welwitschia mirabilis.* **A** Transection through smooth and cylindrical part of axis below the crown, from a 15-year-old plant. **B** Transection of vascular system of crown below the level leaf of meristem; there are two transverse series of bundles. **C** A bundle from **B.** *xp* xylem pole. **D** Coiled vessel isolated from mature plant. (After Bierhorst 1971)

saucer-shaped mass from which vascular traces extend to the leaves, the inflorescence and the single taproot. The cortical parenchyma becomes meristematic, and a phellogen produces an outer corky tissue (Rodin 1963). Within the flanges of the crown are two parallel series of bundles with the xylem on the inside (Fig. 19.3 B, C). Traces from these bundles extend towards the surface, presumably to old reproductive axes. The stem is heavily sclerified.

In early-formed tracheary elements, circular bordered pits are present between the gyres of the secondary wall helix. Vessel elements have single apical or lateral circular pores, oriented transversely or obliquely. In occasional vessel members, the pores are in pairs or, rarely, a foraminate plate has three pores (Bierhorst 1971). The later-formed thick-walled elements have a complex ramiform bordered pit system. A peculiar feature is the occurrence of isolated, unconnected vessels of about 12 cells, which are coiled into tight worm-like masses (Fig. 19.3 D).

Electron microscopic observations indicate that most of the primary and secondary xylem and phloem elements are arranged in radial rows. In the phloem, two types of cells, namely, sieve cells (structure similar to *Ephedra* and *Gnetum*) and parenchyma cells, have been described. The plastids in sieve cells undergo changes such as the formation of starch grains, proliferation of internal membranes, and decrease in the density of the matrix. Osmiophilic globules persist in the plastids (but disappear from the cytoplasm). These plastids, unlike those of *Pinus*, lack crystalline inclusions and fibrillar material at all stages of differentiation. The starch grains present are club-shaped (as in *Gnetum*).

A mature plasmalemma-lined sieve cell contains nucleus, mitochondria, ER and plastids with starch granules. The plastids lack internal membranes and their matrices are hyaline (very clear) unlike the plastids from immature cells. The mitochondria undergo no apparent structural modifications. The sieve cells lack ribosomes, dictyosomes, and microtubules. The vacuoles of young sieve cells and the lumen of mature ones occasionally contain a coarse fibrous substance, similar in appearance to that observed in the vacuoles of parenchymatous cells in the leaf. It should not be confused with slime or P-protein, which is totally absent (see Paliwal 1992). During maturation, the tonoplast, which delimits the vacuolar contents from the cytoplasm in young cells, ceases to be identifiable and the sieve cell appears to contain a single large central cavity (Evert et al. 1973a, b).

An unusual feature of *Welwitschia* is the presence of "spicular" cells (Hooker 1863) throughout the plant body: in leaves, stem, reproductive branches and cone scales. The large, mostly unbranched, sclereids contain numerous crystals embedded in the cell wall. They are closely packed and interlaced, which makes the tissues extremely hard and tough.

The wall of a mature sclereid consists of a thin cellulose primary wall. After the sclereid attains maximal size, calcium oxalate crystals are deposited on its inner wall. The secondary wall layers are formed by further deposition of wall materials, the first layer covering the primary wall and crystals. A mature sclereid becomes slightly lignified.This is followed centripetally by the highly lignified second wall, which is the thickest layer of the secondary wall. The third and innermost layer is cellulosic, the lignified and inner cellulose layers laminated. The lumen is very small. Numerous simple pits are present on the sclereids, the outer and inner extremities of the pits are conically flared (Rodin 1958). At the tips of the sclereid is a branched aperture which perhaps functions like a pit, but is never conically flared.

Leaf.   A mature leaf of *Welwitschia* is isolateral, amphistomatic with sunken stomata. It has large longitudinal and small oblique vascular bundles which anastomose (Fig. 19.4 A, B). The mature leaves have numerous sclereids (Fig. 19.4 D).

A vertical section of a leaf (Fig. 19.5 A) shows:

a) The epidermis is a single layer of cells on the upper and lower surface of leaf. The outer wall of epidermal cells is heavily thickened and consists of a layer of cuticle. The median layer comprises powdery crystals of calcium oxalate, and extends laterally to the subsidiary cells. The innermost cellulosic layer is very thick, specially in a mature leaf. A thin discontinuous layer between the crystal and the inner cellulosic layer contains cutin (Martens 1977b).

The sunken stoma have a stomatal chamber and the development is syndetocheilic.

b) The mesophyll consists of three or four layers of distinct palisade cells on both sides, and a compact central zone of spongy parenchyma (Fig. 19.5 A). The palisade parenchyma is the chief photosynthetic tissue. The chloroplasts in the central mesophyll decrease toward the middle region of leaf. Calcium oxalate granules are present in its cell walls. Hypodermal cellulosic fibres occur in the massive bundles, which may extend to the inner margin of palisade parenchyma. The bundles on the upper surface of the leaf are smaller than those on the lower surface. Secondary walls develop first in the fibres near the epidermis, and maturation proceeds inward. A paradermal section of leaf shows (Fig. 19.4 C) stomata between the bundles of fibres. Occasionally, gum canals occur in mature leaves adjacent to the phloem between vascular bundles. The spindle-shaped unbranched canals are formed by enlargement of irregular chains of mesophyll cells (their walls break down).

c) The vascular system consists of collateral bundles, the xylem faces the upper and the phloem the lower side of the leaf. Fibres surround the vascular bundles, followed by transfusion tissue (Fig. 19.5 B). The latter ensheaths the vascular bundle and strengthens it. Its cells are shorter than the xylem elements, have reticulate or scalariform thickenings, and small simple or bordered pits in the thin-walled areas. In anastomosing bundles, fibres are absent.

Protoxylem tracheids develop from the first procambial strand of the leaf primordium (Rodin 1958). The tracheids have annular or spiral thickenings. The protoxylem is eventually crushed. The metaxylem replaces it, and is the functional xylem. The earlier-formed metaxylem intergrades with the protoxylem and has spiral or annular thickenings with large uniseriate bordered pits. Later metaxylem has vessels and tracheids with spiral, reticulate or scalariform wall thickenings. Large uni- or biseriate bordered pits are common. There is a single perforation on the slanting end wall between two adjacent vessel elements. As many as four bordered pits may also

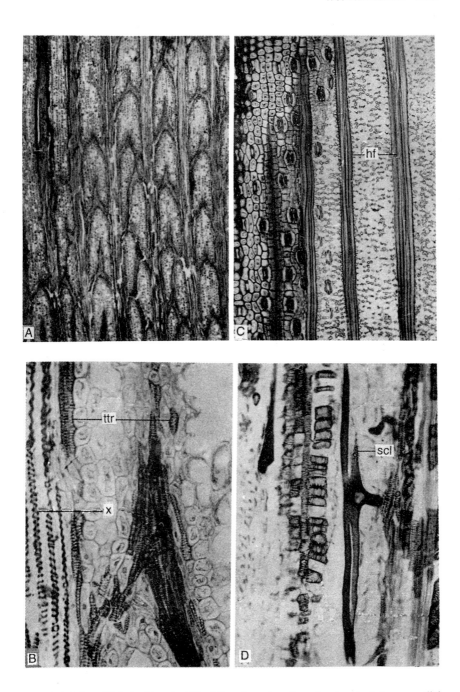

Fig. 19.4 A-D. *Welwitschia mirabilis*. **A** Paradermal section of mature leaf shows parallel longitudinal and smaller anastomosed vascular bundles. **B** Longisection of leaf, the bundles are oblique. *ttr* transfusion tissue. *x* xylem. **C** Oblique paradermal section of mature leaf, note the position of stomata in relation to hypodermal fibre (*hf*) bundles. **D** Longisection of a large, branched crystalliferous sclereid (*scl*) in leaf. (After Rodin 1958)

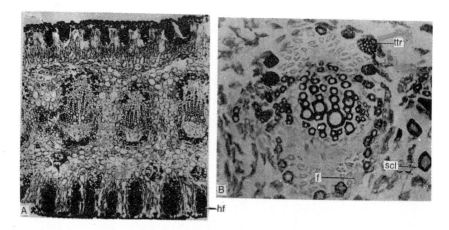

Fig. 19.5 A, B. *Welwitschia mirabilis*. **A** Transection of mature leaf. *hf* hypodermal fibres. **B** Vascular bundle from **A.** *f* fibre. *scl* sclereid. *ttr* transfusion tissue. (After Rodin 1958)

occur on the same end wall. The vessel elements closely resemble the metaxylem tracheids, except for the larger diameter and presence of perforations.

Although tissues resembling secondary xylem and phloem occur in the collateral bundles of the leaf, they are only primary tissues, since a fasicular cambium has not been demonstrated decisively (Rodin 1958).

At the apex of the leaf primordia, toward the tip of the procambial or tracheidal strand, several hypodermal cells become vacuolate, enlarge and become tubular. Their walls thicken and the cells develop into cellulosic fibres, and lignified and crystalliferous sclereids. They form a protective cushion between the epidermis and the apical net of vascular traces (Martens 1977b).

Reproduction

Welwitschia is a dioecious plant. The inflorescences, borne in clusters, arise adventitiously from within the cortical tissues adjacent to a groove from which the leaf grows. They branch in a dichasial manner, each branch terminating in a cone with opposite and decussate scales (Figs. 19.6 A; 19.7 A, B). According to Martens (1959), ontogenetically, both male and female flowers have the same opposite and decussate plan as seen in their floral diagrams (Fig. 19.8 A, B). The cone bracts in both sexes are broad and ovate with tapering thin lateral wings.

Cone Bracts. The male cone bract is thickest at the base and gradually tapers toward the apex, while the female cone bract is quite thin near the base and thick toward the apex. The flattened seed lies in a pocket formed by the thin area, (Fig. 19.7 E), and the mature cone forms a compact structure.

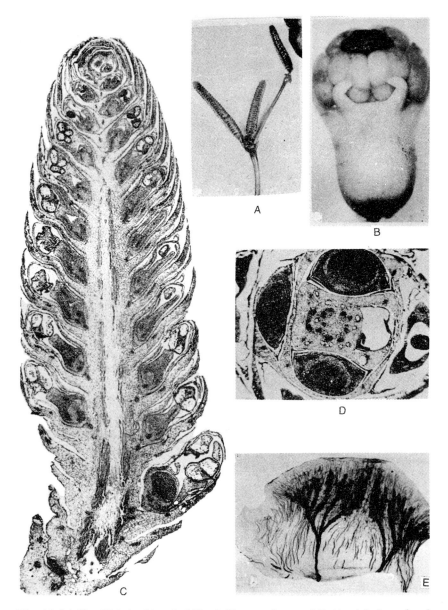

Fig. 19.6 A-E. *Welwitschia mirabilis.* **A** Three male cones. **B** Abaxial view of male flower. **C, D** Longi- (**C**) and transection (**D**) of mature male cone. **E** Cleared male cone bract. (**A, B, D** After Bierhorst 1971, **C, E** after Rodin 1963)

There is a similarity in the venation pattern in the cone bracts of male and female cones (cleared bracts facilitate study). The two vascular bundles are widely separated when they enter the cone bracts. The veins branch dichotomously, with a large number of short dichotomies in the distal

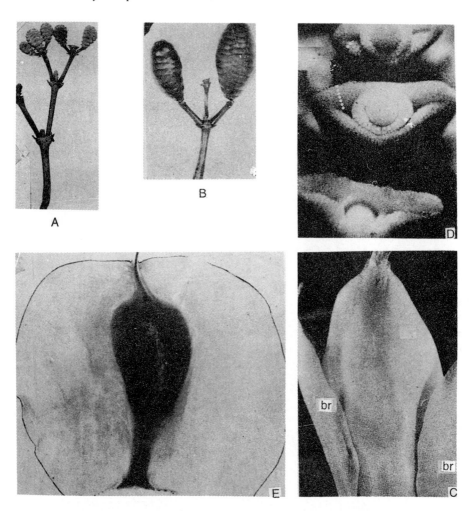

Fig. 19.7 A-E. *Welwitschia mirabilis.* **A** Female inflorescence. **B** Two female cones. **C** Female flower. *br* bract. **D** Successive stages to show initiation of ovule (whole mount). **E** Mature seed. (**A** After Rodin 1963, **B, C, E** after Bierhorst 1971, **D** after Martens 1959).

region (Fig. 19.6 E) where the bract lies exposed. A major part of the female cone bract is winged and devoid of vascular tissue. A large number of fibres and sclereids are present in the distal portion of a mature female cone bract, and nearly obscure the vascular system in cleared material. The cone scales become scarlet red at maturity.

Male Strobilus

A mature microsporangiate strobilus is four-angled, 20–30 mm long and 8 mm wide. It bears broad decussate bracts in four ranks (Fig. 19.6 D) and each bract has in its axil a flower (Fig. 19.6 A). Each flower consists of

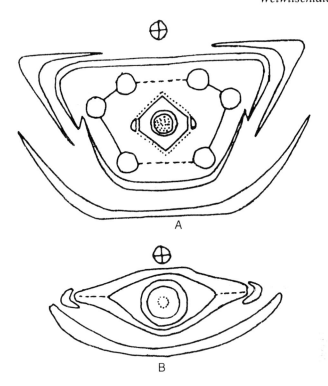

Fig. 19.8 A, B. *Welwitschia mirabilis.* **A, B** Male and female flower, floral diagram. (**A** After Martens 1961, **B** after Martens 1959)

(from outside inward) (Fig. 19.8 A): (a) two pairs of asymmetrically disposed bracts, (b) six microsporangiophores, which are fused basally into a sheath (Fig. 19.6 B); (c) a pair of minute bracts at the base on either side of the centrally located non-functional ovule. The latter has a single integument, which extends upward and flares out at the apex. Each microsporangiophore bears three fused sporangia with their lines of dehiscence radiating from near the point where the septa converge (Fig. 19.6 B). In each flower all the stamens mature at the same time. Flowers at the base of the cone mature first (Fig. 19.6 C). Each stamen in a flower has a single vascular bundle.

Microsporangium

Each microsporangium has an outer layer of cells followed by two tapetal layers enclosing the central fertile tissue.

Male Gametophyte. The precise sequence of development of the male gametophyte has not yet been worked out (Martens and Waterkeyn 1974). A pollen grain at shedding is three-celled; a tube cell, a generative cell; there is a controversy regarding the third cell, interpreted as a prothallial cell or a stalk cell, which aborts even before pollination. The pollen grains are shed through a vertical slit.

Female Strobilus

The ovulate strobilus is comparatively thicker and comprises broad decussate bracts (Figs. 19.7 A, B; 19.8 B). A mature cone is ca. 60 to 80 mm long and 25 mm wide. Each female flower is subtended by a cone scale, and consists of a nucellus enclosed by two envelopes and two small lateral bract-like structures (Figs. 19.7 C, 19.8 B). The inner envelope extends into a long tubular micropyle (Fig. 19.7 C) and is the true integument. The outer envelope is called a perianth.

The ovules arise in the axil of bracts, but the exact morphological nature of meristem which forms the ovule is not known. The outer envelope arises first as the annular hump on the ovular primordium followed by the integument, which is also annular (Fig. 19.7 D). The integument does not cover the nucellus. During later stages of development, the perianth forms a conspicuous laterally extending fibrous wing in the seed (Fig. 19.7 E). The wing has a complex structure which varies according to age, level and the zone (median or lateral) of the seed (Martens 1975). The integument extends apically as a thin micropylar tube.

A discrete epidermis has not been observed in the young nucellus (Fagerlind 1961). The cells of the outermost exposed layer divide periclinally to form the inner archesporial cells and the outer layer. The parietal tissue develops from the (non-functional) archesporial cells.

Megasporogenesis.   A single megaspore mother cell can be distinguished in the third or fourth layer of the nucellus, and during further development its walls become thick (Fig. 19.9 A). Reduction divisions do take place, but the walls are not formed, so that the development of the gametophyte is tetrasporic (Fig. 19.9 A-D). Extrafloral embryo sacs have been observed in the cone axis (Fig. 19.9 E). However, they degenerate at an early stage.

Female Gametophyte.   The spongy tissue is absent in *Welwitschia* (Pearson 1929, Martens and Waterkeyn 1974). The coenomegaspore (Martens 1971) undergoes eight to ten free-nuclear divisions which are synchronous. A central vacuole is absent. The free-nuclear gametophyte shows gradual vacuolation from the chalazal to the micropylar pole (Fig. 19.9 F, G). Walls are laid down by free cell formation (Fig. 19.9 H), and the cytoplasm becomes divided into irregular multinucleate compartments with cell walls. Two zones differentiate; a smaller, fertile micropylar and a larger, sterile chalazal zone (Fig. 19.9 I, J). The micropylar cells are vacuolate with usually two to eight nuclei, whereas the chalazal cells are dense and have 6–12 or more nuclei (Fig. 19.9 J). Although the partitioning of the gametophyte is simultaneous, the walls and vacuoles differentiate more rapidly in the micropylar region (Martens and Waterkeyn 1974). The micropylar cells, along with the contents, form a cluster of embryo sac tubes or prothallial tubes (Fig. 19.10 A, B). These tubes pierce the megaspore membrane and

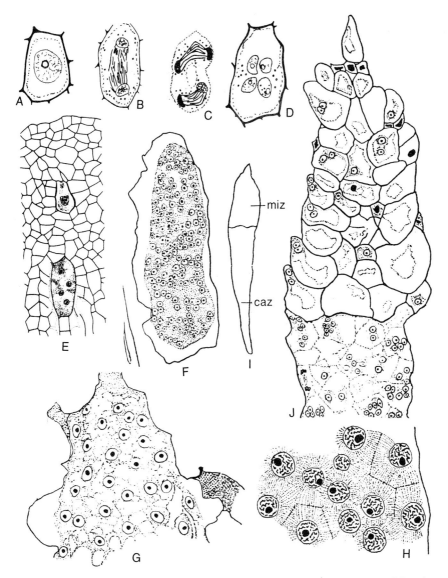

Fig. 19.9 A-J. *Welwitschia mirabilis*, megasporogenesis and development of female gametophyte. **A** Megaspore mother cell. **B, C** Telophase I and II; cell plate is not formed. **D** Four-nucleate coenomegaspore. **E** Longisection of floral axis shows extrafloral megaspore mother cell (above), and coenomegaspore (below). **F** Longisection of free-nuclear gametophyte, the nuclei are scattered all over the homogenous dense cytoplasm. **G** Longisection of upper portion of gametophyte, later stage, cytoplasm vacuolate. **H** Portion of gametophyte shows laying down of walls. **I** Outline diagram for **J;** micropylar (*miz*) and chalazal zone (*caz*) of endosperm. **J** Cellular details of **I.** (**A-E** After Martens 1963, **F-J** after Martens and Waterkeyn 1974)

grow in all directions in the nucellus. Each tube grows independently of the rest and normally ends in an enlarged swelling, called the fertilization bulb

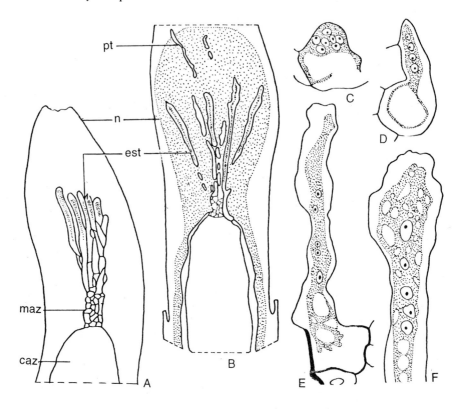

Fig. 19.10 A-F. *Welwitschia mirabilis.* **A, B** Longisection of upper portion of nucelli (*n*) with female gametophyte. *caz* chalazal zone of endosperm. *est* endospermic tube. *miz* micropylar zone of endosperm. *pt* pollen tube. **C-E** Elongated endosperm tube. **F** Tip of endosperm tube shows (swollen) fertilization bulb. (After Martens and Waterkeyn 1974)

(Fig. 19. 10 F), at about half the height of the nucellar cone (Fig. 19.10 A). This level is reached independently of and often before pollination, and eventually exceeds it. Occasionally, the tubes may over-reach the epidermis. The growth zone of the prothallial tube depends on the accumulation of starch in a limited part of the nucellar cone. Branching in the tubes is rare. The nuclei in a tube lie close together during early stages, but later they become arranged in a row (Fig. 19.10 C-F). All the nuclei in the embryo sac tube remain small and are potentially female nuclei. The prothallial tubes obtain their nourishment from the nucellar cells, and thin degenerated remains lie in contact with the tubes.

The haploid nuclei in the polynucleated compartments of the chalazal region undergo mitosis and the daughter nuclei fuse. Finally, a tissue with polyploid nuclei is formed.

The lipid megaspore wall increases in thickness during the growth of endosperm. Later, it becomes thin due to stretching.

Pollination

In the ovule, a honey-like fluid, the pollination drop, is secreted at the tip of the micropylar tube, in the morning hours (around 09.00 h). It persists during the day and disappears before sunset (Pearson 1929). It is probable that the pollination drop may attract insects. No secretion has been observed on the male flowers. The anthers bend over the disc, suggesting the possibility of insect visits. It has not yet been established whether pollination is effected by insects or by wind.

A pollen chamber is formed in the apical papilla of the nucellus by loss of nuclear material and an axial stretching of the hypodermal layers followed by the breakdown of the walls. Finally, a concave chamber is formed (Fig. 19.11 A, B). The pollen germinates in the pollen chamber (Fig. 19.11 B), or at the base of the micropylar tube (Martens and Waterkeyn 1974).

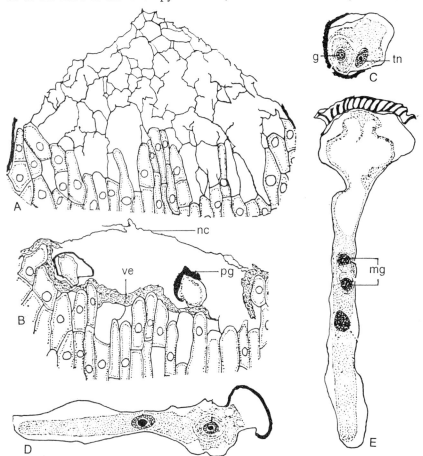

Fig. 19.11 A-E. *Welwitschia mirabilis.* **A, B** Formation of pollen chamber. **A** Nucellar tip with empty cells and ruptured walls; papillate cells at the base of future chamber. **B** Broken cell walls of nucellar tip form a veil (*ve*) on the papillate cells. *nc* nucellar cuticle. *pg* germinated pollen grain. **C-E** Pollen tube. *g* generative cell. *mg* male gamete (nuclei). *tn* tube nucleus. (After Martens and Waterkeyn 1974)

Post-Pollination Development of Male Gametophyte.   The exine splits (into valves). The pollen has a tube nucleus and a generative cell (Fig. 19.11 C). The growth of the pollen tube through the nucellus is inter- and intracellular.   The tube nucleus enters first (Fig. 19.11 D). While two male gametes have been observed (Fig. 19.11 E), the division of the generative cell has not been observed.

Fertilization

The pollen tubes grow downward to the fertilization bulb (Fig. 19.12 A, B), syngamy occurs and a zygote is formed within the fertilization bulb. It is surrounded by a distinct cellulose wall. Thus, zygote formation and its further development occurs in the maternal cytoplasm (Martens and Waterkeyn 1974).

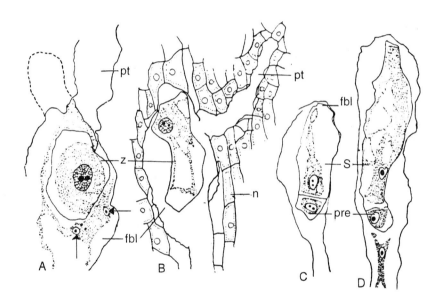

Fig. 19.12 A-D.   *Welwitschia mirabilis.* **A, B** Zygote in the fertilization bulb with binucleolate nucleus, dense cytoplasm, and cellulose wall; two haploid nuclei (arrowed) in the residual cytoplasm of the bulb in **A.** *fbl* fertilization bulb. *n* nucellus. *pt* pollen tube. *z* zygote. **C, D** Division in zygote to form proembryo initial (*pre*) and primary suspensor (*S*). (After Martens and Waterkeyn 1974)

Embryogeny

The zygote has a prominent cell wall (Fig. 19.12 A, B). It elongates, divides inside the fertilization bulb, and gives rise to a long suspensor cell and a small proembryo initial (Fig. 19.12 C, D). the intercalary growth of the suspensor pushes the proembryo towards the endosperm through the broad, free cavity of the endospermic tube. The successive divisions of the proembryo initial produce a terminal group of two (Fig. 19.13 A) and later four cells.

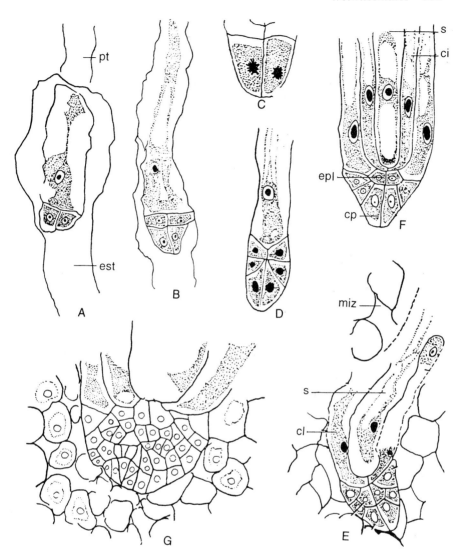

Fig. 19.13 A-G. *Welwitschia mirabilis.* **A** Two-celled apical cell. *est* endosperm tube. *pt* pollen tube. **B** Transverse division of the apical quartet. **C, D** Mitosis (**C**) and oblique division (**D**) in the apical quartet. **E, F** Proembryo in the endosperm; initiation of cortical tier (*ci*) of secondary suspensor. *cp* cap cells. *epl* embryonic plate. *miz* micropylar zone. *S* primary suspensor. **F** Periclinal division in embryonic plate; wall of spearhead has thickened (partially due to an outer incrustation). **G** Hemispherical homogeneous mass of meristematic tissue formed by anticlinal and periclinal divisions in cap cells; two cells form embryonic plate. (After Martens and Waterkeyn 1974)

The terminal cell group gives rise to a horizontal plate of inner cells (Fig. 19.13 B). The latter grows (after the proembryo penetrates the endosperm) peripherally and forms the first cortical tier or ring which covers and strengthens the primary suspensor. A second cortical tier similarly covers the first ring, and the two together form the secondary suspensor.

Further mitoses and oblique divisions (Fig. 19.13 C, D) in the apical quartet eventually form a group of inner and outer cells which divide anticlinally. The components of the outer layer elongate to form the cap cells, and the inner layer forms the embryonic plate (Fig. 19.13 E). The proembryo acquires the shape of a spearhead (Fig. 19.13 E, F). The cap cells are large, distinct, thickened by an external incrustation and better adapted for endosperm penetration than absorption (Martens and Waterkeyn 1974). Further development in the proembryo stops until it penetrates the endosperm and reaches the level beyond the insertion of the integument. Development is resumed, and anticlinal and periclinal divisions in the cap cells produce a meristematic, homogeneous and hemispherical or conical tissue (Fig. 19.13 G). The embryo is still undifferentiated although orientation of cellular rows shows a root centre opposite to the shoot apex, and a coleorhiza. Meanwhile, the peripheral cells of the embryonal mass contribute to the formation of a massive secondary suspensor. In a mature ovule the primary and secondary suspensors are folded in a suspensor sack above the endosperm.

Late embryogeny has been investigated in *Welwitschia*. A mature embryo shows vertical orientation of the two cotyledons, prominent shoot apex, a root centre, voluminous coleorhiza with curved cell rows and the axial rows of the columella. The cotyledons overreach the shoot apex and elongate; a hypocotyl is present. A non-vascular outgrowth of embryo remains embedded in the gametophyte, represented at the mature embryo stage by an annular bulge (collar) in the embryonal axis. As the seed germinates, the collar gives rise to the "feeder" (Fig. 19.14 A; Bower 1881).

The endosperm (gametophyte) develops from a part of the chalazal zone of the gametophyte and undergoes a long period of growth. The outer lateral cell layers and the entire deep zone maintain their meristematic activity during post-fertilization stages. In the micropylar and the central region of the endosperm, there is gradual vacuolation and the cells stretch from the centre to the upper pole. The suspensor sack differentiates in this zone.

After the nuclear fusions in the multinucleate chalazal compartment, the mitoses in the polyploid endosperm cells are followed by wall formation. As the endosperm development advances, the size of cells, nuclei, number of nucleoli and chromosomes in a nucleus, becomes reduced in some areas. This has been interpreted as depolyploidization (Martens and Waterkeyn 1974). It results in the formation of a heterogenous tissue made of large and small cells and nuclei.

The main reserve tissue is relatively poor in starch, but rich in proteinaceous granules; their accumulation compresses and deforms the nuclei. The dermatogen differentiates late and is devoid of inclusions.

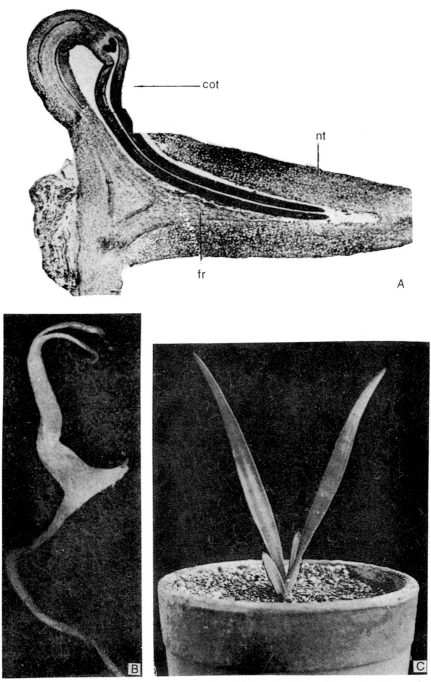

Fig. 19.14 A-C. *Welwitschia mirabilis.* **A** Longisection of germinated seed. *cot* cotyledons. *fr* feeder. *nt* nutritive tissue. **B** Seedling, cotyledons above ground. **C** Eighteen-month-old seedling, cotyledons still healthy. (**A** After Bierhorst 1971, **B** after Rodin 1953b, **C** after Von Willart 1985)

Seed

The seeds are winged. The wing and the seed-coat develop from the perianth (see Female Strobilus). The integument is papery.

There is a double system of vascular bundles in the seed-coat (Martens 1971, 1974). The outer system has two unbranched traces, while the inner system has eight traces which traverse the inner part of the seed and extend up to the base of the integument.

In the chalazal region there is a cupule of spirally thickened tracheids intermixed with thin-walled cells. This extends basally and is not connected with ovular vascular supply. According to Martens (1974), this tracheidal cupule is the relict of nucellar vasculature.

Germination. When the seed germinates, the radicle breaks through the testa (and not through the micropyle), and it is always directed downward (Rodin 1953b). The two cotyledons break through the testa opposite the point where the roots emerge. Figure 19.14 A shows a germinated seed in longisection. The cotyledons are lanceolate and dark red, gradually changing to green within a week. The cotyledons are about 5 mm long and 2 mm broad when they first appear above ground (Fig. 19.14 B), and reach approximately 3 cm in length and 5 mm in width when the seedlings are about 6 months old. They remain healthy for several years. A pair of decussately arranged plumular leaves appear (Fig. 19.14 C), and a meristem becomes active at the base of the leaf at a very early stage.

The venation pattern is similar in both cotyledons and foliage leaves, i.e. longitudinal veins and oblique anastomosing veinlets. Sclereids and vessels are absent in the cotyledons.

Many changes take place at the ultrastructural and biochemical levels when the seed imbibes water for germination (Bornman et al. 1979 a, b, c, Butler et al. 1979a, b, c). The radicle and the compressed hypocotyl are separated by a uniform bulge (collar) which is the future feeder. The embryonic collar cells are activated first after hydration, followed by the gametophytic cells. The latter activates from the embryo outwards, and the rate and sequence suggest the presence of stimulatory factors (in the embryo) which diffuse into the gametophyte and induce and/or enhance metabolic activity therein. Within 36–48 h of hydration, starch, protein and lipid reserves in the collar and developing feeder are degraded, and the rapidly developing seedling becomes dependent on nutritive material in the gametophyte until the plumule emerges. Feeder cells in contact with the gametophytic tissue develop numerous small wall projections invested with plasmalemma and a large number of mitochondria, and act as transfer cells. The collar gives rise to the feeder, which is ca. 5 mm long, at about 36 h following imbibition. The cell walls are mucilaginous and the feeder and gametophyte adhere firmly to each other, which facilitates translocation of nutrients. Butler et al. (1979c) confirms the hypothesis of Bower (1881) that the feeder functions as an absorptive organ. The feeder may have a secondary function to provide the emerging plumule with a firm base.

The reserve material within protein bodies of the embryo and gametophyte is a protein-carbohydrate complex. The immediate digestion of protein body reserves in the embryo and gametophyte interface zone indicates the presence of preexisting hydrolytic enzymes within the protein bodies. However, the enzymes responsible for the breakdown of reserve food in the outer gametophytic tissue are synthesized de novo. Protein hydrolysis precedes lipid digestion, which indicates that some of the resulting free amino acids are used in de novo synthesis of lipases. Lipid bodies, microbodies, mitochondria and amyloplasts are characteristically encircled with ER.

The mean dry mass of the gametophyte decreases by approximately 47% by day 5, and the amount of lipid decrease is 75% and protein 25%. Some of the hydrolyzed fatty acids and amino acids are utilized in the gametophyte, while the major part thereof is probably converted to sugars, which, together with free amino acids, are transported via the feeder to the embryo to be utilized in seedling growth.

Chromosome Number

The somatic chromosome number is 42 (Khoshoo and Ahuja 1963). The karyotype is peculiar and all the chromosomes appear to be rod-shaped with probably terminal centromeres. The chromosomes are asymmetrical with pronounced size difference. The longest pair is 3.25 times longer than the shortest pair, with a gradual transition in size. Two of the chromosomes in the complement are satellited, which coincides with two nucleoli in the metabolic nucleus.

20. Gnetales

Gnetaceae
Gnetum

Gnetum, with more than 30 spp., is confined to the humid, moist regions of the world (see Chap. 1).

Morphology

Gnetum plants are shrubs, trees or climbers with twining stems, and in general habit resemble a dicotyledonous plant (Fig. 20.1). The main stem bears two types of branches: a short shoot of limited growth (internodes present), and a long shoot of unlimited growth (internodes absent). The difference between the types of shoots is not very pronounced. In the climber *G. ula*, the short shoots bear leaves higher up near the branches of the supporting tree. The leaves lie crowded in one plane, in decussate pairs so that the branch looks like a pinnate leaf. The lamina is large, oval, entire with reticulate venation and gives a typically dicotyledonous appearance.

Several *Gnetum* spp. have articulated stems. The joints consist of two parts—one immediately above and the other immediately below the node, the two being separated by an annular groove. The internodes are shed at the joint, and lie in large numbers below the plant, resembling bones.

Anatomy

Root. The root is diarch. The cortex comprises several layers of large polygonal cells filled with starch grains and numerous thick-walled fibres. An endodermis and a five- or six-layered pericycle are present. The secondary growth is normal; the radial walls of tracheids have uniseriate bordered pits with conspicuous bars of Sanio. The vessel elements have smaller multiseriate pits and the bars of Sanio are inconspicuous or absent. The xylem elements in the root are larger than those in the stem, and some of them contain starch grains in later stages. The phloem consists of thin-walled broad rays packed with starch grains, which makes the wood of the root comparatively soft as against that of the stem, where the phloem ray is thick-walled and pitted (P. Maheshwari and V. Vasil 1961a).

Fig. 20.1 *Gnetum gnemon*, a branch with female cones (*fc*). (After P. Maheshwari and V. Vasil 1961a)

Stem. There is a distinct tunica-corpus organization. The single layer of tunica extends over the shoot apex from the axil of the first pair of leaf primordia. The cells at the apex are larger and more vacuolate than those in the flanks. Periclinal divisions occur in the tunica during initiation of the foliar buttress. The corpus comprises two or three layers of subapical initials, central mother cell zone, flanking layers and pith rib meristem.

Immediately following elevation of a pair of leaf primordia, the apex is at its minimum volume. It is a shallow cap of subapical initials and a few central mother cells. Also, a rib meristem may be absent above the first node. It is, however, reconstituted as the apex increases in size.

Transection of a young stem shows a thick cuticle followed by a single-layered epidermis with a thick outer wall, and papillate and rectangular cells. The stomata are sunken. The parenchymatous cortex is 12- to 16-layered. The outer four to six layers are chlorophyllous, followed by a large number of scattered fibres. In older stems there is a conspicuous sclerenchymatous zone (Fig. 20.2 A-C) in the inner part of the cortex. It is formed from large parenchymatous cells which become lignified. These cells have branched and unbranched pit canals. The primary vascular bundles (20–24) are collateral and endarch (Fig. 20.2 A-C). They are arranged in a ring separated by broad medullary rays of considerable height. The wood consists of tracheids and a few vessels, and the phloem comprises sieve cells and phloem parenchyma. The endodermis and pericycle are not discernible. The pith is parenchymatous when young; later, the cells towards the vascular ring become thick-walled, lignified and have numerous pits. Several laticifers occur scattered in the cortex and pith.

Laticifers of *Gnetum* are cellular. They originate as single cells, enlarge and ramify (Martens 1971, Carlquist 1996). Intercellular spaces are associated with their pectic-rich walls.

In the arborescent type (*Gnetum gnemon*), the secondary growth is normal. In climbing species (*G. ula* and *C. africanum*), a new cambium differentiates at various points in the inner part of cortex and gradually become a continuous cylinder. It produces a normally oriented ring of xylem and phloem wedge-shaped bundles separated by medullary rays. The first ring stops growing at the commencement of the outer ring. Successive rings continue to form; some may remain incomplete, which results in an eccentric arrangement of the rings (Fig. 20.2 B), or an eccentric position of the pith. In *G. ula*, the anomalous rings of vascular bundles appear to be extrastelar.

The phellogen arises irregularly in the epidermis and initially spreads unequally. Later, in older internodes, the ring of cork cells becomes complete. The periderm is thin and has lenticels.

The xylem consists of highly tapering tracheids, vessels and xylem parenchyma. The latter are round or elongated with simple pits.

Vessel structure was studied by Muhammad and Sattler (1982) in *G. gnemon* and *G. montanum* by light microscopy (bright field, phase contrast and Nomarski interference optics), SEM and TEM. The tracheary elements earliest to mature are narrow with annular thickenings followed (in the ontogenetic sequence) by longer and wider elements with annular-helical

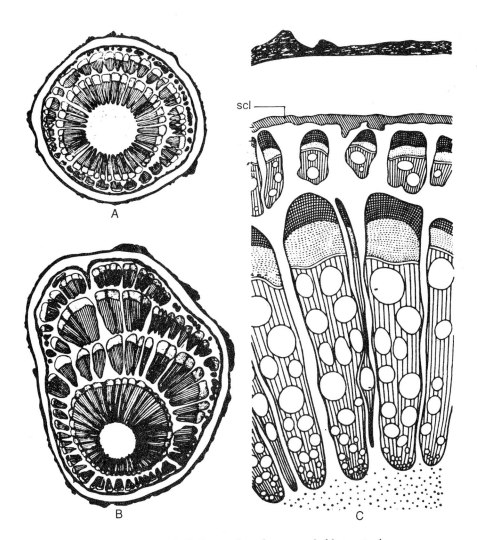

Fig. 20.2 A-C. *Gnetum ula*. **A, B** Transection of young and old stem to show accessory (**A**) and eccentric (**B**) rings of vascular bundles. **C** Vascular bundles from **A**. *scl* sclerenchyma. (After P. Maheshwari and V. Vasil 1961a)

and later only helical thickenings with compound gyres. In annular and helical elements, the gyres are made up of three elementary strands (Fig. 20.3 A). Pits range from slit-shaped to circular, and are formed either between the elementary strands of one gyre or between adjacent gyres. In both species, scalariform pits (Fig. 20.3 B) occur in addition to the circular pits. Larger pits occur in contact with other tracheary elements, while vessel elements in contact with parenchyma have smaller pits. In the nodal region, comparatively large pits with prominent borders occur, especially on the inclined walls of tracheids (Fig. 20.3 C). The pit membrane is thin and does not have a torus (cf. angiosperms), unlike conifers (Fig. 20.4 A). Intervessel

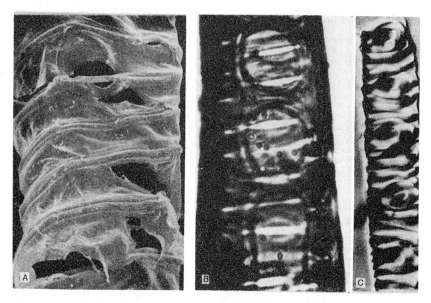

Fig. 20.3 A-C. A, C *Gnetum montanum.* **A** SEM tracheary element from stem shows compound gyre. **B** *G. gnemon,* vessel from petiole with scalariform pits seen through perforation plate. **C** Bordered pits on tracheids from protoxylem. (After Muhammad and Sattler 1982)

pits have well-developed vestures (Fig. 20.4 B), which are absent in tracheids. Both primary and secondary xylem have vessels. Each vessel has two, usually similar, perforation plates at the two ends. The perforation may be scalaroid (vertical row of circular perforations; Fig. 20.5 A), scalariform (row of transversely elongated perforations; Fig. 20.5 B), foraminate (round perforations in alternate or horizontal rows—commonly in the nodal region of stem; Fig. 20.5 C), and simple perforation plate (Fig. 20.5 D). The simple perforation plate is more common in the late metaxylem, especially in the secondary xylem. The breakdown of the wall between the perforations may lead to their fusion horizontally, vertically or haphazardly so that a simple perforation plate is formed. There is no evidence that scalaroid, scalariform, foraminate and intermediate perforation plates are formed in a sequence (Muhammad and Sattler 1982).

Carlquist (1994, 1996) and Carlquist and Robinson (1995) do not agree that perforation plates in *Gnetum* are sealariform or sealariform-like.

Both species of *Gnetum* demonstrate a wide range of variation in the shape of pits and perforations.

The primary and secondary rays are broad, and comprise 10–20 thick-walled pitted and radially elongated cells, which occasionally contain calcium oxalate crystals. In a tangential section the rays appear boat-shaped, and

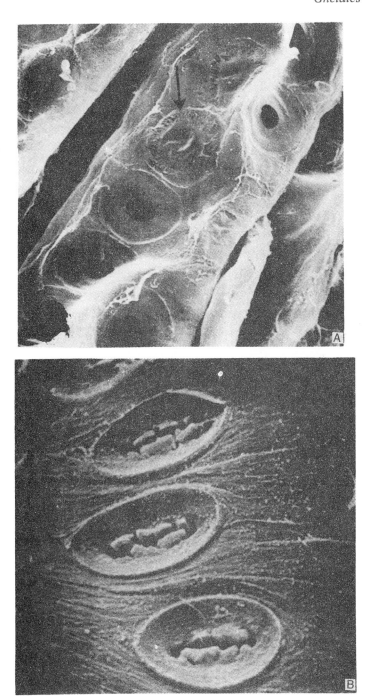

Fig. 20.4 A, B. *Gnetum montanum*, SEM tracheary element from stem. **A** Two pits with intact membranes (arrow). **B** Vessel with vestured pits. (After Muhammad and Sattler 1982)

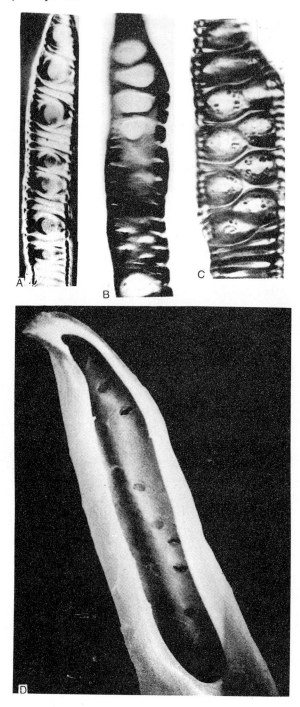

Fig. 20.5 A-D.   *Gnetum montanum.* **A-D** Vessels show scalaroid (**A**), scalariform (**B**), foraminate (**C**) and simple perforation plate (**D**). (After Muhammad and Sattler 1982)

consist of polygonal cells. In addition, uniseriate or biseriate rays have been observed and may be two, three or several cells high.

Paliwal and Behnke (1973) studied the primary and secondary phloem in *G. gnemon*. It comprises only two cell types—the sieve cells and phloem parenchyma (this term is preferred to "albuminous cells" or "companion cells" by the above authors)—which are ontogenetically unrelated and alternate with each other. A regular filing of both the cell types is lost during secondary growth. A sieve cell is larger than a phloem parenchyma cell, and has a parietal layer of cytoplasm, which occasionally clumps together on one side of the cell. The phloem parenchyma cell has dense plasmatic contents.

The secondary growth is initiated in a normal manner. The transection of a young stem shows three to five layers of cambial cells within the vascular bundles. Ray initials are present in the region of medullary rays, but are absent in the phloem tissue itself. The latter is delimited from the cortex by a two-layered parenchymatous sheath followed by sclerieds, and broad rays are present laterally.

Transection of an older stem shows several layers of obliterated crushed and dark phloem cells (*op*, Fig. 20.6 A, B). In none of their preparation could Paliwal and Behnke (1973) recognize the deposition of thickening material in older cells of phloem tissue.

In the active phloem, the parenchyma cells are filled with starch. In the older phloem, they are empty and crushed like the sieve elements. Parenchyma cell differentiation (of all types including those associated with the sieve cells) reveals features comparable to those of the companion cells in angiosperms. The cambial initial—destined to become a parenchyma cell—becomes rich in cellular organelles. The ribosomes are abundantly distributed in the cytoplasm and are deposited on the ER. Microtubules, dictyosomes and mitochondria with well-defined cisternae are also present. Plastids may have ovoid, hemispherical or elongated profiles. These contain thylakoids, a dense matrix, and accumulate varying quantities of starch. The nucleus persists for the entire life of these cells. Cell sap is retained in several small vacuoles, which generally fuse to form a central cavity. They are devoid of proteinaceous contents (see Paliwal 1992).

A differentiating sieve element in *G. gnemon* can be discerned from its neighbouring parenchymatous cell by its very long nucleus. The changes in ER during sieve cell differentiation are significant. The granular ER in the young sieve cells tends to be modified by a swelling of interacisternal spaces. After the loss of ribosomes, the ER assumes a more tubular structure; tubules spread throughout the cell lumen but tend to form aggregates. The ER tubules lie parallel to each other within the aggregates, or may become twisted. Frequently, the ER complexes are associated with the degenerating nuclei or with the plastids. Dictyosomes and microtubules also disappear. Finally, the ER, plastids and mitochondria are the only organelles present

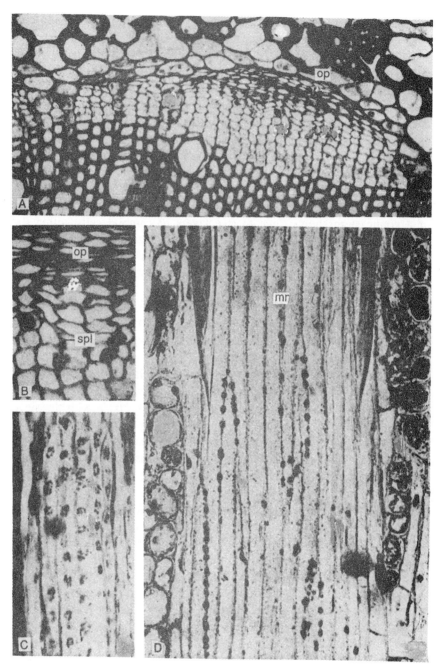

Fig. 20.6 A-D. *Gnetum gnemon.* **A, B** Transection (part of stem) at advanced stage of secondary growth, shows crushed primary phloem (*op*) and thick callose deposits on sieve plates (*spl*). **C** Radial longisection of phloem, the sieve areas with end-plate (arrows). **D** Tangential longisection of phloem between two medullary rays (*mr*); close grouping of sieve areas on the sieve plate, indicated by callose deposition. (After Paliwal and Behnke 1973)

in the "living" protoplast (characteristic condition in gymnosperms). The central cell sap cavity occasionally contains fibrillar material that mixes with and disappears within the protoplast after degeneration of tonoplast. P-protein has not been detected either by light or electron microscopy in young or mature sieve elements (Paliwal and Behnke 1973).

In tangential section of phloem, the sieve plates are arranged on oblique end walls (Figs. 20.6 D; 20.7 A-D). In a radial section each end wall (Fig. 20.6 C) appears to be made up of 7–19 sieve areas. The sieve areas on the lateral walls between two contiguous sieve cells are more widely spaced than those on the end walls.

At maturity, the tapering sieve cells lack cellular contents (Fig. 20.7 A) and plastids occur throughout their length. One to four phloem parenchyma cells remain connected to a sieve cell. They have transverse, oblique or tapering ends. Their lateral association with the sieve elements is concluded from callose plugs which occur on the lateral sieve areas, and take a deep-blue stain with resorcin blue. A well-organized nucleus and other cytoplasmic contents, including plastids, are present even at maturity.

Sieve cells and phloem parenchyma cells are ontogenetically unrelated, but at any time of phloem differentiation, these cells are laterally connected by specialized plasmatic strands.

At no stage was a positive staining reaction observed for proteinaceous substances with mercuric bromophenol blue. This indicates that P-protein (slime) is not produced during the differentiation of sieve cells or phloem parenchyma cells.

Leaf. A vertical section of a leaf shows a well-marked cuticle on epidermal cells with undulate walls on both surfaces, and the mesophyll of a single layer of short palisade cells and a well-developed spongy tissue. Stellately branched sclereids, fibres and latex tubes occur near the lower epidermis, especially in the mid-rib region. The vascular bundles are arranged in a curve. The xylem consists of vessels, tracheids and parenchyma. Below the xylem phloem cells are arranged in regular rows. In *Gnetum ula*, pitted and thick-walled cells form a narrow patch just outside the phloem and towards the protoxylem.

Stomata, in addition to leaves (lower surface), occur on young stems, axes and collars of male and female cones, and on the two outer envelopes of the ovule.

Nautiyal et al. (1976) studied the aerial parts for epidermal structure and ontogeny of stomata in *G. gnemon*, *G. montanum* and *G. ula*.

The aerial parts in the three species show irregularly dispersed stomata and cork warts. The stomatal development is mesoperigenous. The mature stomata are either tetracytic with two mesogene lateral subsidiaries and two perigene polar neighbouring cells, or multicytic with three to seven subsidiaries. The latter may be partly or wholly amphicyclic. Developmental

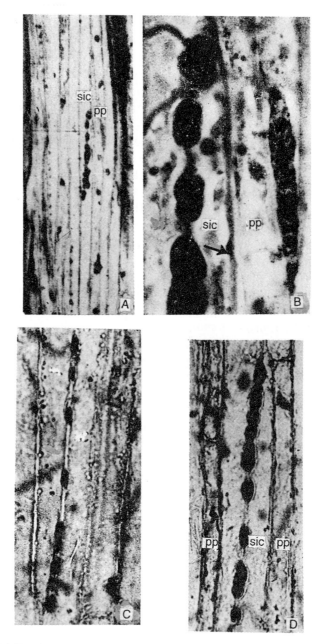

Fig. 20.7 A-D. *Gnetum gnemon*, tangential longisection of phloem. **A** Phloem shows sieve cells (*sic*) with oblique sieve plates and phloem-parenchyma (*pp*) cells. **B** Single sieve plate; association of sieve cells and phloem parenchyma cells is obvious by lateral sieve areas (arrows). **C** Oblique sieve plate between two sieve cells, sieve area dark. **D** Sieve plate. (After Paliwal and Behnke 1973)

studies show that the multicytic stomata are formed by subsequent radial or tangential divisions in the lateral subsidiaries and polar neighbouring cells.

Takeda (1913) reported syndetocheilic development of stomata in the leaves of *Gnetum*. However, P. Maheshwari and V. Vasil (1961b) report haplocheilic development (Fig. 20.8 A-E) in *G. ula* and *G. gnemon..* A few stomata on the leaves show parallel subsidiary cells, which arise from adjoining epidermal cells, and do not seem to have a common origin with the guard cells. Twin stomata (Fig. 20.8 F) also occur, originating from two adjacent stoma mother cells.

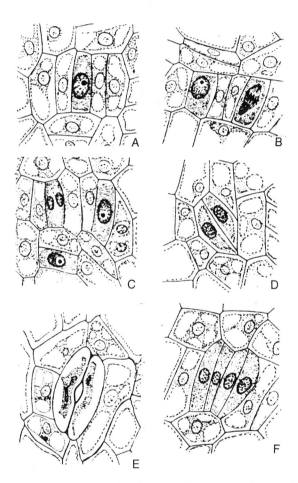

Fig. 20.8 A-F. *Gnetum gnemon,* development of stomata on lower epidermis of leaf. **A-C** Stomatal initials. **D** Guard cells. **E** Older stoma. **F** Twin stomata. (After P.Maheshwari and V. Vasil 1961b)

A wide variety of sclereids develop in the collars of the cones, the perianth of male flowers, and the outer two envelopes of female flowers in *G. ula*

Laticiferous elements are scattered all over the plant in *G. ula* and *G. gnemon*. These are long and, occasionally, branched tubes filled with a latex-like substance. Those present in the pith are initiated from rows of parenchymatous cells during the growth of the embryo.

Reproduction

Gnetum is dioecious. The male and female strobili consist of a stout axis which bears a basal pair of opposite and connate bracts, and usually six to eight superposed cupules or collars. The cupules arise as annular protuberances in acropetal succession (Takaso and Bouman 1986), and an annular rim is formed at the axillary position of each cupule. The cupules are tightly packed in a young strobilus (Figs. 20.9 A; 20.10 A), and the latter do not elongate appreciably during initiation of the annular rims. Later, the internode elongates so that the annular rims separate from the next upper cupule and the axillary position becomes pronounced. According to V. Vasil (1959), however, the annular rim develops as a hump from a few cells lying below each collar. Then this annular meristem comes to lie between the collar which bears it and the one below it, and gives the impression of axillary origin (Fig. 20.10 B-H).

Male Strobilus

Three to six rings of male flowers develop basipetally above each collar (Fig. 20.9 B). A single ring of abortive ovules (Figs. 20.9 B; 20.10 F-H) may occur above the flowers. The flowers in different rings are arranged alternately.

A male flower consists of a stalk bearing two unilocular anthers enclosed in a perianth (Fig. 20.9 D). On maturity, the stalk elongates and pushes the anthers (through an opening in the perianth) beyond the collars of the cone (Fig. 20.9 C, E).

Microsporangium. In a young anther, the hypodermal archesporial cells divide and give rise to a multicelled archesporium (Fig. 20.11 A-C), the outermost layer divides periclinally and differentiates into the parietal and sporogenous cells. The former divide again to produce an outer wall layer and the tapetum (Fig. 20.11 C). The narrow tangentially elongated wall layer cells become compressed during meiosis of the microspore mother cells. The tapetal cells increase in size, have dense cytoplasm and are binucleate (Fig. 20.11 C, G). In *G. gnemon*, the outer tapetal layer is sculptured by the deposition of globules (Fig. 20.11 G), which are 8–110 μm^3 in volume (Carniel 1966). In the beginning, the sculpturing appears coarse, followed by finer globules (Fig. 20.11 H, I), which give the reaction for sporopollenin. In *G. gnemon*, they appear on the outer tangential walls of the tapetum, while in *G. africanum* they appear even on the inner side of the tapetal cells (Waterkeyn 1959). Eventually, the wall layer and tapetum are absorbed and only the epidermis persists in the mature anther (Fig. 20.11 D, F).

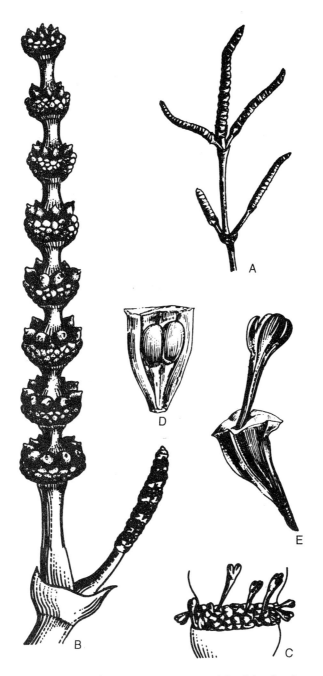

Fig. 20.9 A-E. **A, C-E** *Gnetum ula*, **B** *G. gnemon*. **A** Panicle of male cones. **B** Rings of male flowers and an ovular ring above each collar. **C** Portion of male cone at dehiscence. **D** Longisection of male flowers, two anthers are enclosed within the perianth. **E** Dehisced male flower. (After P. Maheshwari and V. Vasil 1961a)

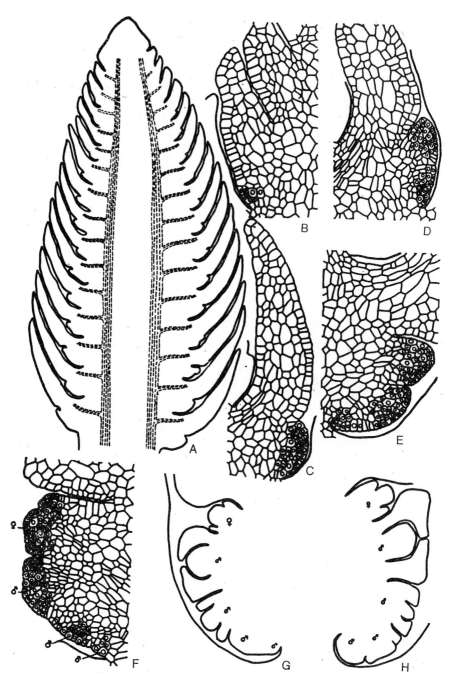

Fig. 20.10 A-H. *Gnetum ula.* **A** Longisection young male cone. **B-H** Portions of longisection to show development of upper ovular and four rings of male flowers. (After P. Maheshwari and V. Vasil 1961a)

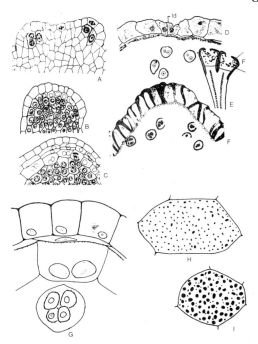

Fig. 20.11 A-I.   **A, D, G-I** *Gnetum gnemon,* **B, C, E, F** *G. ula.* **A-C** Longisection of antherophores. **A, B** Hypodermal archesporium. **C** Differentiation of wall layers, tapetum and sporogenous tissue. **D** Portion of wall with site of dehiscence (*ld*) and two-celled pollen grains. **E** Longisection dehisced male flower. **F** Inset from **E** marked *F* to show epidermal thickenings. **G** Wall layers at tetrad stage, orbicules deposited on the outer tangential tapetal wall. **H, I** Outer tangential wall in surface view to show fine and coarse orbicules. (**A-F** After P. Maheshwari and V. Vasil 1961a, **G-I** after Carniel 1966)

During dehiscence, the epidermal cells become radially enlarged, and the outer wall becomes slightly thickened and cutinized. From the inner tangential walls, fibrous bands of thickenings run upward and outward to the outer wall of each cell (Fig. 20.11 F). The anthers open along a double row of smaller cells, which extend vertically on either side from the tip to some distance below the anther sac (Fig. 20.11 D, E).

Microsporogenesis

The sporogenous cells undergo further divisions, and the microspore mother cells contain dense cytoplasm and prominent nuclei (Fig. 20.12 A). Prior to meiosis, the protoplasts recede and a special mucilaginous wall is secreted between the protoplast and the mother wall. The meiosis is not sychronous even in the same loculus (P. Maheshwari and V. Vasil 1961a).

The reduction divisions are simultaneous (Fig. 20.12 A-E), and cytokinesis takes place by centripetal furrows. The tetrads are tetrahedral (Fig. 20.12 F), isobilateral and decussate. As the microspores enlarge, the callose wall is gradually absorbed, the original wall breaks down, and the young microspores

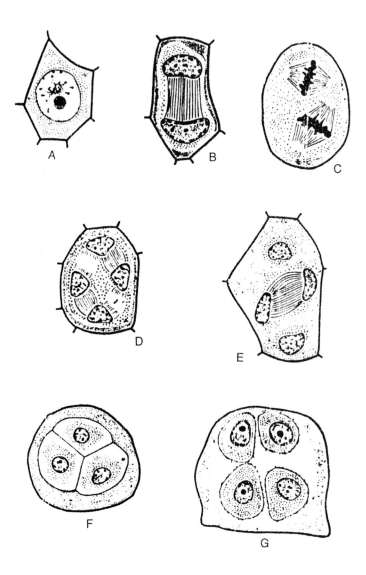

Fig. 20.12 A-G.   *Gnetum ula.* **A-E** Microspore mother cell (**A**), meiosis I (**B**), and
II (**C-E**). **F** Tetrahedral tetrads. **G** Tetrad before liberation of microspores, callose wall
more or less consumed. (After P. Maheshwari and V. Vasil 1961a)

are released. Even before separation, the microspore wall shows minute
spiny protuberances. Gradually, the wall differentiates into a thick, spiny
exine and a thin intine(Fig. 20.12 G). According to Gullvag (1966), the
pollen grains of *G. ula*, *G. gnemon* and *G. montanum* have a single layer
of lamellated exine. This needs confirmation.

Male Gametophyte.   The mature pollen grain is shed at the three-celled stage (P. Maheshwari and V. Vasil 1961a, Martens 1971). The precise sequence by which the three nuclei/cells arise has been investigated by Negi and Madhulata (1957), Waterkeyn (1959), Martens (1971), Swamy (1974) and others. The division of the microspore nucleus results in a small lenticular and a large cell (Fig. 20.13 A-C). The lenticular cell rounds up, does not divide further nor take part in the development of the pollen tube, and degenerates. The nucleus of the larger cell divides again (Fig. 20.13 D, E) and, of the two nuclei, one hyaline with a large nucleolus and is the first to enter the pollen tube. The second nucleus is rich in

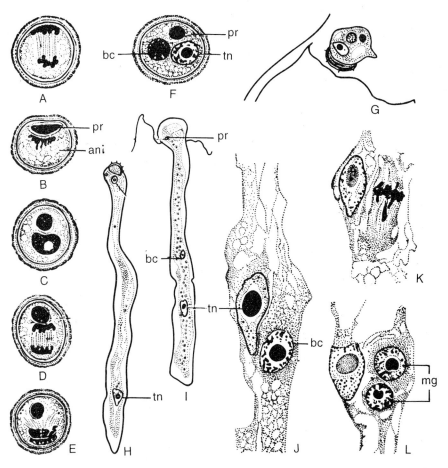

Fig. 20.13 A-L.   **A-F, J-L** *Gnetum africanum.* **A-F** Development of three–celled male gametophyte. *ani* antheridial initial. *bc* body cell. *pr* prothallial cell. *tn* tube nucleus. **G-I** *G. ula.* **G** Germinated (early stage) pollen grain in pollen chamber. **H, I** Pollen tube with body cell (*bc*) and tube nucleus (*tn*). **J-L** Pollen tube with tube nucleus and division of body cell in **K** and two male gametes (*mg*) in **L**. (**A-F, J-L** After Waterkeyn 1959, **G-I** after V. Vasil 1959)

chromatin, has a cytoplasmic sheath of its own (Fig. 20.13 F), and divides in the pollen tube to give rise to two male gametes.

These three nuclei have been variously interpreted. According to Pearson (1912, 1914), they represent the prothallial, tube and generative nuclei. Thompson (1916) presumes that they represent the tube nucleus, and stalk and body cells. Negi and Madhulata (1957) and Waterkeyn (1959) interpret them as the prothallial cell, generative (body) cell, and tube nucleus (Fig. 20.13 F).

Swamy (1974) proposed a new type of gametophyte development in *Gnetum ula*. The microspore divides and forms a small generative (body) cell on one side, and a large cell. The generative cell (which has its own sheath of cytoplasm) eventually lies free in the cytoplasm of the large cell, and then moves into the pollen tube and divides to form two male gametes. The nucleus of the larger cell divides to become a binucleate cell.

Female Strobilus

A female strobilus usually has six to eight collars (Figs. 20.14 A; 20.15 A) and a single ring of ovules above each collar (Fig. 20.14 B, C). Two to eight (usually five or six) ovular primordia differentiate from the annular rim. The latter becomes conspicuous before initiation of the ovules (Fig. 20.15 B). Gradually, the ovules become visible on the upper edge of each collar (Fig. 20.15 C). Generally, the upper few collars have no ovules.

Ovule. The cells in the epidermal and subepidermal layers of the ovular primordium divide actively. The dermal cells undergo both periclinal and anticlinal divisions. Simultaneously, each primordium can be distinguished into an upper region, which bears the ovule, and a lower, cushion-like area. Hairs differentiate from the surface of this cushion and from sterile cells between the ovules (Fig. 20.15 C-E).

Three envelopes arise acropetally on ovular primordia and enclose the nucellus (Figs. 20.15 C-E; 20.16 A-C). The outer envelope (*oe*) is the first to differentiate. It arises laterally to the ovular primordium by peri- and anticlinal divisions in the dermal and subdermal cells (Fig. 20.15 C, D) and is of dual origin. At pollination, the *oe* increases in thickness. About 30 vascular strands are present at the level just above the junction between the inner envelope (*ie*) and the nucellus, and a few of them extend almost to the apex. Laticifer canals (do not form lateral connections) are scattered throughout the envelope, and numerous stomata occur in the outer layer.

After the *oe* is initiated, the apex of the ovule shows a clear uniseriate dermal layer. The middle envelope (me) and inner envelope (*ie*) have a subdermal origin (Takaso and Bouman 1986).

The *me* arises next (Fig. 20.15 E). Around pollination, it grows in thickness, only in the subapical part of the envelope, by periclinal divisions of the subdermal cells which elongate radially. The outer dermal cells remain

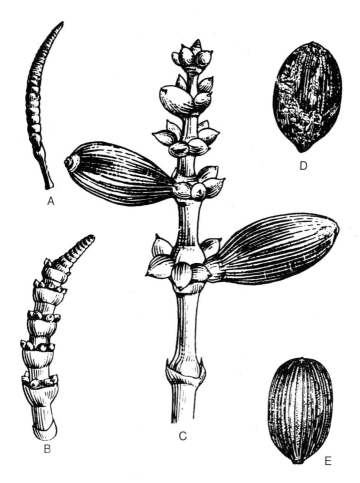

Fig. 20.14 A-E. A, B *Gnetum ula*, young female cone (**A**) with ovules (**B**), **C-E** *G. gnemon*. **C** Cone bearing two seeds. **D, E** Mature seed after shedding (picked up from ground). (After P. Maheshwari and V. Vasil 1961a)

small and bear stomata. Provascular strands extend to subapical region. Laticifers are also present. The *ie* arises last. It develops as a tubular wall around the newly outlined nucellus, grows rapidly and forms several lobes at the apex (Fig. 20.15 A-C). It forms a micropylar tube which extends beyond the rim of the two outer envelopes (Fig. 20.16 C). Soon after pollination, the apical part breaks off. Stomata, sclereids and vascular bundles are absent. However, very few ovules develop into mature seeds (Fig. 20.14 D, E).

In *G. gnemon*, at pollination, the outer epidermal cells of the micropylar tube divide and produce a massive circular structure, called a flange, around the tip of the middle envelope (Figs. 20.16 D; 20.17 A). At a later stage, the outer surface of the flange becomes closely appressed against the outer envelope. After the pollen grains have entered the ovule, the dermal and

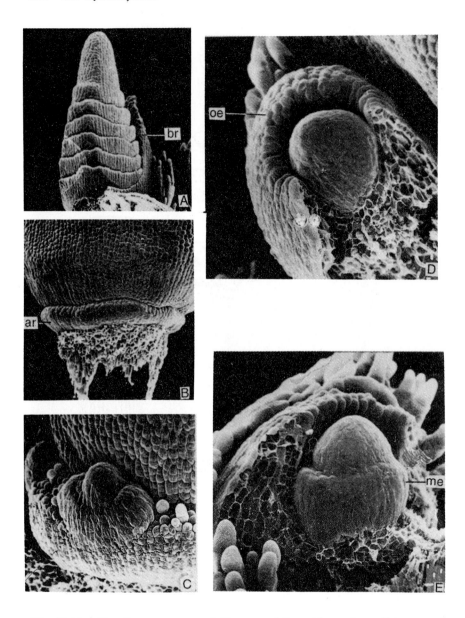

Fig. 20.15 A-E. *Gnetum gnemon.* **A** Young strobilus with one bract (*br*) removed. **B** Axillary region of cupule with annular rim (*ar*). **C** Ovular primordia on the rim, the outer envelope is initiated laterally. Sterile hairs occur on the rim in between the ovules. **D** Ovular apex surrounded by outer envelope (*oe*). **E** Ovule with middle envelope (*me*). (After Takaso and Bouman 1986)

subdermal cells lining the micropylar tube divide to produce a tissue which fills the canal and completely blocks it (Fig. 20.17 B, C). In older seeds of *G. gnemon*, these cells develop thick walls with simple pits (Sanwal 1962).

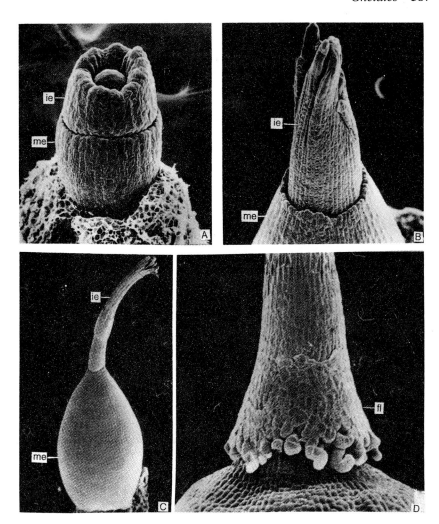

Fig. 20.16 A-D.   *Gnetum gnemon.* **A** Ovule with middle (*me*) and inner envelope (*ie*). **B** Inner envelope shows several apical lobes. **C** Ovule with protruding inner envelope (*ie* inner integument). **D** A flange (*fl*) covers the apical part of middle envelope. (After Takaso and Bouman 1986)

The cells of the upwardly directed flange, the closing tissue and the intermediate tissue of the inner envelope contain a suberin-like substance (Takaso and Bouman 1986). The micropylar tube above the suberized tissue degenerates.

A few cells of the hypodermal layer of the nucellus divide periclinally to form the parietal tissue and the sporogenous cells. Occasionally, the cells of the second and third layer differentiate directly into sporocytes (Takaso and Bouman 1986). The sporogenous cells divide further and produce nearly 12 megaspore mother cells (mgmc). A nucellar cap is formed by

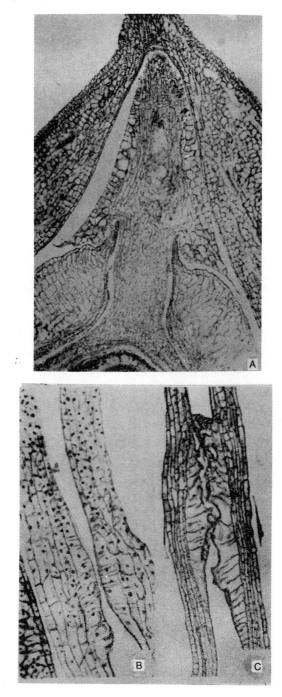

Fig. 20.17 A-C. A *Gnetum gnemon*, longisection upper part of ovule to show the formation of flange from inner envelope. **B, C** *G. ula,* longisection upper portion of micropylar tube. **B** Cells of inner epidermis. **C** Interlocked cells, the central inner epidermal cells extend into the micropyle. (After P. Maheshwari and V. Vasil 1961a)

periclinal division of the apical dermal layer. During the development of the nucellus, a few mgmc degenerate. While these cells are preparing for meiosis I, a fan-shaped tissue, the pavement tissue, of radiating rows of nucellar cells appears at the base of female gametophyte. It has dense cytoplasm, and the developing embryo sac gradually absorbs this tissue.

Megasporogenesis. Meiosis is non-synchronous in the developing megaspore mother cells. Walls are not laid down after meiosis I and II, which results in the formation of a tetranucleate coenomegaspore (Fig. 20.18 A-D). The four nuclei later move to the peripheral cytoplasm surrounding a central vacuole.

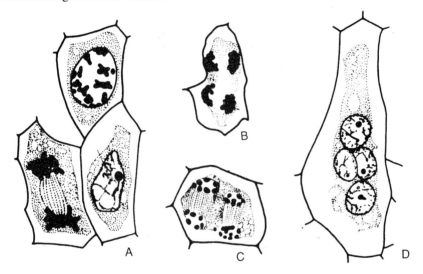

Fig. 20.18 A-D. *Gnetum africanum*, coenomegaspore. **A** Three megaspore mother cells, one in telophase I. **B, C** Meiosis II. **D** Four-nucleate coenomegaspore. (After Waterkeyn 1954)

Female Gametophyte. A number of coenomegaspores develop simultaneously, but only two or three reach the 16-nucleate stage. The nuclei undergo repeated divisions which are synchronous in the beginning. Mitoses proceed in a wave from the base upward, and different divisional stages can be seen in one and the same female gametophyte. In *G. ula* and *G. gnemon*, the divisions are intranuclear, the nuclear membrane disappears at the anaphase. Free-nuclear divisions continue till a definite number of nuclei have been produced, which varies in different species. In *G. leptostachyum* (Thompson 1916) and *G. africanum* (Waterkeyn 1954), there are 512 nuclei, in *G. gnemon* ca. 1000 (Sanwal 1962), and in *G. ula* over 1500 nuclei (V. Vasil 1959). As divisions continue, the gametophyte takes the form of an inverted flask (Fig. 20.19 A, B). The upper part is wide and contains the vacuole, and in the narrow lower part the nuclei aggregate densely. The female gametophyte grows towards the chalazal end.

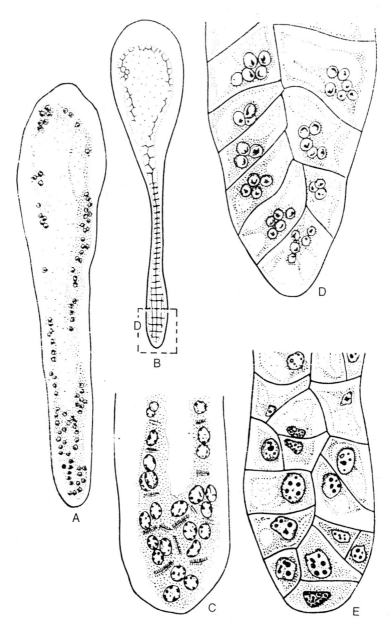

Fig. 20.19 A-E. *Gnetum ula.* **A, B** Longisection female gametophyte. **A** Free-nuclear gametophyte. **B** Cell formation in the lower region of gametophyte, upper portion shows cells and free nuclei. **C-E** Lower region of gametophyte. **C, D** Formation of multinucleate cells. **D** Inset marked *D* in **B**. **E** Uninucleate cells resulting from fusion of nuclei. (After V. Vasil 1959)

Walls are laid down by free cell formation, which begins at the lower end and proceeds upward (Fig. 20.19 B, C). In *G. gnemon*, it is initiated simulataneously at fertilization or immediately after. In *G. ula* and *G. africanum*, the cytoplasm in the upper swollen portion is partitioned before fertilization. It is not regular, and results in groups of either uninucleate or multinucleate cells and packets of free nuclei. In the lower portion, the cells are multinucleate in the beginning, later the nuclei fuse and become polyploid (Fig. 20.19 C-E). These cells divide, the tissue extends upward and fills the micropylar part of the gametophyte, which contains the zygote.

In *Gnetum*, the archegonia are absent. In *G. ula* (V. Vasil 1959), when a pollen tube enters the free-nuclear micropylar region of the gametophyte, one or more group/s of adjoining cells becomes densely cytoplasmic. Only one (rarely two) cells from each group function as the egg. The lower end of gametophyte becomes partly cellular. According to Swamy (1973), the egg in *G. ula* is a uninucleate or coenocytic cell (Fig. 20.20 C). The functional female nuclei are large, and in prophase. In *G. africanum* also certain large and densely cytoplasmic cells in the micropylar region are potential egg cells (Fig. 20.20 A; Waterkeyn 1954).

In *G. gnemon*, one or more nuclei in the vicinity of the pollen tube become conspicuous by their larger size (Fig. 20.20 B). They show striated cytoplasm around them, which later becomes delimited from the rest of the cytoplasm (Sanwal 1962).

In the male cones, the female gametophyte (in the ovules) does not develop beyond the megaspore mother cell, or when the first meiotic division is completed (*G. ula*) or has 64 nuclei (*G. gnemon*). Rarely, only some of the ovules are functional, develop normally and ripen into seeds (*G. gnemon*, *G. africanum*).

Pollination

In *G. ula* and *G. gnemon*, a drop of sugary fluid exudes from the tip of the micropylar tube and catches pollen grains. The slimy pollination drop has appreciable quantities of reducing sugars in *G. gnemon* (Pijl 1953). Surface evaporation causes the contraction of the column of fluid, which brings about the withdrawal of pollen grains down to the nucellus (Lotsy 1899).

Anemophily is probably concerned in the transfer of pollen to the female cone, though entomophily may also be involved. The male inflorescences emit a sweetish odour in the morning. The anthers open between 0.700 and 11.00 h. At about the same time, or earlier, nectar drops are present on fertile and sterile (in the male inflorescences) female flowers. All these characters are suggestive of insect pollination (Pijl 1953).

Kato et al. (1995) studied pollination biology in two *Gnetum* spp. in a lowland mixed dipterocarp forest in Sarawak, Malaysia. A dioecious (shrub species), *G. gnemon* var. *tenerum*, flowered in the evening. Both male and female strobili emitted a putrid odour, that from male strobilus being stronger than that from the female. Pollination droplets are secreted in the evening from ovules on female strobili and from sterile ovules on male strobili.

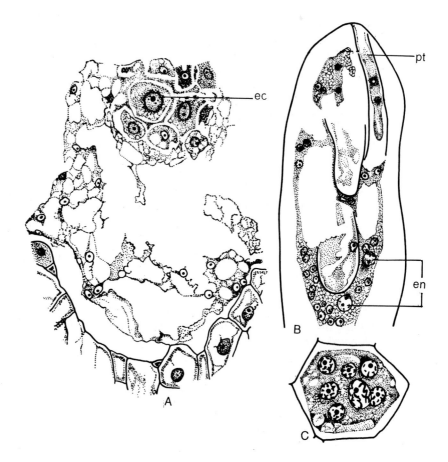

Fig. 20.20 A-C. **A, B** Longisection upper portion of female gametophyte. **A** *Gnetum africanum*, group of cells at the upper end are (potential) eggs (*ec*). **B** *G. gnemon*, two large egg nuclei (*en*) with radiating cytoplasm around them near the pollen tube (*pt*). **C** *G. ula*, multinucleate cell from upper part of gametophyte, the egg nucleus (*en*) is larger than the others. (**A** After Waterkeyn 1954, **B** after Sanwal 1962, **C** after Swamy 1973)

Withdrawal of droplets after pollination has not been observed. The droplets contain sugar at a concentration of 3 to 13%. The sugar concentration is affected by the relative humidity (usually 95% in the evening in the forest floor habitat) of the surrounding air. A very small amount of nectar is also secreted on the outer covers of the microsporangiophores on male strobili. These strobili are visited by nectar-seeking moths of the Pyralidae and Geometridae. The pollen has been observed sticking on the proboscides and antennae of these moths.

G. cuspidatum, a dioecious climber, has cauline strobili on woody stem near the forest floor. Its male strobili (which lack sterile ovules) emit a fungus-like odour (different from that of *G. gnemon* var. *tenerum*) in the evening. Nectar is secreted on collars that subtend flower rings early in the evening (between 17.00 and 18.20 h) with sugar concentration 6 to 10%. The strobili are visited by small flies, Lauxaniidae (Diptera), the pollen sticking to their antenna and legs. In female strobili, the pollination droplet is secreted on the tip of the ovules between 17.30 and 18.00 h and ceases by 21.00 h. The mean sugar concentration of the pollination droplet is 14.7%. During flowering, the female strobili also emit an odour similar to the male strobili.

Evaporation of the exposed droplet/nectar of these *Gnetum* "flowers" seemed to be minimized by nocturnal flowering near the tropical rain forest floor [which, according to Kato et al. (1995), could be a reason why *Gnetum* survived mainly in the tropical rain forests]. The lack of showy "petals" (in gymnosperm) is compensated for by odour.

The two *Gnetum* spp. flower in the understorey of rain forests where the humid habitat is not suitable for anemophily. Therefore the odour and secretion of droplets and/or nectar on exposed parts attract nocturnal pollinators. The sugar concentration is higher than that of anemophilous conifers (*Pinus* 1.2%; Mcwilliam 1958) but is lower than in entomophilous *Ephedra* (72-80%; see Chap. 18). These micropylar secretions also serve to draw pollen into the inner region of the ovule, where pollen germination takes place. According to Kato et al. (1995), the nocturnal pollination system of these *Gnetum* spp. has been derived from unspecialized entomophily, which is presumed to be an original pollination system of the early Gnetales and early angiosperms (Lloyd and Wells 1992). A study of the composition of carbohydrates in pollination droplets and nectars of these *Gnetum* spp. will help in the understanding of evolution of reproductive systems in gymnosperms (Chesnoy 1993).

The ovules are pollinated when the female gametophyte is still free-nuclear.

The pollen grains germinate in the shallow pollen chamber on the nucellar tip (Fig. 20.13 G), and occasionally in the micropylar tube (see P. Maheshwari and V. Vasil 1961a). Germ pores are absent on the exine. During germination, the exine is thrown off, and the intine, nearest to the tube nucleus, grows into a tube (Fig. 20.13 H). The pollen grains contain starch grains. In *G. ula* and *G. gnemon* the pollen tube grows into the nucellus through intercellular spaces. The nucellar cells are full of starch, which nourishes the pollen tubes. The tube nucleus is the first to enter the pollen tube, the body cell moves in after the pollen tube has traversed nearly half the length of the nucellus. The prothallial cell remains in the grain and degenerates in situ (Fig. 20.13 I). The body cell divides to form two male gametes (Fig. 20.13 J-L).

Fertilization

All the pollen tubes do not reach the female gametophyte at the same time, nor do the eggs differentiate simultaneously. As soon as one of the eggs is fertilized, wall formation begins in the upper part of the gametophyte. The latter becomes nearly cellular by the time the second or third egg differentiates.

There are very few observations on fertilization. In *G. ula* the pollen tube lies close to one of the groups of densely cytoplasmic cells present in the upper part of the gametophyte. Only the male nucleus enters the egg; its sheath is cast off outside the egg cell. Both male cells from a pollen tube can function if two egg cells are present close to the pollen tube.

The observations of Swamy (1973) in *G. ula* are somewhat different. The two small male nuclei are discharged into the cell containing the egg nucleus. The functional male nucleus enlarges slightly, comes close to the egg nucleus and enters into prophase (Fig. 20.21 A, B). The second male nucleus stays in the peripheral zone of cytoplasm, and disintegrates. As the male and female nuclei come in to contact, their chromatin becomes more

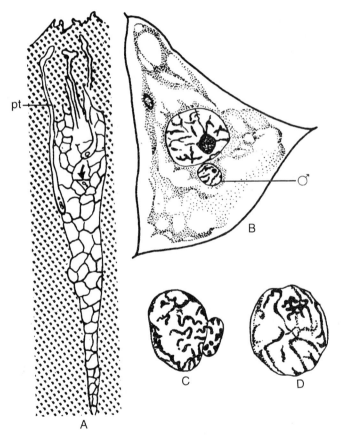

Fig. 20.21 A-D. *Gnetum ula*. **A** Longisection upper part of nucellus with a cellular female gametophyte and pollen tubes (*pt*). **B** Egg cell from **A** (arrow), a functional male nucleus (♂) is in contact with the egg nucleus while the second male nucleus has degenerated. **C, D** Integration of chromosomes of male and female nuclei. (After Swamy 1973)

intense. In the male nucleus the chromatin assumes the form of a short, somewhat curved and stumpy rod-like structure (Fig. 20.21 C). The nucleoli disappear from both the nuclei, and the two groups of chromosomes eventually become indistinguishable (Fig. 20.21 D).

In *G. gnemon*, the two male nuclei from a pollen tube may fertilize two adjacent eggs, as the zygotes often appear in pairs at the tip of the pollen tube. The zygote nucleus is large and hyaline, and has its own cytoplasm and limiting membrane. It is surrounded by many nuclei embedded in dense cytoplasm (Sanwal 1962).

The nature and behaviour of the gametes during fertilization in *G. gnemon* has been studied by Carmichael and Friedman (1995). The pollen tube discharges two male gametes which fuse with nearby undifferentiated female nuclei, within the coenocytic female gametophyte, and form two zygotes. According to Carmichael and Friedman (1995), this is an expression of a rudimentary pattern of double fertilization. The process of fertilization has been studied with light and fluorescence microscopy and the DNA content of various nuclei involved quantified with 4', 6-diamidino-2-phenylindole microspectrofluorometry. It has been observed that both male and female nuclei pass through the S-phase (DNA synthesis phase) of the cell cycle, and double their DNA content from 1C to 2C prior to fertilization. Each of the two zygotes (in association with a pollen tube) contains 4C DNA. Fertilization is therefore dependent on attaining a precise stage within the cell cycle. This reproductive cell cycle pattern is defined as G2 karyogamy (Carmichael and Friedman 1995).

Double fertilization has been reported in *Ephedra* spp. [Two male gametes are released into the archegonium, one fuses with the egg to form the zygote, and the other fuses with the ventral canal nucleus (see Chap. 18)]. In angiosperms the zygote and the primary endosperm nucleus are formed as a result of double fertilization. It has been proposed that double fertilization evolved before the origin of angiosperms. The original manifestation of double fertilization in seed plants may have led to the formation of two embryos (Friedman 1992b, 1994, 1995).

Some angiosperms have diploid endosperm, e.g. *Butomopsis lanceolata* (a primitive angiosperm) (Johri 1936). Of the two male gametes from a pollen tube, one gamete fuses with the egg and the other with the upper polar nucleus (lower polar nucleus is not formed). Thus, the zygote and the primary endosperm nucleus are genetically identical but their products are quite different. Zygote gives rise to the embryo and the primary endosperm nucleus to endosperm—both diploid. There is no such parallel in the gymnosperms.

The term "double fertilization" should be restricted to the fusion of one male gamete with the egg and the other with the secondary nucleus (fused polar nuclei) and not to their products.

Embryogeny

Various investigators report both free-nuclear and nuclear division followed by a wall in the zygote. The details of embryonal development are not very clear. According to P. Maheshwari and V. Vasil (1961a), this confusion is partly due to variation in different species of the genus. Several zygotes are usually formed in the female gametophyte.

In *G. africanum*, successive division in a zygote and its derivatives results in a row of cells (Fig. 20.22 A). Each of these cells gives rise to a suspensor tube devoid of any cross-walls (Waterkeyn 1954).

In *G. gnemon*, the number of zygotes varies from two to four, rarely six. According to Sanwal (1962), the zygote may directly give out a small tubular projection on one side (as also reported by Lotsy 1899), or may divide (Fig. 20.22 B, C) into two cells (as reported by Thompson 1916). Each of these cells may give out a tube or only one of them may "germinate". These tubes have been variously termed proembryonal tubes, embryonal tubes, embryonal suspensor tubes and primary suspensor tubes. Generally, a single tube is given out, rarely two or three (Fig. 20.22 D, E). The zygote nucleus passes into one of the tubes, the other tubes without the nucleus degenerate. Sometimes, two zygotes may "germinate" while they are still in contact (Fig. 20.22 F).

In *G. ula*, the division of the zygote (Fig. 20.22 G) is followed by a wall (V. Vasil 1959). The two daughter cells elongate to form suspensor tubes (Fig. 20.22 H). According to Swamy (1973), the zygote in *G. ula* enlarges and the early divisions are free-nuclear. After the four-nucleate stage, two short processes appear at one pole of the proembryo and grow towards the chalaza. The tubular processes branch, and at the 8- and 16-nucleate stages the nuclei migrate into the tubes. this phenomenon continues until a large number of uninucleate primary suspensor tubes are formed, and all the tubes penetrate the core of the endosperm (female gametophyte) in the chalazal direction.

In *G. gnemon*, the thin-walled embryonal suspensor tube grows downward, while the developing endosperm tissue extends upward. The tubes grow through the intercellular spaces of the endosperm, and become septate to form long uninucleate cells. Above or below a septum, a small protuberance appears, which later elongates into a tubular structure containing a nucleus. Rarely, a branch is given out without any such disposition. The suspensor tubes grow downward into the endosperm and tend to aggregate in the centre, where the cells break down to form a conspicuous cavity. The development of embryo is initiated in only a few suspensor tubes, and a small cell is cut off at the tip. This cell divides transversely and then longitudinally and a quartet is produced. Further divisions result in an embryonal mass, and a long secondary suspensor pushes it into the endosperm. The mature embryo shows two cotyledons and a prominent feeder.

In *G. ula* the suspensor tube becomes coiled in 6 to 7 month-old

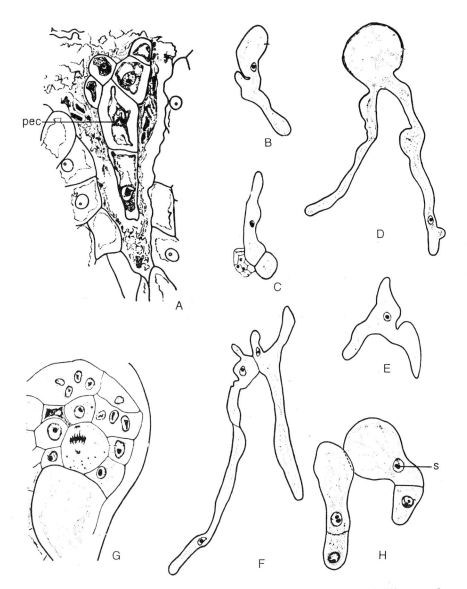

Fig. 20.22 A-H. **A** *Gnetum africanum*. **B-F** *G. gnemon*. **G, H** *G. ula*. **A** Filament of proembryonal cells (*pec*). **B, D, E** Zygote with one, two and three proembryonal tubes, respectively. **C** Two-celled proembryo. **F** Two proembryonal tubes. **G** Division of zygote. **H** Two-celled proembryos, the upper cell elongates and forms suspensor (*S*) tube. (**A** After Waterkeyn 1954, **B-F** after Sanwal 1962, **G, H** after V. Vasil 1959)

proembryos, and ceases to branch. The nucleus divides to form a larger nucleus with poor chromatin, and a smaller one embedded in the dense cytoplasm at the tip (Fig. 20.23 A-C). The latter is cut off from the rest of the tube by a thin wall and forms a pyriform cell, the so-called 'peculiar' cell (Fig. 20.23 C, D). In some tubes it may elongate parallel to the tube and

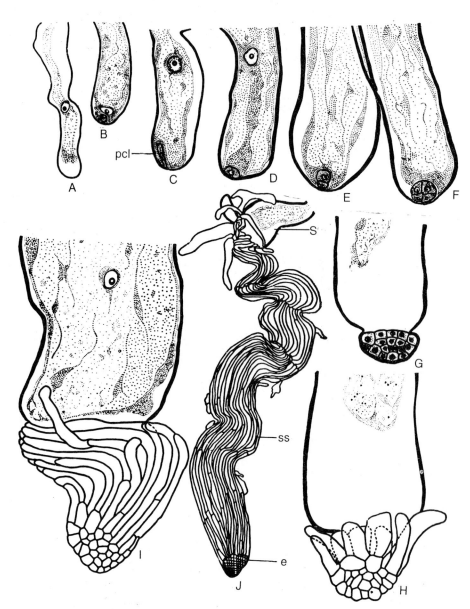

Fig. 20.23 A-J. *Gnetum ula.* **A-D** Formation of peculiar cell (*pcl*). **E-I** Development of embryonal tissue and suspensor. **H, I** Formation of secondary suspensors. **J** Portion of suspensor (*S*), multicelled secondary suspensor (*ss*), and embryonal cells (*e*) at the tip. (After V. Vasil 1959)

remain straight, slightly curved or twisted, and is suspected of degenerating. According to Swamy (1973), due to lack of precise information about its formation and variable shape, this cell has been termed peculiar although there is nothing peculiar in its ontogeny or structure. The pyriform cell ultimately gives rise to mature embryo (V. Vasil 1959, Swamy 1973). Morphologically, it is the embryonal cell. The seed is shed at this stage, and further development takes place on the ground.

The pyriform cell rounds up and divides to produce eight cells (Fig. 20.23 D-F). Further divisions are irregular, and a mass of cells is formed. The cells towards the suspensor tube elongate, divide further and form secondary suspensors (Fig. 20.23 G-I). The cells at the lower end remain compact and densely cytoplasmic, and contribute to the embryo proper (Fig. 20.23 I, J). Gradually, the massive secondary suspensor tubes cease to function and degenerate.

The embryonal mass grows in size, and two cotyledons differentiate which enclose the stem tip. The root apex is initiated at the same time at the opposite end. From the side of the hypocotyl a lateral hump or feeder arises and outgrows the hypocotyl (Fig. 20.24 A-C). The bulk of the feeder results from the cortical cells of the hypocotyl (Sanwal 1962). It has a well-developed vascular supply, is the most prominent part of the embryo, and is longer than the hypocotyl at the mature embryo stage (Fig. 20.24 C). Laticiferous ducts and stomata occur in a mature embryo.

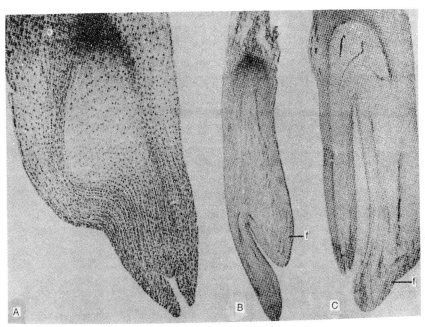

Fig. 20.24 A-C. *Gnetum ula.* **A-C** Longisection of embryo shows initiation, elongation and vascularization of feeder (*f*). (After V. Vasil 1959)

Polyembryony is a regular feature. Many embryos arise by cleavage of the embryonal mass and /or budding of the primary and secondary suspensors. However, only one embryo matures in a seed.

The bulk of the endosperm develops after fertilization, though cell formation may commence before fertilization. In *G. gnemon*, the female gametophyte is free-nuclear at the time of differentiation of the eggs. The lower part becomes cellular when the zygotes appear in the micropylar part (Sanwal 1962). Wall formation is initiated either simultaneously with fertilization, or about that time. In *G. ula*, polyploid cells are produced at the chalazal end of the gametophyte before fertilization, but only uninucleate cells are produced after fertilization (Swamy 1973). With further growth, the endosperm broadens at the expense of adjoining nucellar cells which are consumed laterally as well as at the base. In *G. gnemon* (Sanwal 1962) and *G. africanum* (Waterkeyn 1954) the meristematic activity in the gametophyte is limited to the axial region and a few outer layers, especially at the two poles. Thus, the main mass of endosperm is formed and starch appears first in the outer cells at the lower end, later at the micropylar end, lastly in the inner cells. In older seeds, the cells become gorged with starch grains and may mask the nuclei. Oil droplets have also been detected; in later stages the cell walls show many simple pits (Fig. 20.25 A-D).

Seed

In most species the mature seed is oval and green to red. The endosperm forms the bulk of the seed, and is surrounded by three envelopes. The nucellus is consumed except for a thin strip at the apex.

During the development of (ovule and) seed, the level of insertion of the inner envelope shifts (appreciably) distally. This is due to the development of an ovular tissue between the inner and middle envelopes, accompanied by an intercalary growth in the entire developing seed. The middle and outer envelopes show no such shift in their levels of insertion. A special term, endochalazal, was introduced (Bouman 1984) for such an ovular structure.

Rodin and Kapil (1969) studied the comparative anatomy of the seed coat of *G. gnemon*, *G. ula*, *G. montanum* f. *parvifolium* and *G. neglectum*. The seed coat consists of three layers; (a) outer sarcotesta, (b) middle sclerotesta and (c) inner endotesta (Fig. 20.26). The sarcotesta is free from the base up to apex, and is green and succulent. It is composed of a heavily cutinized epidermis and homogeneous parenchymatous cells. Diverse types of sclereids with lignified walls, and numerous branched fibers and laticifers also occur. In *G. ula*, *G. montanum* and *G. neglectum*, small astroscleriods or brachysclerids occur just below the outer epidermis. Additional astrosclereids are scattered in the deeper tissues. Haplocheilic stomata, as in the leaves, occasionally develop on the outer epidermis.

The sclerotesta forms the protective layer of the seed. It has numerous sclereids of varying shape and may sometimes extend as a basal plate. Depending on the species, at maturity the middle layer may be nearly free from the outer layer, or may be partially or completely fused with it.

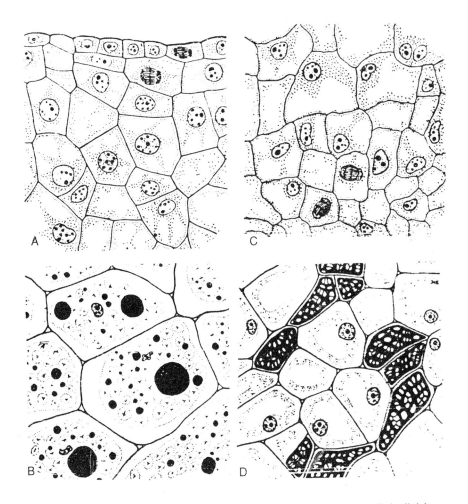

Fig. 20.25 A-D. *Gnetum gnemon.* **A** Endosperm from lower end shows cells in division. **B** Older endosperm shows proteinaceous globules in addition to starch. **C** Endosperm axial region with cells in division. **D** Mature endosperm cells have pitted walls and store starch. (After P. Maheshwari and V. Vasil 1961a)

The endotesta is parenchymatous and membraneous. In a few *Gnetum* spp. it has elongated sclereids (Rodin and Kapil 1969).

All the layers of the seed coat are vascularized (Fig. 20.26). A single ring of vascular bundles enters the base of the seed; each bundle bifurcates, giving rise to an outer series which supplies the outer envelope. The inner series bifurcates again to supply the middle envelope and the peripheral ovular/chalazal tissue which surrounds the endosperm, and extend to the inner envelope. In *G. gnemon*, on an average, 21 bundles are present in the basal ring, while the outer, middle and inner rings contain approximately 62, 18 and 18 bundles, respectively (Rodin and Kapil 1969).

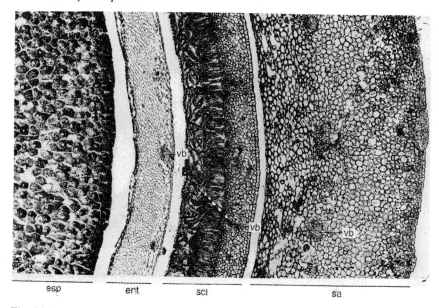

Fig. 20.26. *Gnetum gnemon*, seed coat. Transection of seed shows endosperm (*esp*), endotesta (*ent*), sclerotesta (*scl*), sarcotesta (*sa*), and vascular bundle (*vb*). (After Rodin and Kapil 1969)

Germination. In several *Gnetum* spp., the seed does not germinate immediately on falling to the ground as it is immature and a considerable part of its development takes place on the ground (cf. *Cycas, Ginkgo*). The root with the root cap emerges form the seed first, bends downward, enters the soil, and gives rise to the taproot system (Fig. 20.27 A-C). Next, the upper part of the hypocotyl elongates, pulls the cotyledons out of the seed (Fig. 20.27 D, E) and straightens to bring up the two (occasionally three, Fig. 20.27 K) cotyledons and stem tip in an upright position (Fig. 20.27 F). The cotyledons expand, turn green and resemble a foliage leaf in shape and venation (Fig. 20.27 G-J). The stem tip expands and produces the first pair of plumular leaves, which are opposite and decussate to the cotyledonary leaves (Fig. 20.27 L). The feeder remains inside the seed and occupies the entire space previously filled by the hypocotyl. The endosperm tissue adjacent to the feeder has a lighter colour and appears to be corroded by it. The feeder continues to function as an absorptive organ till a very late stage, and can be seen even after the second and third pair of leaves have developed on the seedling (P. Maheshwari and V. Vasil 1961a).

Chromosome Number

The haploid chromosome number in *G. ula* is n = 22. The bivalents show slight intergradational differences. At diakinesis the configurations observed are in the form of X, L, Y, O and a figure 8; the number of chiasmata in a bivalent varies from one to three. Meiosis is normal. Occasional

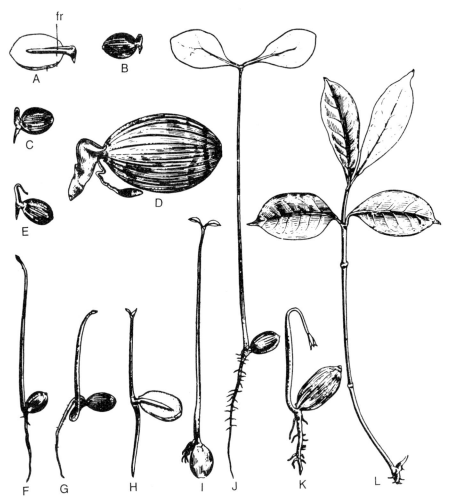

Fig. 20.27 A-L.   A-J, L *Gnetum gnemon,* seed germination. **A-E** Germination of seed.
A Half-endosperm removed after emergence of root. *fr* feeder. **F-K** Development of
seedling. **K** *G. ula,* seedling with three cotyledonary leaves. **L** seedling shows vegetative
leaves. (After P. Maheshwari and V. Vasil 1961a)

diploid grains (inferred from their larger size and form) are observed
(Mehra 1988).

In *G. gnemon,* Fagerlind (1941) noticed n = 22 chromosomes at metaphase
of a single sporocyte (Mehra 1988). Of these, 11 chromosomes are longer
than the rest. During metaphase, the smaller chromosomes are mostly grouped
in the centre of the plate, while the longer chromosomes are arranged
around in a radiating fashion (P. Maheshwari and V. Vasil 1961a). In
G. africanum, the haploid number is between 20 and 25 (Waterkeyn 1959).

Temporal considerations

The life cycle of *G. ula* (growing wild in Peninsular India) takes nearly 18 months from the time of initiation of the flowers to the germination of seeds. The seeds germinate ca. 11–12 months after they are shed from the plant (V. Vasil 1959). In *G. gnemon* the seeds are shed in June/July (in Assam, Eastern India), a few cells have been observed at the tip of a primary suspensor. It takes 2 months for the maturation of the embryo. The seeds germinate in September (Sanwal 1962).

21. Phylogenetic Considerations: Ephedra, Welwitschia and Gnetum

The Gnetopsida has been studied for over a century. The scanty fossil record, restricted to the Tertiary, does not indicate the origin and relationship of this enigmatic group. The discovery of *Ephedripites* pollen (related to modern Gnetopsida pollen, specially *Welwitschia*) in the lower Cretaceous (Trevisan 1980) is significant, as it suggests that at least some member of the group was present at the time of the origin of angiosperms. According to Stewart (1983), based on fossil record, a polyphyletic origin of angiosperms from gymnosperms appears more plausible than previously presumed. As in progymnospermopsida (gymnospermous wood and pteridophytic reproduction), the preangiospermopsida (proangiosperms) could have any one of a several combinations. Those gymnosperms, that combine a few/ some angiospermic characters and form a complex from which evolutionary lines lead to the flowering plants, should be investigated for the origin of angiosperms.

The Gnetopsida ovule has multiple nucellar envelopes, unlike other gymnosperms (which have a single integument only). These have been interpreted as integumentary structures homologous with the outer integument of an angiosperm (Stewart 1983).

Entomophilly in several spp. of *Ephedra, Gnetum* and in *Welwitschia* is rather an evolutionary progress from the primitive gymnospermous mode of anemophilly present in the majority of extinct and extant gymnosperms (*Cycadeoidea* is an exception, where beetles were partially responsible for pollination, Crepet 1972). It is a more economical and probably more effective and reliable form of pollination (Bino et al. 1984). The incidence of an apparently effective insect pollination in gymnosperms is of great interest.

Eames (1952) discusses the affinities of Gnetopsida in considerable detail, and concludes that *Ephedra* is nearer to cordaites and conifers than to *Welwitschia* and *Gnetum*. The ovule and microsporangia of *Ephedra* are appendicular (i.e. terminal on the lateral appendage of a fertile shoot), and not cauline (terminal on the shoot, as in *Welwitschia* and *Gnetum*). According to Eames (1952), this is an important morphological difference, indicating a wide phyletic gap between the three taxa.

Embryologically also, *Ephedra* differs from *Welwitschia* and *Gnetum* (see H. Singh 1978). The male gametophyte has a stalk cell. The female gametophyte is monosporic, and alveoli are formed during wall formation. The archegonia are well developed, there is a distinct and specialized type of proembryogeny and a feeder is absent in the mature embryo. *Welwitschia* and *Gnetum*, on the other hand, lack a stalk cell in the male gametophyte; development of the female gametophyte is tetrasporic, and alveoli are not formed during wall formation, archegonia are absent, the proembryogeny is very different and a feeder is formed in a mature embryo. Thus, *Ephedra* occupies a position isolated from the other two taxa, and may have had a completely different origin (Sporne 1965). *Welwitschia* is also completely isolated with respect to hereditary relationships (Von Willart 1985). Phylogenetically, it must be very old, as there are no recent forms known that relate to other taxonomic groups.

Gnetum has long held a key position in any discussion relating to the origin of angiosperms (see P. Maheshwari and V. Vasil 1961a). Besides the similarities it shares with other members of the group, it has features which show a proximity to the angiosperms.

Leaf.   The external appearance of the broad leaf with netted venation is strikingly angiospermous. The development of the stomata is of the mesoparacytic (Rubiaceous) type (Kausik 1974) which, according to Takhtajan (1969), appears to be the basic type in the evolution of flowering plants. Nautiyal et al. (1976) confirm that the stomata of *Gnetum* differ from those of other gymnosperms (possible exception Bennettitales) and resemble angiosperms such as some Magnoliales and *Casuarina*.

All the six main types of sclereids which are known in angiosperms are also reported in *Gnetum* (Sharma 1975). Outside the flowering plants, the occurrence of laticifers in *G. gnemon* is a rare example (Behnke and Herrmann 1978).

Shoot Apex.   The shoot apex of *Gnetum* is similar to that of angiosperms in the tunica and corpus organization. However, the central mother cell zone is a reminder of gymnosperms such as cycads and *Ginkgo* (Johnson 1950).

There are physiological similarities between *Gnetum* and angiosperms in the chlorophyll synthesis in dark-germinated seedlings (Laudi and Bonatti 1973).

Phloem.   The phloem in *Gnetum* has sieve cells, associated with ontogenetically unrelated parenchymatous cells (in angiosperms sieve tube and companion cells are derived from the same cambial initial and are successive cells in a given row). In *Austrobaileya scandens* (a recognized primitive angiosperm), a tier-by-tier study of the origin of sieve elements and parenchyma (companion) cells led Srivastava (1970) to conclude that true companion cells are absent. Instead, parenchyma cells and sieve elements arise independently from different initials, and are connected to each other by lateral sieve areas. Thus, the phloem of *Gnetum* also closely approaches the angiospermous taxa (see Paliwal and Behnke 1973).

Xylem. Studies on the vessel of *G. gnemon* and *G. montanum* (Muhammad and Sattler 1982) demonstrate a greater range of variation in the shape of pits and perforations than known so far. It consists of typical circular pits (of progymnosperms, pteridosperms and coniferopsids) and non-scalariform perforation plates to scalariform pitting (as in Magnolioideae). Bliss (1921) and MacDuffie (1921) demonstrated striking similarities between perforation plates of *Gnetum* and angiosperms such as *Paeonia, Cydonia* and *Vitis*. Most of the patterns characteristic of and comparable to angiosperms are present, such as: absence of torus in the pit, presence of vestured pits and compound gyres.

Gnetum thus combines both gymnospermous and angiospermous characters. A preliminary survey of 58 characters showed that *Gnetum* shares more than 60% characters with the angiosperms, and ca. 30% with gymnosperms (see Muhammad and Sattler 1982). This hypothesis, however, does not necessarily demonstrate an ancestral relationship to angiosperms, but the possibility that *Gnetum* may have been close to the ancestry of all, or at least some, of the taxa of angiosperms (Muhammad and Sattler 1982). According to P. Maheshwari (1950), it appears best to conclude that, while the angiosperms have probably passed through some such stages as are shown by *Gnetum*, there is no decisive evidence in favour of any close relationship.

22. In Vitro Experimental Studies

Experimental procedures provide excellent opportunity to study growth, development and differentiation of vegetative and reproductive tissues. The gymnosperms—a group of plants between the pteridophytes and angiosperms—have not yet been investigated as thoroughly as the angiosperms.

There are isolated reports of experimental embryology in gymnosperms dating back more than a century (see Norstog 1982). Such studies can be traced to the pioneering work of Professor Carl D. LaRue, University of Michigan, Ann Arbor (USA). He extended his concept of cellular totipotency and developmental potentials in plants to include all aspects of reproduction (LaRue 1954). Despite the easy availability of material, only a few gymnosperms have been investigated, possibly because they are extremely slow-growing plants and less amenable to cultural conditions.

Vegetative Tissues and Organs

Ball (1950) cultured tissues from young adventive shoots growing on the burls of *Sequoia sempervirens*, on diluted Knop's solution with 3% sucrose and 1 mg/l IAA. For the first time, he successfully maintained a continuous culture of coniferous tissue in vitro. Marginal meristems, and cambium-like meristems around groups of tracheids and mature parenchyma cells could be distinguished in the callus mass. The parenchymatous cells occasionally contained tannin. Ball (1955) reported the utilization of various sugars by the callus, the normal chlorophyllous tissue containing glucose, fructose and sucrose. When the callus is grown on any one of these three sugars, the tissue can produce the other two. Maximal growth is achieved on a medium containing sucrose. When either raffinose or galactose is supplied as the source of carbohydrates, sucrose, fructose, glucose, galactose and raffinose are detected in the callus. According to Ball (1955), the enzyme that yields galactose—first from the hydrolysis of excess raffinose—forms raffinose later by condensation of excess galactose and sucrose.

Reinert and White (1956) cultured normal and tumerous (characteristic of the species) tissues of *Picea glauca*, to understand the degree of malignancy of the cells and the biochemical characters of the tumours. The cambial region is excised from the tumorous and non-tumorous portions of tumour-bearing trees and also from normal trees. A complex nutrient medium consisting of White's minerals, 16 amino acids and amides, 8 vitamins and auxins was developed to raise the cultures. The tissues, grown on filter paper moistened with the above medium, turned brown and ceased to grow. This is due to a copper-containing enzyme (phenoloxidase) which appears to be especially active in traumatized tissue. Browning could be prevented by the addition of tyrosine at 40 mg/l, or phenoloxidase inhibitors like diethyldithiocarbamate or sodium ethylxanthate. The tissues continued to grow for some time; with tyrosine the growth was continuous. Additives like 2, 4-D, glutamine and IAA also enhanced growth. The behaviour of normal and tumorous tissues differed; the growth of the tumorous tissues was faster, but they were difficult to establish. The tumorous tissues require vitamin B_{12} while the normal cells do not.

White and Risser (1964) studied conditions for maximal growth of callus derived from tumour tissues of *Picea glauca*. Maximal growth was achieved when 20 mg of callus was inoculated on an agarized substrate (0.5%) at pH 5.5, in square bottles with screw caps. The cultures were maintained at 23°C either in darkness or in diffuse light, and transferred to fresh media every 2 weeks. Under these conditions, the mean increment per passage was about ninefold.

Risser and White (1964) formulated a less complex medium than suggested earlier by Reinert and White (1956). Sucrose at 5% is the most effective carbohydrate source; it can be replaced satisfactorily only by dextrose, while fructose, mannose and raffinose are only half as effective as sucrose. All the 18 amino acids in the medium can be replaced by 1-glutamine (250 mg or more/1). Amongst the auxins tested, the effective growth promoters are 10^{-7}M 2, 4-D, 10^{-5}M IAA and 10^{-3}M β-napthoxyacetic acid; 2, 4-D gives the best result. Adenine, KN, β-alanine, gibberellin and folic acid (in the presence of 2, 4-D) have no promotive effect.

The cells of *Picea glauca* (White 1967) grown in vitro exhibit patterned deposits in the form of strands, furrows or channels, in the cytoplasm, which may act as templates for cellulose and lignin deposition. The cytoplasm is extruded into characteristic papillae (in some cells), which condense to form distinctive extracellular patterns.

Borchert (1967) studied the physiological differences in tissues derived from various explants, or plants of different ages. As an initial step, he determined conditions necessary for in vitro culture of secondary phloem and cambium of *Cupressus lusitanica*. The explants proliferated on a medium

---

Abbreviations: *ABA* abscisic acid, *BA* benzyladenine, *BAP* benzylaminopurine, *BM* basal medium, *CH* casein hydrolysate, *CM* coconut milk, *CW* coconut water, 2, 4-*D* 2, 4-dichlorophenoxyacetic acid, *GA* gibberellic acid, *IAA* indoleacetic acid, *IBA* indolebutyric acid, *KN* kinetin, *MS* Murashige and Skoog (1962), *NAA* napthaleneacetic acid, *TIBA* tri-iodo-benzoic acid, *WB* White's medium, *YE* yeast extract.

containing Heller's minerals and sucrose. When vitamins or growth substances are added to this medium, there is little improvement in growth. However, when they are added to Heller's medium containing higher concentrations of mineral salts, growth is considerably stimulated.

Suspension culture technique is a very useful approach to many problems in cellular biology. Konar (1963b) studied the cell behaviour of callus of *Pinus gerardiana*, grown on Reinert's modified liquid medium. In suspension cultures, the tissues dissociate into single cells, which later divide to form cell aggregates.

Hasegawa et al. (1960) demonstrated the formation of lignin in the cultures of cambial tissue of *Pinus strobus,* using radioactive compounds. Tissues fed with ^{14}C glucose, acetic acid and shikimic acid, and the four fractions obtained (amino acid, acid fraction, phenol and sugar) were analyzed for radioactivity. When the tissues are fed with glucose-1-^{14}C, maximal activity occurs in the acid fraction which contains shikimic acid. The extent of conversion of glucose to shikimic acid is ca. three to five time more than that of acetic acid; the incorporation of acetic acid into shikimic acid is considerably lower. The authors conclude that shikimic acid and lignin are synthesized from glucose in the isolated cambium tissue. They assume that the aromatic nucleus of lignin is formed from sugar via shikimic acid.

Constabel (1965) reports that the extent of tannin production in the cultures of *Juniperus communis* depends on growth, the presence of precursors, and light. With increase in growth rate, there is a decrease in tannin content. When cinnamic, sinapic and ferulic acids are added to the medium to determine whether these compounds function as precursors of tannin, the level of tannin rises. However, these compounds stimulate growth at low concentrations, but have an inhibitory effect at high concentration. Cinnamic aid is definitely a precursor of tannin, whereas ferulic and sinapic acid also act as precursors. There is a significant difference in tannin content of the tissues kept under 14-h light against those kept in dark. In the dark, tissues show higher tannin production.

Al-Talib and Torrey (1959) cultured dormant terminal buds of *Pseudotsuga taxifolia*, and pointed out that isolated dormant terminal buds required oxygen, sucrose, glucose or fructose, and light for growth. They also studied the effect of auxins, gibberellins, kinetin and adenine sulphate on the growth and development of buds. Auxin $10^{-6}M$ had no stimulatory effect on bud development. At a higher concentration ($10^{-4}M$), the development of the leaf is generally retarded. The addition of GA (0.1–1.0 ml/1) in the medium restricts leaf expansion and elongation of the main axis and the buds collapse during later stages. Adenine sulphate reduces bud development and leaf expansion. KN (0.001–1.0 mg/l) has a stimulatory effect on callus growth. Excellent bud development and leaf expansion occur when a Seitz-filtered solution of urea is added to the medium at $10^{-1}M$. Supplements like CM, YE, CH, and watermelon juice are inhibitory to the development of the bud.

A comparative study of in vitro growth and developmental pattern of shoot apices of different gymnosperms would be rewarding (Biswas 1994).

Microspore and Male Gametophyte

In vitro pollen culture of gymnosperms has been successful in obtaining normal or near-normal development. The objectives are to overcome problems of dormancy, viability, storage and longevity. The first success in pollen culture of a gymnosperm resulted by using the hanging drop technique. According to Kuhlwein (1937), 2–20% (lower concentrations give better results) sugar solution is well suited for (in vitro) pollen culture of several gymnosperms. Branscheidt (1930) with the hanging drop method also cultured the pollen of *Taxus* and *Cupressus* in a medium supplemented with sucrose. The sperm nuclei formed in about 20 days. LaRue (1954) cultured pollen grains of *Zamia* on nutrient agar. The development proceeded through several stages until the penultimate division of the generative cell. Blepharoplasts were formed in the sperm mother cells, but not spermatozoids. The tubes remained active (in this stage much longer than under in vivo conditions) and branched.

De Luca and Sabato (1979) reported, the first in vitro spermatogenesis in *Encephalartos altensteinii*. The microsporangia (containing three-celled microspores) were cultured for 8 months. Within 1 week after inoculation, masses of pollen tubes emerged from the sporangia. In 6 months, sperm mother cell formed at the tip of pollen tube. On division, two flagellate sperms were produced, which, however, did not attain maturity.

Pollen of *Ephedra* germinates in a medium containing ovular extracts (Mehra 1938), and forms fully developed pollen tubes with (apparently) normal sperm mother cells. In vivo, the pollen tubes develop rapidly in ca. 10 h (in other gymnosperms ca. 10 months), and are non-haustorial. It may be presumed that specific nutritional requirements in vitro are not as demanding as in vivo.

Male cones of *Pinus roxburghii,* at the microspore mother cell stage, were cultured by Konar (1963c), in White's medium supplemented with nucleic acids. The pollen mother cells show normal development and give rise to mature pollen grains which germinate. The antheridial cell divides to form the stalk and body cell.

The first subculturable callus derived from pollen is that of *Ginkgo* (Tulecke 1953). The mature pollen is cultured at the shedding stage (Tulecke 1953, 1957, 1960). The gametophyte develops normally till the formation of flagellated spermatozoids, but motility has not been observed. The pollen tube forms the usual haustorial structure (see Chap. 12). Several abnormalities have been observed: formation of septate pollen tubes; divisions in the tube nucleus (forms large coenocytic vesicles, each containing nearly 50 nuclei) and, less frequently, divisions in the prothallial, stalk, body and (occasionally) even in the sperm cells to form a cluster of cells (Fig. 22.1). Such clusters

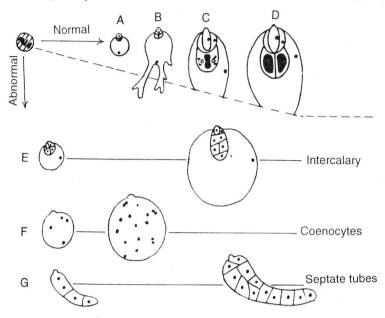

Fig. 22.1. *Ginkgo biloba.* Diagrammatic representation of normal and abnormal development of male gametophyte under in vitro conditions ; note embryo-like structures in E, G. (After Tulecke 1957)

from several pollen tubes give rise to a tissue of meristematic parenchymatous cells, which is white and friable, and separates readily into groups of adherent cells. This tissue is basically haploid and forms a subculturable callus. Tulecke and Nickell (1960) maintained this tissue (in vitro) continuously for a number of years, without cell differentiation. Tulecke et al. (1965) devised a setup, the Phytostat, for obtaining continuous suspension cultures of *Ginkgo biloba* pollen tissue.

Biochemical and physiological studies have been conducted to determine the differences in the free and protein amino acids, sugars, and non-volatile organic acids in pollen and in tissue derived from it. A strain of pollen tissue was isolated which requires arginine (1000 mg/l) for its growth (Tulecke 1960). Canavanine, an antimetabolite of arginine, inhibits the growth of the arginine-requiring strain. Ammonia could, to a certain extent, replace arginine. Similar subculturable callus has been obtained from the pollen of *Taxus* (Tulecke 1959), *Torreya* (Tulecke and Sehgal 1963), *Ephedra foliata* (Konar 1963d) and *Thuja* (Rao and Mehta 1969). This tissue is haploid and remains undifferentiated.

Razmologov (1973) observed that the generative cell of the pollen grain of *Cupressus* produces a callus-like tissue in vitro. Bonga and McInnis (1975) cultured micro-sporophylls of *Pinus resinosa*. The first division of the immature pollen produces equal-sized cells. Cold treatment results in more pollen with such cells, and an increase in callus initiation. When cold treatment is followed by centrifugation, callusing increases significantly.

However, none of the pollen (or microspore) cultures, or the resultant tissue, regenerates buds, roots or embryos. This may be due to a higher capacity of gymnospermous tissue for proliferation or that the correct stage of culture for induction of embryos has not been ascertained. According to Norstog (1982), the culture of microspores and microgametophytes, unlike angiosperms, do not produce the hoped-for clones of haploid tissue and embryos. However, the production of subculturable callus indicates that improved techniques may result in cloning and diploidization, as is now possible in some angiosperms.

The in vitro production of mature microgametophytes of cycads and *Ginkgo* provides much additional information about spermatogenesis. There is scanty information, especially at the ultrastructural level, concerning the development of microgametophyte and sperm cells of other gymnosperms. Rohr (1973b) made an ultrastructural study of sperm cells formed within the cultured pollen tubes of *Taxus baccata*. In some of the electron micrographs, ultrastructural aspects of well-defined sperm cells have been observed.

Female Gametophyte

The production of plants through vegetative growth of prothallial tissues (apogamy) is of considerable scientific interest. It demonstrates the ability of cells to manifest almost the entire range of morphogenetic potential of the species (totipotency). It is valuable to geneticists, horticulturists and foresters as a means of rapid vegetative propagation. The haploid female gametophyte can be used as an alternate source (pollen grains are readily available but with numerous genotypes) for the production of haploids and, being massive, is more amenable than the male gametophyte. This may be due to its larger size, sterile location, and a distinct meristematic phase.

Du Chartre (1888) described root formation by the female gametophyte of *Cycas*, which is the earliest record of regeneration by megametophyte of a gymnosperm. Coulter and Chrysler (1904) made a similar observation in *Zamia*.

LaRue (1948, 1954) successfully maintained the female gametophyte (excised at about the time of fertilization) of *Zamia floridana* and later *Cycas revoluta* on simple media, and obtained regeneration. Both root and shoots in *Zamia* (Fig. 22.2) and *Cycas*, as well as adventive embryos (in *Zamia*, see below), have been reported. The leaves on the shoots are small and closely resemble those of the seedling.

Norstog (1965b, 1967b) and Norstog and Rhamstein (1967) made a detailed study of the initiation of callus, adventive embryos, roots and leaves from the gametophytic tissue of *Zamia integrifolia* and *Cycas circinalis*. The megagametophyte of *Zamia* responded variously to additions, deletions and changes in the concentration of auxins, cytokinins and amino acids (Figs. 22.3; 22.4). Without these additives, only callusing occurred. The

Fig. 22.2. *Zamia floridana.* Regeneration of roots and shoots in cultured megagametophyte. (After LaRue 1948).

addition of 2, 4-D and KN induces callusing followed by rooting. The addition of glutamine and asparagine to the auxin-cytokinin medium results in a considerable increase (from 48 to 92%) in the number of megagametophytes which exhibit morphogenetic response. In addition, there is regeneration of shoots as well as roots (Fig. 22.3). The organization of the gametophyte influences the orientation of regenerated roots and shoots (Norstog 1965b). These organs tend to form at the archegonial end of gametophytes on media without an auxin. When auxin and KN are both present in the media, this polarizing effect is largely offset and roots and shoots form in other regions as well. The presence of an organized mass of cells (megagametophyte) appears to be essential in the direct regeneration

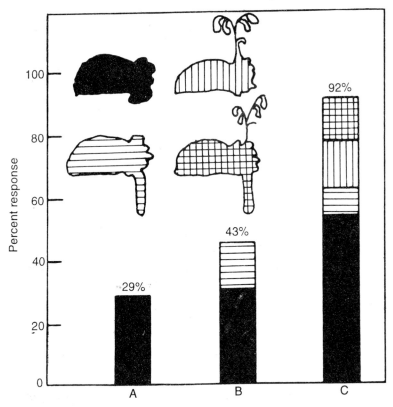

Fig. 22.3. *Zamia integrifolia*, megagametophyte cultured on three different media. **A** Minimal basal mineral medium. **B** Basal medium + 2, 4-D + KN. **C** Basal medium + 2, 4-D + KN + glutamine + asparagine. *Black* callus. *Horizontal lines* callus + root. *Vertical lines* callus + bud. *Cross-hatched* callus + root + bud. (After Norstog 1967b)

of roots and shoots. The roots originate from peripheral areas of meristematic cells, and retain the haploid chromosome number.

There is a different response on a medium with different concentrations of auxin and cytokinin (Norstog and Rhamsteine 1967). The megagametophytes are cultured on a high-auxin medium (10 mg/l 2, 4-D) in shake cultures. When the resulting callus is transferred to a medium with lower levels of auxin and cytokinin (1 mg/l each of 2,4-D and KN), it could be maintained and subcultured without any organ regeneration. When the cells are transferred to a medium without auxin or cytokinin, embryoids (embryo-like structures; see H. Singh 1978) differentiate. This sequence repeatedly gives the same result. Similarly, the megagametophytic tissue of *Cycas circinalis* produces subculturable callus which does not form roots, shoots or embryoids.

The megagametophytes from mature seeds of *Ceratozamia mexicana*, *Cycas revoluta* and *Encephalartos umbeluziensis* were cultured on modified White's media (according to DeLuca et al. 1979) with 2% sucrose—with

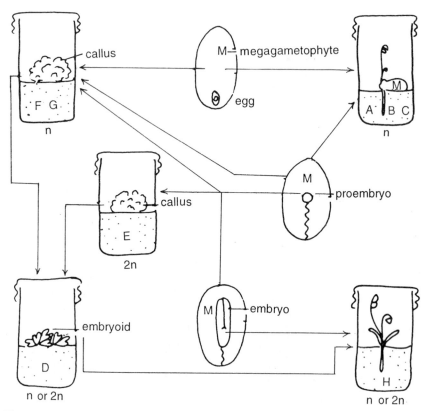

Fig. 22.4. *Zamia integrifolia*. Megagametophytes and embryos grown on eight different media: **A** Basal mineral medium without 2, 4-D and KN; **B** Basal mineral medium + 2, 4-D + KN (1.0 mg/1 each); **C** Basal mineral medium + 2, 4-D + KN (1.0 mg/l each) + asparagine + glutamine (400 mg/l each); **D** Medium containing alanine (100 mg/l) + glutamine + asparagine (400 mg/l each) + adenine (10 mg/l) + NH$_4$-malate (100 mg/l); **E** Medium same as in D + 2, 4-D + KN (1.0 mg/l each); **F** Modified Murashige and Skoog medium + alanine (100 mg/l) + glutamine (400 mg/l) + 2, 4-D + KN (1.0 mg/l each); **G** Medium same as E +2, 4-D (10 mg/l); **H** Basal medium + sucrose, vitamins not added. All media from A-G contain 2% sucrose and vitamins. (After Norstog 1967b)

or without 2, 4-D and KN. Callusing occurred in all the taxa but regeneration only in *Ceratozamia* and *Cycas*. The callus of *Ceratozamia* (on BM) showed differentiation of only adventive embryos, but later circinate leaves developed. On a medium enriched with amino acids, embryos and haploid roots were also formed. The *Cycas* megagametophytes, when callused on an enriched medium, produce a large number of pseudobulbils (interpreted as root primordia by the authors). Coralloid roots are formed when the megagametophytes are cultured on a medium containing 2, 4-D and KN. These roots lacked the endophytic blue-green alga but are anatomically comparable to in vivo roots (DeLuca and Sabato 1980).

Tulecke (1964, 1967) reported that *Ginkgo* megagametophyte produces a subculturable callus but, unlike *Zamia*, there is no regeneration of roots, shoots and embryoids. Rohr (1977) cultured the female gametophyte of *Ginkgo*, 4-6 months after fertilization. The superficial tissues turned deep green and produced a variety of outgrowths. A tissue isolated from the gametophyte and subcultured continues to grow even after several transfers on the same medium (see Norstog 1982).

Borchert (1968) reports callus formation from the megagametophytes of *Pinus lambertiana, P. resinosa, P. nigra* and *P. mugo*. Only non-subculturable proliferations of callus is obtained from the female gametophytes of *P. resinosa* (Bonga and Fowler 1970). After 5-7 weeks of culture, 10% megagametophytes of *P. nigra* and about 14% of *P. mugo* produce a bright green callus capable of subculture (Bonga 1974). In *P. nigra*, the callus comprises haploid cells, and in *P. mugo*, predominantly haploid with a few diploid cells.

Konar and M.N. Singh (1979) cultured the female gametophyte excised from mature seeds of *Ephedra foliata*. Initiation of callus (Fig. 22.5 A, B) occurs on BM (MS + 2% sucrose + 10% CM) + 2, 4-D (2 mg/l); stock cultures were maintained on a lower concentration of 2,4–D (1 mg/l). The subcultures on BM + 2 mg/l KN (auxin-free medium) result in the formation of multiple shoots and roots (Fig. 22.5 C). Transfer of callus with shoot buds and roots on KN-free medium produced better growth of shoots only (Fig. 22.5 D). According to Konar and M.N. Singh (1979), 2, 4-D stimulates callus formation. However, extensive proliferation and formation of subculturable callus is obtained only when coconut milk is added. In the absence of CM, the tissue neither regenerates nor yields a subculturable callus.

M.N. Singh et al. (1981) excised female gametophytes from ovules at the archegonial stage and cultured them on BM (MS + 2% sucrose + 10% CM). Growth and morphogenetic responses of the explants to auxins (at different concentrations) and their interactions with cytokinins has been studied. BM + 2, 4-D (2 mg/l) induces profuse callusing which subsequently produces roots. NAA (4 mg/l) is optimal for callus growth and rooting. Combinations of 2, 4-D and KN are more effective in inducing roots and shoot buds, than a combination of 2, 4-D and BAP. The addition of BAP (0.05 mg/l) to a medium containing optimal concentration of NAA (4 mg/l) results in the formation of a large number of roots. A high concentration of BAP (8 mg/l) stimulates shoot bud formation. The root tips and callus have haploid number of chromosomes (n = 7).

Bhatnagar and M.N. Singh (1984) cultured female gametophyte of *Ephedra foliata* for inducing haploids. The age of the explant is important to determine patterns of regeneration. Explants cultured before archegonia formation do not respond.

Longitudinal halves of the female gametophyte have been used as explants. Transverse halves (to ascertain the role of the archegonium on morphogenetic responses) when inoculated on the BM (MS+2% sucrose+10% CM

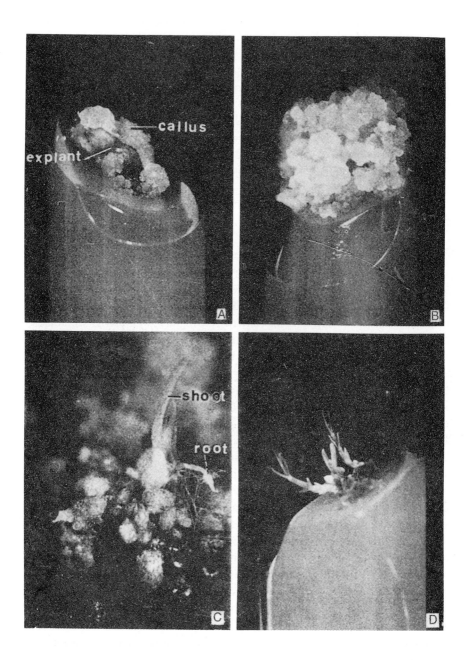

Fig. 22.5 A-D. *Ephedra foliata,* female gametophyte. **A, B** 2- and 4- week-old culture on BM + 2 mg/l 2, 4-D. **C** Callus with shoot and root on BM + 2 mg/l KN, 4 weeks after subculture from BM + 1 mg/l 2, 4-D. **D** Shoots differentiate from the callus. (After Konar and M.N. Singh 1979)

+2 parts 10^{-6} NAA showed no significant difference (between the micropylar and chalazal halves) in callus growth and roots differentiated in both halves. As compared to longitudinal halves, in transverse halves there is much less callus growth and fewer roots.

In *Ephedra*, 2, 4-D causes callus formation, regeneration does not occur. With the addition of CM (10%) the female gametophyte produces callus and roots differentiate. However, CM—in the absence of sucrose, even though 2, 4-D at 2 parts 10^{-6} is present in the medium—supports scanty callus growth. CM has a promotory effect on growth only when the medium has been supplemented with essential growth substances (Bhatnagar and M.N. Singh 1984).

The regeneration of roots depends upon the addition of NAA (substitute for 2, 4-D). NAA at 4 parts 10^{-6} is optimal for the growth of callus and differentiation of roots per culture of the female gametophyte at both archegonial and mature embryo stages.

Callus growth and morphogenetic development are strongly influenced by the ratio of NAA and BAP, and the developmental stage of the gametophyte. Root regeneration depends on (the presence of) NAA, while BAP has a modifying effect. When lower concentrations of BAP (0.05–0.5 parts 10^{-6}) is incorporated in the medium containing (0.05–4 parts 10^{-6}) NAA (explants at the archegonial stage), BAP promotes the rhizogenic effect of NAA. The roots differentiate in 91% cultures (Fig. 22.6 A), and the average number of roots in the explant is higher than in the explants at the mature embryo (stage), which show roots in only 33% cultures. At higher concentration of BAP (6.0 parts 10^{-6} and 8.0 parts 10^{-6}), the explants at the archegonial stage differentiate shoot buds (Fig. 22.6 B), while those at the mature embryo stage do not respond at all. Callus on explants at the archegonial stage produces the maximal number of roots at 1 part 10^{-6} BAP, and the mature embryo stage at 0.5 part 10^{-6}.

Callus initiation, growth and regeneration are optimal at 4 parts 10^{-6} NAA. The addition of 0.05 part 10^{-6} BAP in the medium (containing optimal concentration of NAA) increases the rate of callus growth, percentage of cultures forming roots, and the number of roots. BAP (above 6 parts 10^{-6}) does not stimulate growth of callus, and does not effect root differentiation. However, a higher (2–4 parts 10^{-6}) concentration of BAP (along with NAA 0.5 part 10^{-6}) improves the differentiation of shoot buds (but the roots are fewer) at the archegonial stage (Fig. 22.6 C, D).

In interaction studies of 2, 4-D and KN, the cultured female gametophyte showed significant differences in growth and regeneration of root and shoot buds. Shoot buds regenerated at lower concentration of 2, 4-D (0.05–2 parts 10^{-6}) and different concentrations of KN (6 parts 10^{-6}), and formed normal shoots (Fig. 22.7 A). Further growth is inhibited at 0.5 part 10^{-6} and 2 parts 10^{-6} 2, 4-D. When grown on BM without 2, 4-D and KN, growth is resumed, and new roots are also formed. This response indicates that

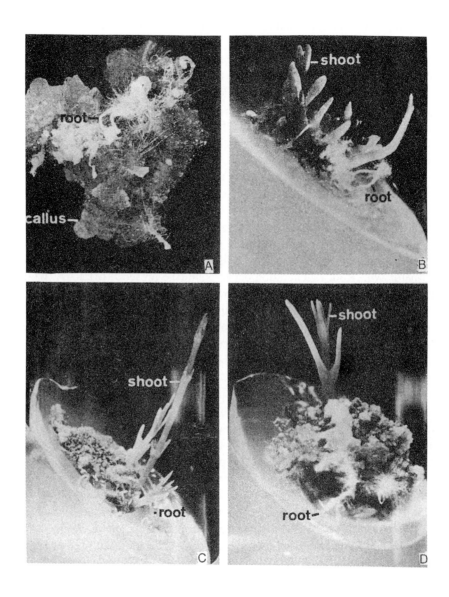

Fig. 22.6 A-D. *Ephedra foliata,* female gametophyte at archegonial stage. **A** 4-week-old culture on BM + 0.05 part 10^{-6} NAA + 0.05 part 10^{-6} BAP shows callus with roots. **B, C, D** 6-week-old culture. **B** On BM + 0.05 part 10^{-6} NAA + 6.0 parts 10^{-6} BAP shows formation of root and shoot buds. **C** On BM + 0.5 part 10^{-6} NAA + 2.0 parts 10^{-6} BAP; callus mass with roots and shoots. **D** On BM + 0.5 part 10^{-6} NAA + 4.0 parts 10^{-6} BAP shows differentiation of roots and shoots. (After Bhatnagar and M.N. Singh 1984).

2, 4-D at relatively high concentrations, together with KN, has an inductive effect on shoot bud initiation; its continued presence is not necessary for further growth. 2, 4-D (0.05–2 parts 10^{-6}), when combined with lower concentrations of KN (0.05 part 10^{-6} and 0.5 part 10^{-6}), generally favours root formation. A combination of 2, 4-D (0.05 part 10^{-6}) and KN (4 parts 10^{-6}) is optimal for the induction of root and shoot buds. Higher concentrations of KN (6 and 8 parts 10^{-6}) inhibit rooting (Fig. 22.7 B).

The callus cells and root tips have haploid chromosomes (Fig. 22.7 C); rarely, a few cells are diploid (Fig. 22.7 D).

According to Gauthert (1966), a differentiated cell can undergo dedifferentiation and provide new cells that are likely to exhibit various other types of specializations under in vitro conditions. The formation of an organized structure begins with meristemoids (Torrey 1966). The meristemoids are small, and comprise densely cytoplasmic compact groups of cells, which have prominent nuclei (Fig. 22.8 A). They respond to directional stimuli (within the culture) to form a primordium (Fig. 22.8 B, C), which can give rise to root, shoot bud or an embryo-like structure. The meristemoids in the female gametophyte cultures of *Ephedra* are located on the surface, or embedded in the callus tissue (Bhatnagar and M.N. Singh 1984). The deep-seated meristemoids organize only into root primordia, but the peripheral ones give rise to root as well as shoot bud primordia. Initially, there is no vascular connection between the shoot bud and the callus, but it is established later (Fig. 22.8 D, E).

Nagmani and Bonga (1985) report embryogenesis in subcultured callus of *Larix decidua*. The megagametophytes are excised from immature ovules. The entire (coenocytic or prearchegoinal cellular) gametophyte is sown on the modified nutrient medium (MS 1962 or Litvay et al. 1981, either at full strength or with all macro- and microelements—except $FeSO_4$ and Na_2 EDTA—reduced to half-strength). Gametophytes at an advanced stage of development are cut longitudinally (archegonia, proembryos/embryos and suspensors are removed) and each half is placed with its cut surface in contact with the medium. After a small gametophytic callus has been formed, it is transferred to hormone-free medium. Thereafter, the callus is subdivided and subcultured (on the same medium) every 2 weeks. After 2 to 3 weeks, a large number of embryoids with suspensor differentiated. The embryoids comprised small isodiametric cells with dense cytoplasm, large nuclei and several large nucleoli. Most embryoids remain small, but some develop into plantlets (the ploidy could not be determined). After six to eight subcultures, the number of embryoids gradually declined.

In gymnosperms, haploid cultures have been established more often from megagametophytes than from microsporophylls, microsporangium or pollen.

The cultures of female gametophytes may prove to be a useful means of achieving haploid and doubled-haploid trees of gymnosperms incorporating desirable genetic characters. In cycads, megagametophyte culture demonstrates a considerable capacity for regeneration. Further investigations are necessary.

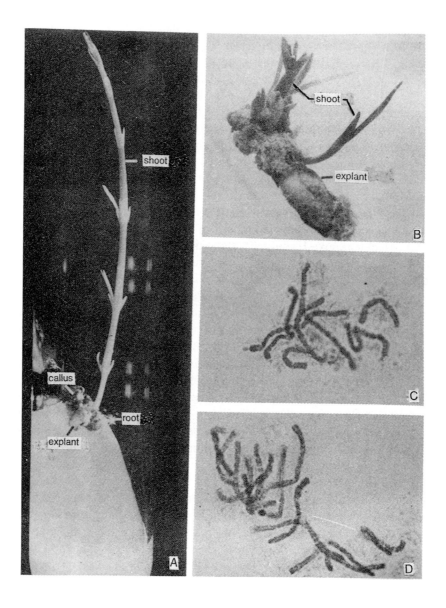

Fig. 22.7 A-D. *Ephedra foliata*, female gametophyte. **A, B** 6-week-old culture. **A** On BM + 0.05 part 10^{-6} 2, 4-D + 1.0 part 10^{-6} KN shows shoot and root formed from callus. **B** On BM + 0.05 part 10^{-6} 2, 4-D + 6.0 parts 10^{-6} KN; large number of shoot buds. **C, D** Acetocarmine squash of p-dichlorobenzene-treated root tip of in vitro-differentiated root from callus shows n = 7 haploid (**C**) and 2n = 14 diploid (**D**) chromosomes. (After Bhatnagar and M.N. Singh 1984)

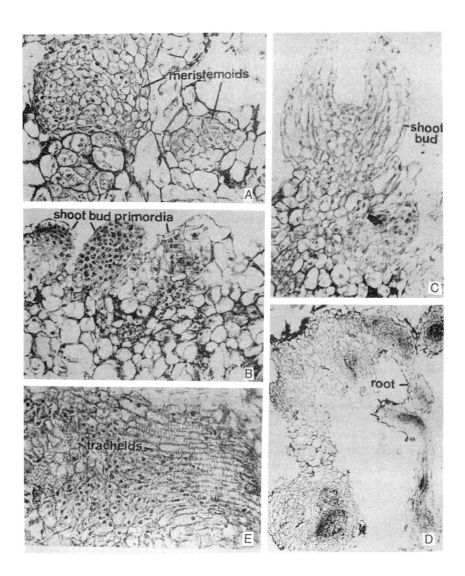

Fig. 22.8 A-E. *Ephedra foliata*, transections of callus. **A** Meristematic zones. **B** Shoot bud primordia along the periphery. **C** Shoot bud with apical meristem and two leaf primordia. **D, E** Vasculate nodules and root. **E** Inset from **D** shows vascular connection between root and callus. (After Bhatnagar and M.N. Singh 1984)

Embryo

Studies on embryo culture are of great importance in determining the factors influencing the development of embryos for overcoming dormancy, and for rearing hybrid embryos.

The earliest report of embryo culture in gymnosperms was by A. Schmidt in 1924 (see Durzan and Campbell 1974). He cultured embryos of *Pinus sylvestris* in organic nutrient solution. Li (1934) and Li and Shen (1934) cultured embryos of *Ginkgo* and demonstrated that the excised embryos of gymnosperms can germinate and produce plantlets, on relatively simple nutrient media, which replace the female gametophyte. LaRue (1936) cultured the excised embryos of a number of plants—*Pinus resinosa, Picea canadensis, Tsuga canadensis* and *Pseudotsuga menziesii* (*taxifolia*)— and pointed out that the immature embryos (2–4 mm in length) grow and develop into plantlets. Subsequent investigations have shown that the excised mature embryo is capable of giving rise to a seedling independently of the megagametophyte.

Radforth and his associates (Radforth 1936, Radforth and Bonga 1960, Radforth et al. 1958) cultured proembryos of *Ginkgo* and *Pinus nigra*. One of their objectives was to assess the developmental potentialities of the embryo itself. It was observed that the embryo develops radially and forms callus-like masses (suspensor is not formed). In *Ginkgo* (Radforth et al. 1958) the archegonia-containing free-nuclear proembryos (without suspensor formation) are excised along with the minimum amount of circumjacent gametophytic tissue, and cultured on nutrient agar. Cell divisions occur in the cultured proembryos but there is no suspensor formation. The absence of polarity in the embryo is perhaps due to its uniform accessibility to nutrients (Fig. 22.9 A, B).

Sterling (1949b) cultured immature embryos of larch (*Larix* sp.). In some cultures, the proembryonic mass cleavaged, but none of the units completed normal development. Similar results were reported by Loo and Wang (1943), Radforth and Pegoraro (1955), Berlyn (1962) and Konar and Oberoi (1965). This is partly due to suboptimal culture media. Thomas (1970) cultured the undifferentiated proembryos of *Pinus* on Heller, Halperin, and Murashige and Skoog (1962) media. The proembryo survived up to 48 h on all three media, but up to 8 days only on Halperin medium. However, there was neither cell division nor cellular elongation.

Norstog and Rhamstine (1967) observed that in *Zamia integrifolia* the embryo cultures with suspensors formed normal dicotyledonous (occasionally polycotyledonous) embryos. Normal or near-normal differentiation occurs on media without auxin and KN. In a relatively low concentration of 2, 4-D (0.01–0.1 mg/1), and KN (0.05–0.5 mg/1), a subculturable callus is formed. The embryos differentiate on transfer to an auxin-free medium. Cells and tissues of *Zamia* (apparently) are much more plastic in their morphogenetic responses, than other gymnosperms. This plasticity was elucidated after a trial of 78 different media, of these 8 showed marked selective effects on growth and development (Norstog 1967b, Norstog and Rhamsteine 1967).

V. Vasil (1963) obtained callus from the embryos of *Gnetum ula* cultured

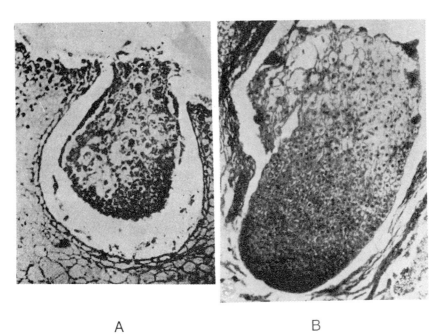

A   B

Fig. 22.9 A, B   *Ginkgo biloba*. Embryos of same age in vitro (**A**) and in vivo (**B**). Note absence of suspensor (formation) and polarity in embryo in **A**; the embryo has a distinct polarity in **B**. (After Radforth et al. 1958)

at the embryonal cell (the so-called peculiar cell) stage, on White's basal medium supplemented with (500 mg/l) CH and YE. The callus comprised thin-walled cells of various size and shape. In a subculture, after 4 weeks the callus formed uniseriate filaments, or small buds, or proembryos. Attempts to induce differentiation of cotyledons in these proembryos were unsuccessful.

In gymnosperms, the culture of very young embryos (proembryos) has not been successful. Normal patterns of embryogenesis do not occur in vitro even with partial or complete growth. The cultured embryos usually exhibit radial symmetry/asymmetry rather than axial organization.

Using partially mature *Ginkgo* embryos, Ball (1956a) observed that the primary root originates endogenously between the base of the hypocotyl and the hypocotyl cap. He concluded that the hypocotyl meristem is not synonymous with the root meristem. Further studies (Ball 1956b) demonstrated the developmental potential of the root initials, and a longitudinal split generates two new complete apices (Fig. 22.10 A, B).

The root and shoot of *Ginkgo* embryos have different requirements with respect to sugars (sucrose, glucose, raffinose, galactose, levulose), IAA, glutamine, and CM. The uptake of these substances occurs only through the cotyledons [Ball (1959); Fig. 22.11)], which indicates that the cotyledons may not only take up and distribute growth regulators but also mediate the morphogenesis of root and shoot.

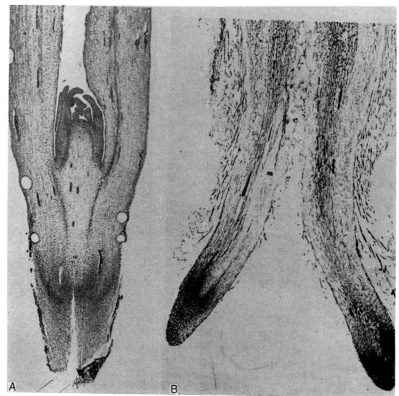

Fig. 22.10 A, B. *Ginkgo biloba.* Effect of a longitudinal incision at the tip of the hypocotyl of embryo on root development in vitro **A** Longisection embryo immediately after longitudinal incision at the tip of hypocotyl. **B** Same, 9-day-old culture, each half of the hypocotyl has grown into a normal root. (After Ball 1956b)

Ginkgo embryos could be grown in vitro, with or without cotyledons. However, the growth of cotyledons, hypocotyl and root ceased after about 15 days (Li 1934). According to Bulard (1952), normal growth of embryos occurred only when the cotyledons were in contact with the nutrient medium.

Brown and Gifford (1958) demonstrated the role of cotyledons in affecting root growth in *Pinus lambertiana* embryos grown in vitro. They devised a double-medium technique in which the embryos were suspended with the cotyledons embedded in agar, and the hypocotyl tip remained immersed in a liquid medium contained in an inner vial. Excellent root growth occurs if the roots remain immersed in a sugar-free medium. However, root elongation occurs only in the presence of sugar.

In *P. lambertiana* Greenwood and Berlyn (1965) reported physiological polarity in sections of embryos. The hypocotyl of the dormant embryos were cut into 2-3 mm segments and cultured in inverted tubes on Knop's medium containing agar (0.8%) and sucrose (4%). When the segments were inoculated with their morphological basal end downward in air, roots emerged at this end. On reversing the orientation, root did not differentiate.

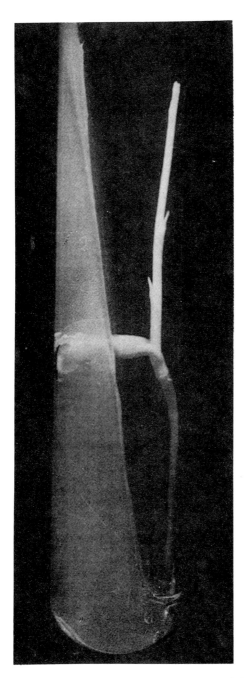

Fig. 22.11. *Ginkgo biloba*, embryo cultured with only the cotyledons embedded in medium. There is a well-developed root and shoot. (After Ball 1959)

Berlyn and Miksche (1965) studied the in vitro growth behaviour of mature embryos of *P. lambertiana.* The embryos show good growth even without cotyledons. They point out that the removal of shoot meristem (of the embryos) suppresses root development. The orientation of the excised embryo in relation to gravity and light alters embryo growth. On placing the culture tubes (containing the embryos) in a vertical position in light, the growth was much better than in the tubes placed horizontally. In the dark, the cultures in a horizontal position showed a very high growth rate. However, this differential effect of light or orientation occurred only in excised embryos; the embryos with gametophytes showed no such response. The authors conclude that the gametophyte inhibits tropic responses of the embryo.

Normally, the germinating embryos of gymnosperms develop chlorophyll in the dark, while excised embryos on nutrient media do not. However, a germinating *Pinus* embryo (with the female gametophyte removed) may maintain chlorophyll synthesis for a short period in the dark (Bogorad 1950, Sacher 1956). In cultures, when the embryos are suspended with the cotyledons embedded in agar media containing yeast, malt and female gametophyte extracts, Engvild (1964) observed that apparently a specific chlorophyll synthesis-stimulating molecule is not involved; chlorophyll synthesis could be promoted by sucrose, B-vitamin, urea and a mixture of amino acids.

Le Page-Degivry (1968, 1970, 1973a, b, c) and Le Page Degivry and Garello (1973) investigated the dormancy of the embryo of *Taxus baccata*. Immature embryos cultured on agar media grew very slightly. When cultured in a liquid medium first and transferred to agar media (after 8 days), germination occurred. According to these investigators, the inhibitors for germination leach out of the embryos in liquid culture. An analysis showed that the excised uncultured embryos, as well as the liquid medium (in which the embryos had been cultured for 15–20 days) contained substances with properties of ABA; the extracts of the cultured embryos lacked ABA. Following leaching, treatment of embryos with gibberellic acid, or chilling (at 4°C), broke the dormancy. In *Ginkgo* also, gibberellic acid promotes in vitro germination of excised embryos (Bulard and Le Page-Degivry 1968).

Excised embryos of *Pinus strobus* were cultured on MS medium supplemented with different growth hormones (Minocha 1980). On BM containing 3–6% sucrose, the embryos developed into plantlets; with 1–2% sucrose, the root primordia did not grow satisfactorily. The addition of GA_3 (0.01–20 mg/1) suppressed the growth of root primordia; while auxin (NAA/2, 4–D/IBA—each 0.01–2 mg/l) led to callus formation. It showed no organogenesis. IBA (1–5 mg/1) induced adventitious shoots from the hypocotyl of some embryos. Similar shoots were formed at the tip of cotyledons when TIBA (0.5–1 mg/l) was added to the medium (Minocha 1980).

The excised embryos of *Peseudotsuga menziesii* reared on MS medium, with or without the female gametophyte or its extract, exhibited striking differences in size. Obviously, the female gametophyte contains some substances which have a growth-promoting effect on the embryo. It is thermostable and diffusible in agar. The extract also has a synergistic effect with CM in promoting morphogenesis in suspension cultures of *Pseudotsuga* (Mapes and Zaerr 1981).

Sommer et al. (1975) cultured the embryos of *Pinus palustris* on Gresshoff and Doy (1972) medium (modified) for the first 4–6 weeks. Numerous adventitious buds along elongating cotyledons were formed. The buds, when excised and grown on modified Risser and White (1964) spruce medium, formed roots and vigorous plantlets.

The excised embryos of *Pinus ponderosa* formed callus and initiated a few buds from the tip of cotyledons when placed on medium devised by Sommer et al. (1975) and Cheng (1975). When they were placed on Reilly and Washer's (1977) medium (full or half-strength dilution); or on Cheng (1975) medium (half-strength dilution), buds differentiated along the entire length of those cotyledons in contact with the medium. For bud initiation, BA must be present in the medium; the addition of auxin increases the yield of excisable cotyledonary buds (Ellis and Bilderback 1984).

Konar and Oberoi (1965) cultured the embryos of *Biota orientalis*, at different stages of development, on White's medium with various supplements such as 2, 4-D (1–10 mg/l), IAA (0.1–5.0 mg/l), CM (10–20%), CH (1–5 mg/l) and YE (500–1000 mg/l). Other media were also tested: Crone (see Radforth and Pegaro 1955), Butenko and Yakovleva (1962) and Tepfer et al. (1963). Sucrose (2%) was used as the carbon source and the cultures were maintained at 25 ± 2 °C with ca. 8 h of light.

Embryos cultured before the initiation of cotyledons showed occasional proliferation; those with cotyledons and shoot apex developed rapidly. Within 3–4 days, the cotyledons mature and turn green; the hypocotyl swells and a short root develops at the basal end. On WB + CM (10 mg/l) + 2, 4-D (1 mg/l) + IAA (0.1 mg/l), the embryos develop into small seedlings in 15 days. On WB + IAA (0.1 mg/l), the hypocotyl proliferates and forms a massive friable green callus. In these embryos, the root and the shoot are initiated but their further growth is arrested. In the absence of initiation of root and shoot, the hypocotyl fails to callus.

In Crone's medium, the root develops and reaches 15 mm in 20 days; several well-developed secondary roots are also formed. The development of shoot is, however, suppressed. In Butenko's medium the cotyledons in 90% of embryos mature within 1 week. Concurrently, a short taproot is formed at the radicular end, followed by slight callusing of the hypocotyl and appreciable swelling of the cotyledons (Fig 22.12 A). In these cultures, the shoot and root failed to grow further. The 8-week-old cultures had five to seven small swellings on the inner surface of the cotyledons and become prominent within another week (Fig. 22.12 B, C). These swellings gradually differentiated into embryo-like structures (embryoids) with two prominent

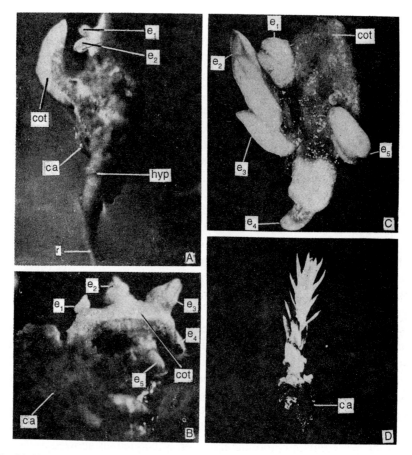

Fig. 22.12 A-D. *Biota orientalis.* **A** Young seedling with swollen hypocotyl (*hyp*) and cotyledons (*cot*) bearing embryoids (*e*). *ca* callus. *r* root. **B, C** Embryoids. **D** Young shoot from cotyledonary embryoid. (After Konar and Oberoi 1965)

cotyledons. Finally, they germinated and formed shoots with normal leaves. When the embryoids are subcultured with a small amount of cotyledonary tissue, normal shoots are produced with spirally arranged leaves (Fig. 22.12 D). In none of the media did the embryoids develop a root. However, the two cotyledonary leaves and the dome-shaped apex compare well with a normal plantlet (seedling). This is probably the first report of induction of embryoids on the cotyledons of a gymnosperm.

Konar and M.N. Singh (1980) reported continuous callus cultures and regeneration of shoots from hypocotyl, cotyledons and callus obtained from mature embryos of *Pinus wallichiana*.

The excised segments of hypocotyl, when cultured on modified MS containing a very low concentration of salts supplemented with 0.1 mg/l NAA, give rise to a continuously growing callus. Higher concentration of NAA (1, 3 mg/l) inhibits callus growth. Explants grown on BM + different concentrations of BAP (0.1, 0.5, 1, 3 mg/l) show a general increase in size within 12 days. At 0.1 mg/l callusing occurs within 24 days—the callus is green and its growth slow. Better growth results with increase in BAP (0.5 mg/l). At 1 mg/l, regeneration of shoot buds with interphase of callus has been observed but the percentage of regeneration is low. At a higher concentration of BAP (3 mg/l), callus growth is prolific.

In mature embryos excised and cultured on BM + 0.1 mg/l NAA, the cotyledons turn light green and elongate five times in 7 days. Callusing occurs after 12 days and complete dedifferentiation of embryos after 3 months. This medium was supplemented with different concentrations of ammonium nitrate (200 mg/l to 1400 mg/l) to determine the optimal concentration to support maximal callus growth. The lowest concentration (200 mg/l) brings about faster growth. The increased concentration brings about a slow decline in callus growth. Callus grown on BM + 0.1 mg/l NAA, when subcultured on BM + 1 mg/l BAP, shoot buds differentiated after about 30 days (Fig. 22.13 A). Callus with shoot buds when transferred to MS + 2% sucrose + 10% CM, forms shoots (Fig. 22.13 B).

The embryos cultured on BM + 1 mg/l BAP show a general increase in size, and numerous shoot buds differentiate along the surface of the hypocotyl (Fig. 22.13 C) after 24 days. On transfer to MS + 2% sucrose + 10% CM, shoots develop. On BM + 1 mg/l BAP + 2 mg/l NAA, the cotyledons ceased to elongate, and appeared "succulent". The cotyledons and hypocotyl form callus when they are in contact with the medium. The green and compact callus can be subcultured continuously on the same medium. Within 3 weeks, primary leaves and shoot buds become visible along the surface of the cotyledon and hypocotyl. Excised shoot buds, cultured on MS + 2% sucrose + 10% CM, differentiated into shoots, and the tissue in contact with the nutrient medium dedifferentiated and formed callus (Fig. 22.13 D).

Bhatnagar et al. (1983) made preliminary investigations on organ differentiation in tissue cultures of *Cedrus deodara* and *Pinus roxburghii*. In *C. deodara*, callus could be induced from mature female gametophyte and mature embryo. Adventitious shoot buds formed on the hypocotyl of embryos cultured on MS + KN 2 mg/l. The embryos of *P. roxburghii*, cultured on a medium with BAP 1 mg/l and NAA 2 mg/l, differentiated primary needles and shoot buds on the cotyledons.

In *Welwitschia mirabilis* the growing point of the stem is extremely short-lived; further growth is from a meristem at the base of each leaf which persists throughout the life of the plant (see Chap. 19). To obtain differentiation of shoot apices, embryo, somatic tissue and single-cell cultures give best results (Button et al. 1971)

The seeds of *W. mirabilis* contain inhibitors for germination (see Button et al. 1971). However, the excised embryos from surface-sterilized seeds grow very well on an agarized substrate containing MS + organic additives (mg/l glycine 5.0, inositol 50, calcium pantothenate 0.03, thiamine

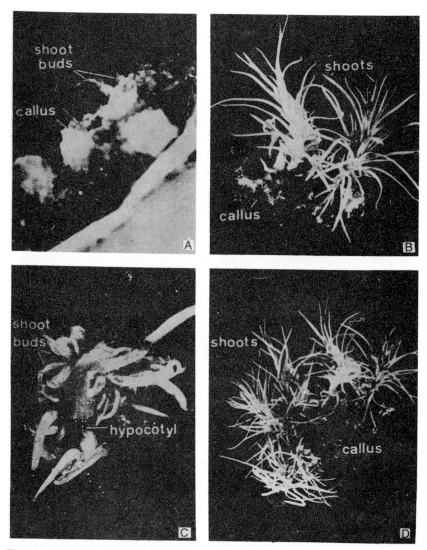

Fig. 22.13 A-D. *Pinus wallichiana*, subculture of callus from excised embryos.
A 40-day-old culture on BM + 1 mg/l BAP shows shoot bud. **B** 4-month-old culture on
MS + 2% sucrose + 10% CM transferred from **A** shows shoot formation. **C** 30-day-old
culture on BM + 1 mg/l BAP, shows shoot bud (formation) along the surface of hypocotyl.
D 4-month-old culture on MS + 2% sucrose + 10% CM — transferred from BM
+ 1 mg/l BAP + 2 mg/l NAA, — shows shoot differentiation; tissue in contact with the
medium has dedifferentiated from callus. (After Konar and M.N. Singh 1980)

hydrochloride 0.3, niacin 1.0, pyridoxine hydrochloride 0.05) + CH 400 +
sucrose 4×10^4 + NAA 1.0 or 5.0 mg/l. There is prolific callus development
from the hypocotyl-root axis. It is friable and consists of abnormally large
(75–200 μm in diameter) and variously shaped cells. Several cells showed
conspicuous nuclei, nucleoli and well-developed vacuoles traversed by

numerous thick cytoplasmic strands which showed active cyclosis. These cells are (potentially) totipotent.

Small segments of this callus are transferred to 500-ml culture flasks containing 10 ml liquid medium (as above) but containing CM 10% (v/v) instead of NAA. The flasks (attached to a culture wheel) are rotated at 1 rpm. The alternate flooding and aeration of the callus (in the flasks) promoted disintegration of the callus and cell multiplication. The cell suspension (mounted on microslides in buffered methylene blue) was examined under the phase-contrast microscope. The cells were spherical, cylindrical or like fungal hypha. The majority of the cells were live, did not take the stain, and showed cyclosis. Other cells were irregular (in shape) and took the stain. The cells proliferated by equational division, and by budding from papilla-like outgrowths on the callus.

Conifer Biotechnology
The practice of silviculture and tree improvement is important in forestry. Renewal of forests is obligatory to ensure future crops. The yield depends on the quality and vigour of the trees. However, there may be qualities/traits which are not easy to acquire or resolve. In such areas, biotechnology has significant impact on tree improvement. The object of conifer biotechnology is plant propagation and improvement (as in conventional methods).

Somatic Embryogenesis
Embryogenesis that does not begin with the zygote is asexual or somatic. It is the process by which somatic cells develop into differentiated plants through characteristic zygotic stages.

There are two types of somatic embryogenesis: (a) Direct somatic embryogenesis—the embryo develops directly from the explant tissue. (b) Indirect somatic embryogenesis—the embryo develops from callus or cell suspensions.

Cells which can develop into somatic embryos have embryogenic competence. The selection of the correct developmental stage of an explant, appropriate media and environmental conditions, and repeated transfers are generally necessary for successful embryogenesis. Immature zygotic embryos are the best explants. Embryos from mature stored seeds have also been used (such seeds are available throughout the year). However, the recovery of plantlets from somatic embryos is poor. Somatic embryos occur in a repressed state of development in an induction media and require a differentiation media for maturation.

Hakman et al. (1985) studied the development of somatic embryos initiated from immature (at various developmental stages) embryos of *Picea abies*, on basal medium gelled with 0.5% agar and supplemented with 2, 4-D (10^{-5}M) and BA (5×10^{-6}M) for tissue cultures, and with activated charcoal

(1% w/v) for embryo development. The immature embryos produced a highly friable and embryogenic callus which could be maintained in subcultures. Polarized and organized somatic embryos, with long highly vacuolate cells (suspensor) at one end, and a group of meristematic cells (embryonal) at the other end are formed. These structures are comparable to the early stages of normal zygotic embryogeny.

In subculture, the embryos form a bipolar shoot-root axis with an independent and closed vascular system. Frequently, either the shoot or the root meristem fail to differentiate. Embryogenic tissue obtained on agarized media could be transferred to liquid media and maintained in subcultures for at least 6 months. Differentiation of somatic embryos has also been observed in liquid cultures. This investigation shows that cultures with a high potential for initiation of somatic embryos have been obtained from immature embryos of *P. abies*. Similar results were also obtained from immature embryos isolated from seeds; the cones had been stored at 4°C for up to 6 months. This procedure significantly extends the period of availability of proembryos and immature embryos.

Gupta and Durzan (1986a, b) obtained callus in *Pinus lambertiana* from the radicle of mature zygotic embryos. However, immature embryos showed a greater degree of competence. About 1–2% of the embryos developed from the callus, and produced plantlets. Lelu et al. (1987) reported production of somatic embryos from seedling cotyledons of *Picea abies*. NAA with BA gave the best results. Becwar et al. (1987) developed a plating-counting method for *Picea abies*. Nagmani et al. (1987) compared somatic embryogenesis in *Picea glauca* and *P. abies*. Hakman et al. (1987) studied it in *Picea glauca* using light and electron microscopy. Gupta and Durzan (1987a) used the term polyembryogenesis to describe the somatic embryo cultures of *Pinus taeda*. A four-nucleate free-nuclear stage occurs, followed by segregation of the nuclei and delineation of a proembryonal cell. According to these authors, the pine somatic embryogenesis system truly represents zygotic polyembryony.

It is possible to cryopreserve somatic embryo-competent cells of *Picea alba* and *Pinus taeda* in liquid N_2 (−196 °C) and recover viable cultures (Gupta et al. 1987). This is also true for *Picea glauca* (see Dunstan 1988).

For encapsulation the somatic embryos (*Picea abies, Pinus taeda* and *P lambertiana*) are mixed with 1.5% sodium alginate and dropped singly into a 50–100 mM solution of $Ca(NO_3)_2$. A coating of calcium alginate formed around the embryos and hardened to form capsules during a 20–30-min incubation. Encapsulated somatic embryos have been stored for 6 months at 4 °C without loss of viability. The plants from encapsulated embryos develop on vermiculite, perlite, sand or peat, but at very low frequencies (Gupta 1988). Somatic embryos of *Picea glauca* can be maintained for extended periods with repeated subcultures in liquid suspensions. With the use of bioreactors, it would eventually be possible to produce large quantities of elite genotypes on a regular schedule.

The advantages of somatic embryogenesis are (a) A large number of plantlets can be produced. (b) Cell suspensions which are amenable to bioreactor technology can cut down labour, time and cost. (c) Encapsulated somatic embryos with controlled dormancy could form an efficient storage and delivery system, as in natural seeds. (d) Cell-suspension somatic embryogenesis system is better suited for genetic engineering techniques (Gupta 1988).

The development of a method of embryogenesis in vitro for conifers is an important breakthrough and a rapidly growing culture system is available (Dunstan 1988).

A culture system comparable to somatic embryogenesis is gametophytic embryogenesis. It originates from haploid gametophytes, gynogenesis from female gametophyte (see Nagmani and Bonga 1985, described on p 421) and androgenesis from male gametophyte (microspore).

The culture of gametophytic tissue is of interest to tree breeders for use as homozygous diploid (doubled haploid) plants to study inheritance.

The first report of conifer protoplast (described below) regeneration to somatic embryos is in *Pinus taeda* and *Picea glauca* (Gupta and Durzan 1987b). Somatic embryo-competent cell cultures were used as the starting material; 95% protoplast viability was obtained by using 4-day-old suspension cultures of somatic embryos of *Pinus teada*.

The regeneration of somatic embryos from protoplasts of individual cells of conifers is useful for propagation and provides a method for genetically improving species by somatic hybrization.

Somatic Embryogenesis in Some Conifers

Species	Explants	Investigator/s
Abies alba	Immature zygotic embryos	Schuller et al. (1989)
Larix decidua	Megagametophytes	Nagmani and Bonga (1985), von Aderkas et al. (1987), von Aderkas and Bonga (1988)
L. eurolepis	Immature zygotic embryos	Klimaszewska (1989), von Aderkas et al. (1990)
Picea abies	Immature zygotic embryos	Hakman et al. (1985), Hakman and von Arnold (1985)
	Mature zygotic embryos	Jain et al. (1988) Verhagen and Wann (1989), von Arnold and Hakman (1986), von Arnold (1987)
	Mature zygotic embryos and 3–7-day-old seedlings	Krogstrup (1986) Lelu et al. (1987)
	Megagametophytes	Simola and Santanen (1990)
P. glauca	Immature zygotic embryos	Attree et al. (1989), Hakman and Fowke (1987), Lu and Thorpe (1987)

Contd

Species	Explants	Investigator/s
P. glauca	Mature zygotic embryos	Tremblay (1990)
	12–30-day-old seedlings	Attree et al. (1990)
P. sitchensis	Immature zygotic embryos	Krogstrup et al. (1988)
	Mature zygotic embryos	von Arnold and Woodward (1988)
Picea mariana	Immature zygotic embryos	Hakman and Fowke (1987), Tautorus et al. (1990)
	Mature zygotic embryos	Tautorus et al. (1990)
	12–21-day-old seedlings	Attree et al. (1990)
Pinus caribaea	Immature zygotic embryos	Laine and A. David (1990)
P. elliottii	Immature zygotic embryos	Jain et al. (1989)
P. lambertiana	Mature zygotic embryos	Gupta and Durzan (1986a)
P. radiata	Immature zygotic embryos	Chandler et al. (1989)
P. strobus	Immature zygotic embryos	Finer et al. (1989)
P. taeda	Immature zygotic embryos	Gupta and Durzan (1987a), Becwar et al. (1990, 1991)
Pseudotsuga menziesii	Immature zygotic embryos	Durzan and Gupta (1987)
Sequoia sempervirens	7-day-old seedlings	Bourgkard and Favre (1988)

Genetic Transformation

Genetic transformation is the incorporation of native or modified gene in a recipient organism. The "foreign" genes can be inserted into the conifer genome in vitro by two methods: (a) Direct gene transfer. (b) *Agrobacterium*-mediated transfer. In the first method, the transfer is by microinjection (see Miki et al. 1987). It requires a micromanipulator and microsyringe to deliver small volumes of plasmid directly into the recipient nucleus. Microinjection reduces the regenerative capacity of protoplasts but transformation frequency is high.

Agrobacterium is a naturally occurring bacterium which is capable of transforming a wide variety of plants by introducing DNA from its tumour-inducing plasmid. It has a wide host range and has been reported to be infective in gymnosperms (De Cleene and De Ley 1976). Conifer cultures that lend themselves to *Agrobacterium* transformation are somatic embryos and protoplast-derived cells. Scanty work has been reported in conifers (see Dunstan 1988).

Protoplast Culture

Inbreeding for tree improvement requires a long time to produce successive generations; many trees can not be easily inbred due to problems of self-incompatibility and in-breeding depression. Isolation and fusion of protoplasts have stimulated considerable interest, specially in the development of new conifer hybrids. The protoplast is plasmalemma-bound without a wall. It is useful for propagation, for plant breeding by somatic hybridization or genetic transformation, and is also of use in resolving questions of basic plant biology. The protoplast is obtained by removing the cell wall constituents enzymatically in an osmotic solution, which slightly plasmolyzes the cell. However, improved techniques need to be developed for isolating protoplasts in high frequencies; for separating viable protoplasts from intact cells, cellular debris, and the cell wall-degrading enzymes used to isolate the protoplasts; for fusing the protolplasts; to distinguish between hybrids and non-hybrids among the fusion products, and for regenerating whole plants from the protoplasts.

The most frequently studied coniferous material is plantlet cotyledon-derived protoplasts: *Pseudotsuga menziesii* (Kirby and Cheng 1979), *Pinus pinaster* (A. David and H. David 1979), *P. coulteri* (Patel et al. 1984). Other sources of protoplasts are (haploid) pollen (Duhoux 1980), root (Faye and A. David 1983), and primary needles of plantlet buds of *Pinus pinaster* (H. David et al. 1986).

Micropropagation

Micropropagation is the foremost area of plant biotechnology in terms of commercial application. Micropropagation through tissue cultures is a means of producing a new plant and a large number of plants can be obtained from a very small amount of tissue from an elite species. The multiplication potential of this technique is immense. It offers a new dimension and unparalled opportunity for forest tree improvement (see Karnosky 1981). The utilization of tissue culture in the propagation and genetic improvement of conifers promises several benefits, such as faster development and multiplication of selected and improved genotypes, and the transfer of genes between non-fertile parents. However, improved techiques have still to be developed. Also, genetic differentiation of cells in culture creates uncertainty as to the fidelity of reproduction by micropropagation (Berlyn et al. 1986).

Horgan and Holland (1989) developed a reliable method of rooting micropropagated shoots from mature *Pinus radiata*. There was difficulty in achieving consistent results in the beginning, but finally a high percentage (78%) of rooting was achieved. Shoots that had been cold-stored prior to rooting did not survive. Maximal rooting occurs in a free-draining peat-pumice-perlite medium in propagation trays maintained under controlled environmental conditions. A well-managed watering regime and gradual conditioning of the shoots from the day of setting enhanced survival and rooting. Removal of callus from the base of the shoots prior to an auxin treatment significantly improved survival (30–70%) and rooting (16–56%). It was further improved by a 5-week prerooting treatment in nutrient medium containing 6% sucrose. The shoots formed vigorous plantlets, which could be successfully acclimatized to full sunlight within 4–6 weeks of transfer to root trainers. Plantlet survival ranged from 90 to 100%.

The number of roots and their percentage in tissue-cultured *Pseudotsuga menzieii* is influenced by the rooting substrate, sucrose concentration and boron in the rooting medium, shoot height and shoot generation. Peat-perlite is a better substrate (produces 70% rooted shoots) than agar (0% rooted shoots). In the former, organized cell divisions occur and are associated with tracheidal nests, while in agar the proliferation is unorganized and not restricted to the tracheidal nests. Optimal sucrose concentration is 4% to produce nodular or rooted shoots. With the addition of 3 mg/l boric acid, 100% of the shoots rooted and the mean root number was 11. The number and percentage of roots is higher with shoots 3 cm long; shoot responses are more (80%) in third and fourth generation than in second generation (36%) shoots (Mohammad et al. 1989).

A number of factors have been identified which affect the rooting and acclimatization of conifers. The quality of shoot, age of the donor, clone, temperature and substrate influence the production of root. Biochemically, root initiation is a high-energy process which requires a continuous supply of free sugar from the medium. One of the most important factors is auxin, its type, concentration, mode of applications and duration of exposure. When used alone, it is usually one of the synthetic auxins NAA (0.05–0.5 mg/l) or IBA (1–10 mg/l). Combinations of NAA, IBA or IAA (10–20 mg/l) have been more effective with *Pinus radiata, P. rigida, Picea abies,* and *Pseudotsuga menziesii* (see Mohammad and Vidaver 1988). The type of auxin used affects the formation of callus at the shoot base, and the number of roots per rooted shoot.

During acclimatization, it is important to minimize water stress; new growth is necessary for the plantlets to become fully autotrophic. The quality of the root system is critical, a well-branched fibrous system is essential for good performance in the field. It is influenced by the rooting substrate, root pruning and the container. Further work is needed; the immediate areas are: role of sucrose in rooting and subsequent plantlet survival, the potential role of ectomycorrhizal associations prior to or during transfer stages, and the cause(s) of early maturation following transfer to the greenhouse (Mohammad and Vidaver 1988).

Some gymnospermous forest tree species from which entire plants have been regenerated through tissue cultures (based on Karnosky 1981).

Species	Explant	Investigator/s
Araucaria cunninghamii	Shoot tip	Haines and de Fossard (1977)
Cryptomeria japonica	Hypocotyl, stem	Isikawa (1974)
Picea abies	Megagametophyte	Bonga (1977)
P. glauca	Hypocotyl	Campbell and Durzan (1976)
P. sitchensis	Embryo, needles, shoot tips	Webb and Street (1977)

Contd

Species	Explant	Investigator/s
Pinus elliottii	Cotyledon	Sommer and Brown (1974)
P. palustris	Cotyledon, embryo	Sommer et al. (1975)
P. pinaster	Cotyledon, hypocotyl	H. David et al. (1978)
P. radiata	Cotyledon, hypocotyl	Reilly and Washer (1977), Smith et al. (1982)
P. rigida		
P. sabiniana	Cotyledon	Brown and Sommer (1977)
P. virginiana		
P. taeda	Cotyledon	Sommer and Brown (1974)
P. strobus	Embryo	Minocha (1980)
Pseudotsuga menziesii	Cotyledon	Cheng and Voqui (1977)
Sequoia sempervirens	Juvenile shoot	Ball (1978)
Thuja plicata	Cotyledon, shoot tip	Coleman and Thorpe (1977)
Tsuga heterophylla	Cotyledon	Cheng (1976)

To meet the ever-increasing demand on the forests, therefore, traditional improvement and silvicultural practices along with developing biotechnologies are required.

23. Economic Importance

The gymnosperms are predominantly woody plants. They are frequently used in landscaping of parks and gardens because of their evergreen habit and symmetrical appearance.

General Aspects

Various *Cycas* species are cultivated in the garden as palms.

Ginkgo biloba is grown in groups or as avenue trees. The male trees are preferred, as the ripened ovules (on female plants) have a disagreeable odour, like that of rancid butter. However, the male and female plants cannot be distinguished in the vegetative state. The plant is exceptionally resistant to attacks of insects and fungi and can be grown successfully in modern cities.

The conifers are much valued, and several taxa are grown wherever the climate permits; *Thuja, Biota, Araucaria, Juniperus* and *Pinus* are quite popular. *Cupressus funebris* is commonly planted in China and India in cemeteries. *Cryptomeria japonica* lends beauty to the groves and gardens of Japan. It forms the famous avenue which leads to the temples and tomb of Icyasu (founder of Tokugawa dynasty) at Nikko. The avenue, laid out in the 17th century, has been maintained by replacing dead or damaged trees. Much attention was paid to forest administration by the Tokugawa Government. Tree felling was restricted. Certain trees, such as *Chamaecyparis obtusa, C. pisifera, Sciadopitys verticillata, Thuja standishii* and *Thujopsis dolabrata*, the celebrated "five trees of Kisco" have been strictly preserved. *Agathis alba* is grown as an ornamental tree in Java. *Juniperus horizontalis*—the creeping juniper—spreads over the ground, and is used for covering exposed banks. Several horticultural shrub varieties of *J. virginiana* have been developed for ornamental plantings (Hemmerley 1970). They vary widely in foliage colour which may be "white", gold-tipped or blue-green, and turn purple in winter. The shape of the plant may be pyramidal, spreading or globose. They need abundant sun, but can grow in sandy or dry soil where other shrubs will not grow. *Chamaecyparis obtusa* var. *compacta* is a dense-growing dwarf with very short branches and branchlets. *C. obtusa* var. *nana* is a spreading green-leaved bush. Both the varieties are suitable for rock gardens. *Cephalotaxus* is a spreading bush which makes excellent screens in the garden; it withstands pruning well and is used as a hedge plant. *Taxus* spp. are widely planted in gardens. It has several erect branches from the base, stands clipping well, and is one of the best hedge plants, used extensively in topiary work.

Various species of *Abies, Picea, Pinus, Pseudotsuga* and *Juniperus* are popular as Christmas trees. In Europe *Abies alba*, and in the USA *A. balsamea* are preferred (they hold the needles longer than other species). Before World War II, *Juniperus virginiana* was the traditional Christmas tree (Hemmerley 1970). Bonsai is an ancient Japanese art form (planting of trees in shallow trays or pots). The trees or shrubs are grown indefinitely in proportionately small containers and treated with certain miniaturization/ dwarfing techniques to impart in them an appearance of age. *Cycas revoluta*, various species of *Pinus* and *Juniperus* are very popular for growing bonsai. In Japan, *Chamaecyparis obtusa, Pinus densiflora, P. parviflora, P. thunbergii* and *Podocarpus nagi* are largely used for dwarfing. However, almost any conifer will make a dwarf plant.

Fresh or dried leaves of various cycads are utilized as decorative material. The foliage of *Cycas revoluta* is used for making wreaths, floral decorations and/or artificial palm trees for window display. The leaves are boiled in water, soaked in a preservative solution for several days (Thieret 1958), or plunged in boiling seawater. The leaves become leathery and do not break on drying.

The gymnosperms are economically important for their timber, in the manufacture of paper and board, resins, tannins, essential and fatty oils, as food supplements and pharmaceuticals.

Woods

Most of the commercially important gymnospermous woods are obtained from conifers and taxads. Anatmoically, these woods comprise tracheids, xylem parenchyma and xylem rays. They lack xylem fibres for the most part, and the high cellulose content imparts a softer texture than the wood of angiosperms (see P. Maheshwari and H. Singh 1960). In *Larix decidua, Agathis* and *Taxus*, the wood is very hard. In conifers, the heartwood and sapwood are not always well defined. The distinction between spring and autumn wood is usually better marked in trees from temperate regions than from warmer regions. A few coniferous woods have a definite taste, for example bitter (*Dacrydium colensoi*) or astringent (*Agathis australis*), while others are greasy to the touch (*D. colensoi, Taxodium distichum*).

Most coniferous woods are straight-grained, light-coloured, light-weight and strong in comparison to their weight. They can be easily worked, have good nail-taking properties, take a fine finish with sharp tools, and polish and paint well. These qualities make them suitable for a wide range of work where strength and durability are not essential: furniture, cabinet making, flooring, interior decoration, for general carpentery, building purposes, joinery, for boats and shipbuilding, poles, posts, railway sleepers, cooperage, veneers, etc. Some of the taxa have scented wood or are beautifully figured and are, therefore, specially valued for making chests, cabinets, and for panelling. All species in a genus usually have the same odour, except *Chamaecyparis nootkatensis* which is different from other species.

Araucariaceae. The wood of *Agathis* is strong, durable, pale yellow to yellow-brown with a silky, even texture and lustrous surface. *A. australis* is the principal timber tree of New Zealand. The wood has several uses, including making piano parts, artificial limbs, and reproduction of antique furniture, especially in France (Whitmore 1980 a, b). Occasionally, the wood of *Araucaria araucana* has dark brown fixed knots which contrast strikingly with the rest of the wood. Logs of *A. angustifolia* and *A. cunninghamia* are used for plywood manufacture.

Cupressaceae. The wood of *Callitris* is fragrant and shows distinct heartwood and sapwood. It is sometimes beautifully figured. Due to the presence of phenol and other chemical compounds, the wood is resistant to white ants. It is valuable as timber, and mostly the wood of *C. glauca* is used. In *Chamaecyparis,* the wood is durable, finishes with a glossy surface, and has a pleasant lasting spicy odour which repels moths and insects. *C. lawsoniana* is one of the most valuable timbers of North America. The fragrant essential oil present in the wood is a powerful diuretic, which makes it necessary for workers in the saw mills to occasionally change to other woods (Dallimore and Jackson 1966). The wood is also used for aeroplanes, canoe paddle, and organ pipe making. *C. nootkatensis* is considered to be one of the best woods for battery separators. *C. obtusa* has white-, straw-coloured or pink wood and is popular in Japan. It is used as a base for superior lacquer ware. Due to its straight and even-grained nature, the wood can be cut up into very thin shavings and plated into hats. The bark is used for roofing, as it is decay-resistant. *C. pisifera* is used for bentwood articles. *Cupressus* wood is very durable, and has a spicy-resinous odour which is insect-repellent. *C. sempervirens* (gopherwood) wood is reputed to be one of the four woods used in the construction of the cross upon which Christ was crucified (Baumann 1960). The wood lasts indefinitely under water. It has been a favourite wood for coffins ever since the Egyptians introduced it. It is considered an excellent furniture wood in France and Italy. The doors of St. Peter's in Rome were made from this wood, and lasted for nearly 1000 years (P. Maheshwari and H. Singh 1960). The timber of *C. torulosa* is often used in India in temples. When burnt as incense, it forms very little ash. The wood is suitable for second-grade pencils, battery separators, and in aircraft manufacture.

Juniperus wood is fragrant, reddish or reddish brown, and rarely damaged by insects. It is very durable and especially resistant to weather conditions. *J. virginiana* has a cream-coloured sapwood with projections into the dark red heartwood which gives it a distinct appearance. It is durable even when in contact with the soil, and is resistant to microbial decay. It was the standard wood for pencils until it became scarce. *J. macropoda* wood is suitable for pencils in India; it is also burnt as incense. *J. recurva* and *J. wallichiana* wood are also burnt in Buddhist temples. The wood of *Libocedrus* is reddish brown, resinous, fragrant, durable and easily worked. The heartwood of *L. decurrens* contains a volatile oil and small quantities of carvacrol and hydrothymoquinone—the last two highly fungicidal (Anderson 1967).

All the high-grade woods are used in the manufacture of lead pencils, and venetian blinds. The wood of *Thuja* is fragrant, non-resinous, very durable, and yellowish or reddish brown. The heartwood of *J. plicata* is very durable, and contains tropolones (thujaplicins), which are extremely toxic to wood-destroying fungi (Anderson 1955). The wood is used for glasshouse construction, beehives, and outhouses. In the USA, native Americans used it for canoes by hollowing out the wood, and splitting the trunks for their totem poles.

Pinaceae. The colour of wood in *Abies* varies from white-yellowish to reddish brown. It is light and used for packing cases, matchwood, wood wool, aircraft work, light camp furniture, and plywood. The odourless wood is much in demand for butter, lard and grocery boxes. *Cedrus* wood is very durable, oily, fragrant, insect-repellant, and rot-resistant. The wood of *C. libani* was extensively used for coffins and mummy cases by the ancient Egyptians (see P. Maheshwari and Biswas 1970). In India, *C. deodara* is a general utility wood and is the strongest of the Indian coniferous woods. Due to the presence of oil, the seasoned heartwood is very durable and rarely attacked by white ants or fungi. It is, however, difficult to glue or polish, and unsuitable for painting and varnishing, as resin oozes out from the knots even after seasoning.

The wood of *Larix* is one of the heaviest, strongest and toughest of the softwoods. It is durable even in contact with the ground. The wood is extensively used for garden furniture, and outdoor feeding racks and troughs for cattle. The plant has a naturally curving lower part which is ideal for boat- and shipbuilding. Some of the important species are *Larix decidua*, *L. laricina*, *L. leptolepis* and *L. occidentalis*.

Picea wood is soft, odourless, long-fibered, white or pinkish, black or dark-brown, works well, and finishes with a satiny surface. It is resonant and is prized for making sounding boards of pianos and bodies of violin and organ pipes. The timber of *P. abies* has a natural lustre, and is extensively used for toys, carving, wood chips for hats, and fruit baskets. The wood from Romania is excellent for making sounding boards for musical instruments. It is popular as dairy and kitchen table-top and dressers as the wood can be washed and scrubbed to give a clean appearance. *P. sitchensis* wood is the most valuable of all spruce woods because of its combined qualities of strength and lightness. It is used for plywood for special laminates in aeroplane and glider construction.

Pinus timber is commercially important and valuable. The soft pines have a straight-grained, uniform, easy-to-work wood, comparatively free of resin. The hard pines have resinous, heavy, hard, strong and durable wood; heartwood and sapwood are well defined. The group is heterogenous, each species has distinctive wood. It is highly inflammable and can be used as a torch. *P. lambertiana* yields large quantities of unblemished, high-quality timber. The wood of *P. palustris* has a world-wide reputation. It is very hard, strong, durable, resinous, and occasionally has decorative markings. It is esteemed for heavy construction work such as bridge and naval architecture. *P. roxburghii* timber is moderately hard, resinous, light and non-durable. It is extensively used for packing cases, cheap furniture and charcoal. *P. sylvestris* wood is superior, and has excellent lasting qualities. The treated best-grade wood, when laid on a good concrete foundation, withstands very heavy wear, as shown by blocks which remained in use for over 13 years on Westminster Bridge (London) where a continuous and heavy traffic flows (Dallimore and Jackson 1966). Some other notable timber-yielding species are: *P. ponderosa*, *P. resinosa*, *P. taeda*, *P. strobus* and *P. wallichiana*.

Pseudotsuga wood is variable in character. The heartwood is reddish- or brownish yellow and sapwood creamy yellow with a pronounced difference in spring and summer wood. Resin ducts are usually present in groups. The tracheids have spiral thickenings, otherwise superficially the wood resembles *Larix*. The tall tree gives a clear timber which can be used for heavy construction work. The wood is especially suited for bending. In the USA, most veneer and plywood is made from this plant (P. Maheshwari and H. Singh 1960). The brownish wood of *Tsuga* normally lacks resin ducts; the heartwood is darker than sapwood. The wood is tenacious and holds nails well, hence its value as a box wood. The important species are *T. heterophylla* and *T. canadensis*.

Podocarpaceae. *Podocarpus* is commonly called yellow wood because of its colour; occasionally it may be brown or reddish. The wood is not economical for cheaper manufactures as it takes nails only when it is bored and is more expensive to work. Several species are important timber trees in their native countries. *P. ferrugineus* and *P. totara* wood resists teredo (shipworm molluscs) and is used for marine work, such as piles for bridges and dock. The yellow or reddish wood of *Dacrydium* is resinous, durable, and occasionally beautifully figured. *D. biforme* wood is durable even on contact with the ground. *D. colensoi* wood is of low specific gravity, very strong, elastic and does not shrink much on seasoning. *D. intermedium* wood is strong, durable, resinous, and highly inflammable.

Taxodiaceae. The wood of *Cryptomeria japonica* is fragrant, strong, durable, coarse and variably grained, and is not readily attacked by insects. It is a widely used timber in Japan. The heartwood of *Sequoia sempervirens* is a rich dull red, and sapwood yellowish white. It is soft, fine-grained, easy to work, and durable when exposed to weather or soil. It does not warp or shrink readily. The wood has many uses, since it is available in large dimensions free from defects. The plywood can also be used in exposed situations. *Sequoiadendron gigantea* wood has a limited use. Both the trees are preserved for their scarcity. *Taxodium distichum* wood is soft, straight-grained, non-shrinking, and has a characteristic sour odour. The heartwood varies from red to almost black, and the sapwood is white or yellowish white. It is very durable in wet places, resistant to insects, and immune to white ants. It is used for construction of water tanks, water pipes, and is the best all round timber for chemical tanks and vats, since it does not impart any colour or taste to the products. It is particularly suitable for greenhouse construction, cooling towers and roof planks of dye houses.

Taxaceae. The wood of *Taxus* is strong, oily, elastic, close-grained and very durable (tertiary spiral thickenings are present in the tracheids). It is the heaviest of the softwoods, with good mechanical properties; it works well, and finishes with a smooth glossy surface. The heartwood of *T. baccata* is reddish brown, toning with age and exposure, the sapwood is pale yellow-white. It has irregular growth rings, which give it a decorative value. Transections of the base of old trees may show several distinct "hearts", which are due to erect shoots growing from near the base, thickening and being overgrown by the main trunk. Such sections are prized as tabletops. In ancient times, the wood was popular for bows and it is still used in archery. The wood of *Torreya nucifera* is lustrous, and durable under water. It is used for water pails and Japanese chess men.

Ginkgoaceae. *Ginkgo biloba* wood is light and brittle. In China and Japan it is used for chess boards and chess men.

Paper and Board

Paper can be made from any natural fibrous material depending on the amount, nature, softness and pliability of the cellulose present in the cell wall. At present, nearly 95% of the paper produced in the world comes from wood, and in the USA and Canada (the largest paper producers), 85% of the wood pulp is made from coniferous wood. Conifers are important because of the greater average length of the fibres and larger percentage of long fibres per given volume of wood.

The wood is converted into pulp or fibrous mass by various processes: (a) Ground wood or mechanical process, (b) Chemical processes.

In the mechanical process, the pulp is produced by pressing the logs against revolving pulpstone in the presence of water. The resins, lignin and other undesirable materials are not removed, and they resist bleaching agents, which causes the paper to turn yellow. The fibres do not felt readily and the pulp is poor in quality. Usually, the mechanical pulp is blended with a varying percentage of chemical pulp to give the paper the necessary strength. Ground-wood pulp is mostly used for newsprint (blended with 20% sulphite pulp).

In chemical processes, the wood chips are subjected to chemical treatment/cooking which removes most of the lignin and other materials, and nearly pure cellulose fibres remain. For pulping softwoods, the chemical processes such as sulphite and sulphate or kraft are used. The sulphite process has an acidic reaction. Bleached sulphite pulp is used in the manufacture of higher grades of paper. The sulphate or kraft process operates under alkaline conditions. This method is adopted for coniferous wood, especially pines, which have a high resin content. The digesting solution dissolves resins, wax and fat from the wood. The pulp is mostly used in the manufacture of strong grades of paper such as kraft paper, used for wrapping, paper bags and paperboard. The important conifers used extensively for the manufacture of paper and board are: *Pinus* sp. for kraft paper; *Picea*, *Abies* and *Tsuga* for higher grades of writing and printing paper; *Pinus*, *Picea* and *Pseudotsuga* for thermal and sound insulation boards. From *Picea* and *Tsuga* a number of derived products are also obtained, such as transparent films, photographic films, artificial sponges, sausage casings and lacquers. The bast fibre of *Sequoia sempervirens* is formed into sheets for battery separators (Isenberg 1956). Other conifers used in the industry are: *Thuja plicata*, *Agathis palmerstoni* and *A. microstachya*; *Araucaria angustifoila*, *A. bidwilli* and *A. cunninghami*; *Larix laricina*, *L. occidentalis* and *L. decidua*.

Resins

Resins are complex plant exudates which vary in their chemical composition, and are related to the terpenes or essential oils. They are insoluble in water but are soluble in vegetable oils and organic solvents such as alcohol, ether and carbon bisulphide. When heated, resins melt with progressive distillation of volatile oils as the temperature increases. The conifers are one of the major resin-yielders in the world. Resins harden gradually as their oil evaporates, which makes them invaluable in industries like paints and varnish, lacquers, paper-sizing and medicine. They are dissolved in solvents and painted on the given surface; after the oil and solvent evaporate a thin waterproof coating remains on the surface. Resinous substances have long been used as waterproof coatings. The Egyptians varnished their mummy cases with them. In most conifers, the resin remains mixed with either abundant essential oil (oleo-resin) or very little of it (hard resins). In others, the resin may be mixed with gum (gum resin) as in *Araucaria* and it is not possible to separate them.

Hard Resins

The hard resins are usually solid, more or less transparent, and brittle. They are an excellent source of varnish as they have a low oil content, and dissolve readily in alcohol. Some of the uses are: (a) Printing inks. Resins are ingredients of speciality inks which include gloss, non-rub, non-scratch food carton, candy wrapper, cellophane, soap wrapper, gravure inks, etc. These inks are applied to surfaces other than paper. (b) Adhesives, (c) Pyrotechnics (fireworks industry), and (d) Linoleum.

Copal. The term copal is of Mexican origin, and is applied to a large group of resins obtained from a variety of plants. It is of varying hardness, and has relatively high melting point. The softer copals are almost completely soluble in cold alcohol, and used as such for spirit varnishes. Harder copals need heat treatment or "running" to render them soluble in drying oils. There are three principal copals, two are from conifers.

a) **Kauri Copal**. This is obtained from *Agathis australis,* the most important tree of New Zealand. The copal is chiefly fossil in nature and is collected from sites of present or former kauri forests. It is dug up from ridges and swamps which furnish the bulk of the supply, and the pieces may weigh as much as 45 kg. The range gum is the best grade of kauri. An inferior bush gum is obtained by tappping living trees. A small amount of resin exudes naturally and forms green gum. The resin is extremely valuable for varnish and especially suited for marine and outdoor work. It has a high depth of gloss combined with elasticity and durability. For a long time, this resin held a premier position in varnish trade. An inferior grade of kauri copal is also used for making linoleum.

b) **Manila Copal.** This is obtained from *Agathis alba* distributed throughout Malaysia. The first shipments of this copal were from Manila, and the name has persisted. In Malaya, the trees occur only at higher elevations. The tapped resin does not harden there for some unknown reason (Mantell 1950), and is known as syrup copal. The resin exudes naturally; during high winds and storms, cracks and fissures are formed in the tree. The resin flows out from them and accumulates as large lumps up to 18 kg in weight, in the forks of trees from where it is collected. Resin is also obtained by systematic tapping. The resin is present in the bark; pieces of it are removed and the exposed surface is cleaned. After every collection, bark is removed from upper edge of the wound for a new opening. Resin also occurs in a fossilized state and is collected from the earth. It consists of large, irregular milky pieces with a yellowish interior. In Borneo, a semi-fossil grade known as Pontianac is obtained and is the hardest variety of copal. The resin is used in oil and spirit varnish and paints, lacquers and in making linoleum. It is used as a sizing material and has numerous other applications such as preparation of plastics, driers, adhesives, oil cloth, printing inks, and waterproofing compositions.

Amber. This is a fossil, water-insoluble tree resin which has attained a stable state after various changes such as loss of volatile constituents, oxidation and polymerization processes and a prolonged burial period. Amber is a general word for resins which are heterogeneous and differ in chemical and physical properties. It occurs throughout the world in widely separated localities such as Burma, Japan, Alaska, North, Central and South America, and many European countries as well as near-Eastern countries. When one speaks of amber, it is primarily the fossil resin which is found chiefly on the shores of the Baltic sea, and it is the most important commercial amber. Chemically, it is a resin from a number of extinct conifers of which *Pinus succinifera* is the principal source. These plants flourished during the Eocine on the shores of a former sea.

Amber is exceedingly hard and brittle, the color varies from yellow to brown or even black. It occurs in several forms, the most important is succinite. When rubbed, it takes a high polish, becomes negatively electrified, and emits a characteristic aromatic odour.

Amber was highly prized by the ancient peoples; the Greeks and Romans for jewellery, beads and other ornamental purposes. Certain magical properties were attributed to it. At present, it is used for the mouthpiece of pipes, and holders for cigars and cigarettes. It is also used in medicine and X-ray therapy. Human blood does not coagulate if kept in amber containers. The darker grades yield a valuable varnish, but it is too expensive to be of much use. Scientifically, amber is of interest because remains of insect or plant material are occasionally embedded in it. These existed when the tree was alive and the resin was fluid. This helps in the study of evolution of life on this earth.

Sandarac. Sandarac is a pale yellow resin obtained from *Tetraclinis articulata* (Vahl) Mast. (African Sandarc), or various species of *Callitris* (Australian Sandarc). The resin is formed between inner and outer layers of the bark and exudes as small "tears". The hard white, rather brittle, spirit varnish is particularly adapted to the finishing of metal objects, to which it gives a good lustre. One of its largest applications is as a primer for metal surfaces, where it provides a high degree of adhesion. It is used as paper and leather varnish, for glass and procelain cements, spirit varnishes and lacquers for photographic work, and as a constituent of incense. Many of its uses are in combination with mastic (resinous exudation of trees of

Pistacia, Anacardiaceae) and elemi (oleo-resins of different origin belong to members of the Burseraceae). Mastic sandarc has been used for the preservation of old paintings (Mantell 1950). An alcoholic solution of sandarc on cotton wool is used as a temporary filling for teeth.

Oleoresins

The oleoresins have a substantial amount of essential oil which makes it almost liquid in nature, and they have a characteristic aroma. The two components can be separated by distillation. Turpentine is an oleoresin obtained exclusively from conifers.

Turpentine.   The pine oleoresin is called pine gum, pine pitch or turpentine. It is a viscous, honey-like liquid which is obtained, for commercial use, by tapping living trees. On distillation the turpentine yields the essential oil (spirits of turpentine) and rosin (also called colophony). Both of them are immensely useful, and important industries have grown around them. The turpentine or Naval Stores industry is one of the oldest forest industries. In the 17th century, the wooden sailing vessels used large quantities of oleoresin and gave the name naval stores to the industry. The United States leads in production, followed by France, whose products are of the highest quality. Spain is third, followed by the European countries, India and East Asia.

In the USA, the major turpentine pines are *Pinus palustris* and *P. caribaea*; some of the other species of importance are *P. taeda*, *P. ponderosa*, *P. lambertiana* and *P. contorta*. In Europe, *P. pinaster*, *P. halepensis*, *P. nigra*, *P. pinea* and *P. sylvestries* are the common yielders. In India, *P. roxburghii*, in the East Indies, *P. merkusii*, and in the Phillipines, *P. khasya* are turpentined.

The trees are tapped for oleoresin. The method of tapping varies in different countries although the basic principal is the same. In the USA, in early days, tapping methods were very destructive and injurious to the tree. Cavities known as "boxes" were cut near the base of the tree and the oleoresin collected in them. Improved methods of tapping, called the Henry system, which involves shallow chipping and the use of several types of cup-and-gutter system followed. In a tree for turpentining (exceeding 23 cm in diameter), basal incisions are made in the trunk and metal gutters are slipped into it. These guide the oleoresin into a metal cup suspended below the gutters. A strip of bark and wood (approximately 1.75 cm deep) is removed just above the gutter for a distance of one-third of the circumference of the tree. This stimulates the flow of resin and induces the formation of new ducts. At present, the recommended practice is bark chipping and acid treatment. The surface is prepared by removing a strip of bark only by using especially designed scrapers, and no wood is removed. The fresh streak is then sprayed with 50% H_2SO_4. The acid stimulates the flow of resin by not allowing the wound to heal quickly (Anderson 1955).

A similar effect is shown by inoculation of *Fusarium lateritum* f. *pini* (Clapper 1954). Such streaks produce 50–100% more oleoresin during the first week, and continue more than normal yield during the second and third week. The interval between chipping can be increased from the usual 1 week to 2/3 weeks. This technique leaves the tree base in better condition for subsequent utilization as plywood or lumber.

Much turpentine and rosin are also obtained from old pine stumps and logging waste, by steam and solvent processes.

The oleoresins collected in the cup are emptied into larger vessels and transported to the factory, where they are washed and cleaned before distillation. The distillate is collected in barrels, where the oil of turpentine rises and is run off for storage. The residue (rosin), while still hot, is run through a series of screens to remove impurities and is transferred to a cooling vat. It is then put into slack barrels where it completely hardens within 24 h.

The oil of turpentine has many uses. Due to its properties as a solvent, it acts as a thinner in the paint and varnish industry. It is useful for printing cloth, especially cotton and wool, in perfumery, pharmaceutical and allied industries, and as a solvent for rubber and gutta-percha.

Rosin is even more important in industry. Depending on grade and quality, it has several uses, such as in the manufacture of soap, varnishes, paint driers, oil cloth, linoleum, sealing wax, adhesives, plastics, etc. Superior grades are used for paper sizing, enamels and ointments. Palustric acid has been discovered from rosin which is useful in paper sizing.

Canada Balsam. This oleoresin is obtained from *Abies balsamea* (balsam fir), the most widely distributed conifer in North America and Canada. The oleoresin is located in elongated, schizogenous ducts or blisters on the bark. It is collected by using a pot with a spout, cut at an angle, which is forced into the blisters, and the oleoresin drains out. Most of the collection is made in the province of Quebec in Canada. Each tree yields approx. 230–285 g in a year. Canada balsam is a viscid, yellowish or greenish substance which does not granulate or crystalize on drying. It is transparent, and has a high refractive index, approximating that of glass. This property makes it extremely suitable as a mounting medium for microscopic objects, and as cement for lenses in optical work. It is also used as a fixative for soap and perfumes and as a component of collodion and several plasters. *Pseudotsuga taxifolia* and *Tsuga canadensis* also have oleoresin with similar properties and uses.

Venice Turpentine. This is obtained from *Larix decidua*. The resin ducts are located in the heartwood of the tree. To collect the resin, a hole is bored into the trunk; one single hole may suffice for the whole life of a tree. The trees are tapped in spring. The oleoresin is a yellowish/greenish liquid and has a characteristic odour and taste. It is used in varnishes, lithographic work, veterinary medicine and histology.

Tannins

Tannins are organic compounds, glucosidal, have an acid reaction, and are astringent. They are useful because of their ability to unite with the proteins of animal skin to form a strong, flexible, resistant insoluble substance known as leather. Tannins also react with salts of iron to form dark blue or greenish black compounds which form the basis of common inks. They are useful in medicine because of their astringent nature. Tannins are used in the petroleum industry as a dispersant to control the viscosity of mud in oil-well drilling. The bark of *Tsuga canadensis* has 8–14% of tannin and has been in use in the USA as the chief domestic source. In Europe, *Larix decidua*, *Picea abies*, and in New Zealand *Phyllocladus trichomanoides* are used. The latter is utilized for glove leather, as it contains a bright orange-yellow dye in addition to tannin, which gives the kid gloves their particular shade.

Conifers do not form very important tannin yielders but small industries develop wherever large quantities of the bark are available.

Essential Oils

Almost all conifers with resin ducts can yield essential oils, but they are not always commercially important. Steam distillation of the young branches and adherent leaves, wood and sawdust yields the oil, which is used extensively in preparations of deodorants, room sprays, disinfectants, preparations of bath salts, perfumery and medicine. Some other uses of the oil are: Himalayan cedarwood oil (*Cedrus deodara*) and red cedarwood oil (*Juniperus virginiana*) are used for clearing tissues in histological work, and for use with the oil immersion lens of the microscope. Oil of juniper (*J. communis*) is used as an essence for flavouring several European liquors, such as gin, which owe their characteristic aroma to it. Oil of cade (*J. oxycedrus*) is widely used in the treatment of chronic eczema, other skin diseases, preparation of medical soaps, and healing of cuts and cutaneous diseases of animals. Oil of savin (*J. sabina*) was once used as an antirheumatic, and vermifuge but has lost much of its importance due to its disagreeable odour, irritating effect and toxicity. Huonpine wood (*Dacrydium franklinii*) oil is highly germicidal, and is used as a preservative of casein and other nitrogenous products; it exterminates powder pests and furniture borers, and is an effective insect-repellant; also used in scenting transparent soaps.

Fatty Oils

The seeds of several gymnosperms have fatty oils, mostly not available for oil-milling, as the seeds are eaten either raw or roasted. The fleshy layers of seeds of *Macrozamia* contain a bright orange oil, which resembles palm oil in its physical and chemical properties. *Cephalotaxus drupacea* seeds yield a fatty oil which is used as an illuminant, and is of local importance in Japan. The small brown seeds of *Torreya nucifera* yield an oil called kaya-no-abura, which is used traditionally for cooking in Japan (Burke 1975). It is also used for paints and other technical preparations (Eckey 1954). The seed kernels of *Gnetum ula* yield 14.2% fixed oil. It is used in South India for massage in rheumatism, as an illuminant, and for edible purpose.

Tall oil is a by-product of the sulphate pulp industry. The waste liquor from pine pulp mills, after treatment, results in the crude tall oil which is refined by steam distillation. The composition of the crude oil is approximately 20–60% fatty acids, 10–60% resin acid and 5–24% unsaponifiable material, mainly sterols (Hildititch 1956). It has innumerable uses. Products made with the oil are asphalt emulsions, wetting agents, binders, cement addition agents, waterproofing agents, boring oils, cutting oils, sulphonated oils, mould lubricants etc. During World War II, for the quickly constructed landing fields laid out in swampy land to be strong enough for bombers, a small amount of tall oil was used in the paving, which induced the wet earth to dehydrate and stick to the asphalt. Tall oil is the lowest-priced of all organic acids and also a source of components such as abietic acid in unpolymerized form, oleic and linoleic acids and sterols.

Food Supplements

Young succulent leaves of various *Cycas* spp. (*C. circinalis*, *C. pectinata*, *C. revoluta*, *C. rumphii*, *C. siamensis*), and *Gnetum* (also the inflorescence) are cooked and eaten as vegetables in their native countries.

The starch present in the pith and cortex of the stem and endosperm of the seeds of cycads is extracted and used as a food. Most cycads yield a large amount of starch (from the stem) at about 7 years of age; the best time for extraction is prior to a flush of new leaves. The male plants have more starch than female, and the starch content also varies from season to season. The plant is felled and the innermost cylindrical axis of the stem is removed. It is sliced into thin, oval or circular discs, spread upon mats and sun-dried. When crisp, it is pounded into flour, sifted and mixed in water and poured into a vessel, allowed to stand till the starchy substance is deposited at the bottom. The clear liquid is drained off, and the precipitate rolled about between boards until spherical pellets called sago are formed. Various grades such as bullet sago and pearl sago are formed. The majority of the manufacturers make an amorphous flour, which is stored. The starch is, however, a poor man's food, or used in times of scarcity in several areas in southern and southeastern Asia, New Caledonia, Indo-China, Malaya, India, Burma, Sri Lanka and Fiji.

It is more economical to extract starch from the seeds than from the stem; for the latter the entire plant has to be destroyed. A *Cycas* plant produces annually about 550 seeds, which yield nearly the same amount of starch (ca. 2.26 kg) as a 1.25-m-long stem (Thieret 1958).

The various parts of the plants and the starch extracted from cycad seeds and stems often contain a toxic principle, which should be removed before using it as food.

The pith of *Encephalartos* is used to make kaffir bread by the aborigines of South Africa. The pith is scooped out and buried in the earth to rot for 2 months. It is taken out, kneaded with water, made into cakes and baked in embers under the ashes. Dutch colonists of South Africa gave the name kaffer brood boom, hottentot brood boom or simply brood boom (bread tree) to several *Encephalartos* spp. These names are still in use.

Roasted seeds of *Ginkgo* are eaten at feasts in China and Japan to promote digestion and diminish the effects of drinking wine (Dallimore and Jackson 1966).

Pine seeds are rich in fats and proteins. The seeds of several species are large enough, have a good flavour, and are edible. Some of the species are *P. pineae* (Europe), *P. cembra* (Europe, Siberia), *P. armandi* (China), *P. gerardiana* (India, Afghanistan), and *P. cembroides*, *P. edulis*, *P. monophylla*, *P. sabiniana*, *P. parryana* and *P. counteri* (North America). *P. pineae* has been used as a food item in the northern Mediterranean region for over 2000 years. The nut is referred to as pignolia (England), pinone (Italy) and pignon (France).

In Italy, the kernels are largely used for making confectionery. Chocolate manufacturers mix them with cocoa (*Theobroma cacao*) and the product is a great delicacy. Traditionally, the kernels are used in soups. Raw or roasted kernels are also eaten as dessert. In North America, the processing of nuts has been mechanized. The kernels are dried or roasted in oil or manufactured into nut-coffee and other sweets and candies. The seeds of *Pinus edulis* were a staple diet of the Navajo Indians.

The Aborigines in Queensland depend on the ripe seeds of *Araucaria bidwilli* for food. They travel long distances to the groves or forests and feed on the seeds, which are very fattening. There is a restriction by the government upon felling these trees in one tract of hilly country, where several trees are reserved for the natives. They apportion the trees among themselves, so that each tribe has its own trees, which are again divided amongst families. The trees are thus handed down from generation to generation (Dallimore and Jackson 1966). *A. araucana* seeds are similarly used by the natives of Chile, and *A. angustifolia* in Brazil (Howes 1948).

The seeds of *Torreya nucifera* are an important article of food in Japan. The seeds of *Gnetum ula* and *G. gnemon* are eaten roasted or cooked. The seed kernel is mashed, moulded into cakes or biscuits, dried in the sun and fried in boiling oil (P. Maheshwari and V. Vasil 1961a).

A sugary exudation with cathartic properties is obtained from the heartwood, particularly from charred and wounded trees, or from sawed wood of *Pinus lambertiana* (sugar pine). It is sometimes used as a substitute for sugar.

Spruce Beer. This is a fermented liquor made from an extract of twigs and leaves of *Picea abies*. It is then mixed with treacle or other sugary substances (Dallimore and Jackson 1966).

Pharmaceuticals

Ephedra, known as ma-huang, has been a common medicine in China for over 5000 years (see Trease and Evans 1983).

The chief sources of the drug are *E. sinica, E. equisetina, E. intermedia, E. gerardiana* and *E. major*. The ephedras contain ca. 0.5–2.0% of alkaloids, which varies with species, seasons and age of the plant. The maxium is reached when the plants are ca. 4 years old, and still flowering. The alkaloids present are 30–90% ephedrine (and its isomers) and pseudoephidrine, which were isolated in 1887, although their pharmacological value was discovered only much later, and have been in extensive use during the present century. Physiologically and chemically, ephedrine resembles epinephrine (adrenaline), a hormone-like substance with a stimulating action on the sympathetic nervous system. The roots also contain a number of macrocyclic alkaloids (ephedradines) which have hypotensive properties. However, the medical properties of root and stem are in opposition to each other (Shiu-Ying Hu 1969). The root is prescribed for checking excessive perspiration in a weak patient.

Green branches of *Ephedra* are collected in autumn when it has the highest level of ephedrine. They are dried completely, first in shade and later in the sun. The best material is dry, thick, light green, solid with a bitter taste and little odour.

Ma-huang is hot, bitter and warming. It is prescribed for typhoid, bad colds, fever without sweat, pain all over the body, in joints, swelling of ankles, short breath, etc. The anti-inflammatry action is due to a recently discovered oxazolidone related to ephedrine (Trease and Evans 1983). Among the main clinical applications of ephedrine are for bronchospasm, as a nasal decongestant, and in certain allergic disorders.

Taxol. Taxol is a drug obtained from the dried inner bark (phloem-cambial tissues) of *Taxus brevifolia* (Pacific yew). It has unique therapeutic qualities against ovarian cancer, breast cancer, non-small-cell lung cancer, melanoma and colon cancer. Taxol is a complex diterpene—a 20-carbon taxane containing a rare oxetane ring and an amide side-chain with antineoplastic properties (see Edgington 1991). Wani et al in 1969, discovered and named it 'Taxol' (Wani et al. 1971, Edgington 1991). It is considered to be the most important novel natural anticancer drug after 15 years search by the National Cancer Institute (NCI) in USA. It is regarded as the prototype of a new class of cancer chemotherapeutic agents (Cragg et al. 1993, Chen 1990). The drug has been approved for clinical treatment of ovarian and breast cancer by the Food and Drug Adminstration (FDA) in the USA (see Zhong 1995).

Taxol catalyzes rapid microtubule formation, stabilizing them against depolymerization. This gives the drug two tumour fighting mechanisms:

(a) It freezes the mitotic spindle, prevents the depolymerization that pulls the chromosomes into the two halves of the rapidly dividing tumor cells. This suspends the cells at G_2 or M phase, ultimately killing it.

(b) Taxol inhibits cell migration, which may prevent the spread of metastatic cancer cells (Edgington 1991).

Taxus is a slow-growing plant of restricted geographical distribution. The amount of taxol in the bark is relatively low - ca. 0.01% of the dry weight. Ca. 7000 kg bark is required to produce 1 kg of taxol by current bark extraction procedures (Cragg et al. 1993). To produce 1 g of the substance, 3 or 4 trees (at least 60 years old) are required. *T. brevifolia* is the only commercial source for the drug at present. However, other species of *Taxus* (or *Austrotaxus* - see Zhong 1995) also produce taxol and are potential sources (Fett-Neto et al. 1992).

Total synthesis of taxol has not been achieved; semi-synthetic methods have been developed. The tissue of *T. brevifolia* has been successfully cultured to produce taxol, related alkaloids and alkaloid precursors (Fett-Neto et al. 1992). Tissue and cell cultures of other species of *Taxus* have also been established (see Zhong 1995) - *T. cuspidata, T. canadensis, T. baccata, T. yunnanensis, T. chinensis, T. chinensis* var *mainei, T. globosa, T. wallichiana, T. floaridana*. Most of them contain taxol or taxol-like compounds (Kang and Hou 1993). In in vitro cultures, callus can be induced from any tissue from any part of the plant - bark, cambium, needle, stem, seed, aril and roots. For optimum yield of taxol, the bark or cambial tissue is preferred (Fett-Neto et al. 1992). Due to the slow growth of *Taxus* spp., tissue and cell cultures of *Taxus* are also recalcitrant and require continuous effort for the production of taxol. The possibility of commercializing large-scale 'yew' cell culture is being considered as the demand for taxol is continually increasing (*Taxus* species may become extinct if its exploitation continues).

Scientists are also exploring alternative sources for taxol. *Taxomyces andreanae*, an endophytic hyphomyceteous fungus associated with the phloem of *Taxus brevifolia*, also produces taxol. The fungus is cultured in semi-synthetic liquid medium and produces taxol and related compounds. Biotechnology may ultimately enhance the production of taxol by *Taxomyces andreanae* (Babu et al. 1993).

In *Cycas* a mucilaginous and transparent gum exudes from the stem when it is wounded. Later, it hardens and turns light brown. The gum swells when placed in water, becomes colourless and transparent. It produces rapid supuration when applied to malignant ulcers. The gum is also used as an antidote for snake and insect bites. Other gum yielders are *Dioon*, *Encephalartos* and *Macrozamia*.

The gymnosperms are indeed very useful in our daily life.

24 Concluding Remarks

Seasonal Changes in Secondary Growth. A comparative comprehensive study of the seasonal variations in anatomy would be very useful for uses of the timber (Sharma 1994). Cambial growth research is a productive field for future studies. The mechanism by which the vascular cambium originated in early plants is not fully understood. Factors such as rainfall, temperature and photoperiod directly affect the ontogeny of tissues. The influence on the development of vascular cambium (meristem that produces secondary vascular tissues) and its derivatives is closely related to the effects of internal influences—physiological, structural and genetic. Because of this interaction between external and internal influences, investigations of vascular tissues of the plant as a whole are necessary (see Munting and Willemse 1987).

Little is known about the influence of external factors on the development of secondary phloem. The phloem and xylem may differentiate simultaneously, or phloem differentiation may precede or follow the differentiation of xylem. Functional sieve cells are present throughout the year in the secondary phloem of various Pinaceae (Alfieri and Evert 1973). The last-formed sieve cells overwinter, and remain functional throughout the next growing season. According to Alfieri and Mottola (1983), the differentiation of phloem may follow xylem differentiation. The secondary phloem affects cambium activity and, through the cambium, the secondary xylem. Internal influences on the secondary phloem are more pronounced than external. The external and internal factors have a direct influence on sieve tubes and sieve cells, during development as well as at maturity. Therefore further investigations are necessary to understand the formation of secondary phloem. According to Denne (1976), the major seasonal variation in tracheid size and thickness of wall in seedlings of *Picea sitchensis* is likely to be associated rather with changes in the production of hormones, than with variation in substrate availability.

Population and Wood Requirement. The world population is growing fast and the non-renewable resources are dwindling. The demand for wood and wood products is continuously rising, through direct burning of fuel, production of charcoal, alcohol (methane), and through the use of cellulose by the plastic industry. Cell and tissue culture techniques can introduce new genetic variation and reduce barriers to crossability amongst species, and thus assume a role supporting biotechnology (Mohammad and Vidavar 1988).

Pollution. There have been several studies on the harmful effects of air pollutants on plants. Inhibition of seed germination, retarded tree growth, and damage to leaves are some of them (see Renzoni et al. 1990). *Pinus pinea* is very sensitive to pollution, and it may be used as an environmental indicator of air pollution. In vitro germination responses have been investigated from two localities in Italy (San Rossore National Park, and Pisa Town Centre) characterized by different types and levels of pollutants. The pollen from San Rossore showed a higher rate of germination (80%) while those from Pisa showed only 40%. Also, pollen from Pisa had several morphological anomalies, such as reduced size and wrinkled grains. The data suggest that there is higher pollution in Pisa than in San Rossore. This has an adverse/disturbing effect on sporogenesis, microspore maturation and development of pollen grains. The reduction in pollen fertility caused by pollutants may lead to a decrease in seed production (Renzoni et al. 1990). Similar investigations in other taxa would be fruitful.

Symbiosis. The study of coralloid roots in cycads is fascinating. Since its discovery in *Cycas revoluta* by Reinke (1872), little work has been done on this aspect (see Chap. 8), and knowledge about coralloid root growth and development is inadequate. There is no information on the transition stage between precoralloid and coralloid roots. The identification of specific stages of development based on distinct morphological characteristics would provide a more accurate basis for further anatomical and ultrastructural investigations. The study of the processes involved in the transition of apices from the precoralloid to coralloid condition, the mechanism involved in initial invasion by algae, colonization and establishement of symbioses in mature precoralloids, and continued invasion of active coralloid apices would be interesting. There is also urgent need for a detailed study of the various dermal tissues of precoralloids, coralloids and normal lateral roots in order to determine the identity of the papillose sheath (root cap or epidermis) and the algal zone (epidermis or specialized cortical tissues).

Symbiosis between N_2-fixing Cyanobacteria and cereal crops has been attempted as a possible alternative to reliance on fertilizers (see Ahern and Staff 1994). Further investigation of the process of invasion and establishment of symbiosis in cycad coralloid roots may provide significant results.

Reproductive Biology. The study of the reproductive biology of gymnosperms requires a good deal of attention; only a few taxa have been studied. Detailed investigations are necessary to fill up the gaps in our knowledge of anatomy, reproductive biology, cytology, genetics and breeding behaviour. Physiological, biochemical, cell and nuclear studies have been more or less completely ignored and deserve immediate attention. Significant knowledge (so far unreported because of the exclusive use of the light microscope) will emerge by employing new techniques—histo- and cytochemistry, electron (TEM, SEM), fluorescence and interference microscopy.

Pollen Biology. Intensive investigations (using modern tools and techniques) on pollination biology—the role of the pollination drop, anemophily/entomophily, pollen viability, origin and role of pollen tube, wall/alveoli formation in female gametophyte, archegonium etc.—are necessary.

Life History. The life history of several gymnosperms and their temporal correlations in the reproductive cycle needs to be investigated. Wild gene resources can be identified and superior genetic variation can be multiplied or bred under control conditions through the knowledge of reproductive biology. It can be used for tree improvement programmes.

An ethnobotanical survey for the use of gymnosperms would be of great use (Sharma 1994).

Finally, our understanding and knowledge of the gymnosperms is far from complete. Detailed investigations (structural, physiological and biochemical) have so far been confined to only a few taxa. Therefore, there is urgent need to intensify these studies.

The possibilities of studying this fascinating group of plants—Gymnosperms: the last group of archaegoniates—are endless.

References

Ahern CP, Staff IA (1994) Symbiosis in cycads: The origin and development of coralloid roots in *Macrozamia communis* (Cycadaceae). Am J Bot 81: 1559–1570

Aldrich HC, Vasil IK (1970) Ultrastructure of the postmeiotic nuclear envelope in microspores of *Podocarpus macrophyllus*. J Ultrastr Res 32: 307–315

Alfieri FJ, Evert RF (1973) Structure and seasonal development of the secondary phloem in the Pinaceae. Bot Gaz 134: 17–25

Alfieri FJ, Mottola PM (1983) Seasonal changes in the phloem of *Ephedra californica* Wats. Bot Gaz 144: 204–246

Alosi MC, Alfieri FJ (1972) Ontogeny and structure of the secondary phloem in *Ephedra*. Am J Bot 59: 818–827

Al-Talib KH, Torrey JG (1959) The aseptic culture of isolated buds of *Pseudotsuga taxifolia*. Plant Physiol 34: 630–637

Anderson AB (1955) Recovery and utilization of tree extractives. Econ Bot 9: 106–140

Anderson AB (1967) Silvichemicals from the forest. Econ Bot 21: 15–30

Andre D (1956) Contribution à l'étude morphologique du cône femelle de quelques gymnospermes (Céphalotaxacées, Junipéroidées, Taxacées). Natur Monsp 8: 3–35

Andrews Jr HN (1961) Studies in Paleobotany, pp 1–487. John Wiley, New York

Archangelsky SA (1965) Fossil Ginkgoales from the Tico Flora, Santa Cruz Province, Argentina. Bull Br Mus Nat Hist, Geol 10: 121–137

Arnold CA (1948) Classification of gymnosperms from the viewpoint of paleobotany. Bot Gaz 110: 2–12

Arnold CA (1953) Origin and relationships of the cycads. Phytomorphology 3: 51–65

Arnold CA (1967) The proper designations of the foliage and stems of the Cordaitales. Phytomorphology 17: 346–350

Attree SM, Dunstan DI, Fowke LC (1989) Initiation of embryogenic callus and suspension cultures, and improved embryo regeneration from protoplasts of white spruce (*Picea glauca*). Can J Bot 67: 1790–1795

Attree SM, Tautorus TE, Dunstan DI, Fowke LC (1990) Somatic embryo maturation, germination, and soil establishment of plants of black and white spruce (*Picea mariana* and *Picea glauca*). Can J Bot 68: 2583–2589

Audran JC (1965) Contribution à l'étude de la structure de la paroi du grain de pollen chez *Dioon, Stangeria, Ceratozamia, Cycas* et *Encephalartos*. Ann Univ ARERS (Reims) 3: 130–144

Audran JC (1971) Contribution à l'étude la microsporogénése chez les Cycadales: Sur la présence de systéms á functions autophagiqus dans les cellules-méres primordiales durant leur phase de multiplication chez le *Ceratozamia mexicana* (Cycadacees). Ann Univ ARERS (Reims) 9: 106–121

Avanzi S, Cionini PG (1971) A DNA cytophotometric investigation on the development of the female gametophyte of *Ginkgo biloba*. Caryologia 24: 105–116

Babu CR, Joshi A, Sardesai N (1993) Biodiversity: A Genetic Treasure. Indian Farming 43: 25–36

Baird AM (1939) A contribution to the life history of *Macrozamia reidlei*. J Proc R Soc West Aust 25: 153–175

Baird AM (1953) The life history of *Callitris*. Phytomorphology 3: 258–284

Bakshi BK, Kumar D (1968) Forest tree mycorrhiza. Indian For 94; 79–84

Ball E (1950) Differentiation in a callus culture of *Sequoia sempervirens*. Growth 14: 295–325

Ball E (1955) Studies of the nutrition of the callus culture of *Sequoia sempervirens*. Ann Biol 31: 281–305

Ball E (1956a) Growth of the embryo of *Ginkgo biloba* under experimental conditions: 1. The apex of the first root of the seedling in vitro. Am J Bot 43: 488–495

Ball E (1956b) Growth of the embryo of *Ginkgo biloba* under experimental conditions: 2. Effects of a longitudinal split in the tip of the hypocotyl. Am J Bot 43: 802–810

Ball E (1959) Growth of the embryo of *Ginkgo biloba* under experimental conditions: 3. Growth rates of root and shoot upon media absorbed through the cotyledons. Am J Bot 46: 130–139

Ball EA (1978) Cloning in vitro of *Sequoia sempervirens* (Abstr), p 163. In: Thorpe TA (ed) Frontiers of Plant Tissue Culture 1978. Proc Fourth Int Congr Plant Tissue and Cell Culture. Calgary (Canada)

Bamber RK, Summerville R, Gregory J (1978) Unusual cells in the mesophyll zone of the leaves of *Araucaria*. Aust J Bot 26: 177–187

Barnett JR (1974) Secondary phloem in *Pinus radiata* D. Don. 1. Structure of differentiating sieve cells. N Z J Bot 12: 245–274

Baumann BB (1960) The botanical aspects of ancient Egyptian embalming and burial. Econ Bot 14: 84–104

Baylis GTS, McNabb RFR, Morrison TM (1963) The mycorrhizal nodules of podocarps. Trans Br Mycol Soc 46: 378–384

Beck CB (1960) Connection between *Archaeopteris* and *Callixylon*. Science 131: 1524–1525

Beck CB (1960) The identity of *Archaeopteris* and *Callixylon*. Brittonia 12: 351–368

Beck CB (1962) Reconstructions of *Archaeopteris*, and further consideration of its phylogenetic position. Am J Bot 49: 373–382

Beck CB (1970) The appearance of gymnospermous structure. Biol Rev 45: 379–400

Beck CB (1976) Current status of the Progymnospermopsida. Rev Palaeobot Palynol 21: 5–23

Beck CB (1981) *Archaeopteris* and its role in vascular plant evolution In: Niklas KJ (ed) Paleobotany, Paleoecology and Evolution, Vol 1: pp 193–230. Praeger Publ, New York

Beck CB (1985) Gymnosperm phylogeny: A commentary on the views of S.V. Meyen. Bot Rev 51: 273–294

Becwar MR, Noland TL, Wann SR (1987) A method for qualification of the level of somatic embryogenesis among Norway spruce callus lines. Plant Cell Rep 6: 35–38

Becwar MR, Nagmani R, Wann SR (1990) Initiation of embryogenic cultures and somatic embryo development in loblolly pine (*Pinus taeda*). Can J For Res 20: 810–817

Becwar MR, Blush TD, Brown DW, Chesick EE (1991) Multiple patternal genotypes in embryogenic tissue derived from individual immature loblolly pine seeds. Plant Cell, Tissue and Organ Culture 26: 37–44

Behnke HD, Paliwal GS (1973) Ultrastructure of phloem and its development in *Gnetum gnemon* with some observations of *Ephedra campylopoda*. Protoplasma 78: 305–319

Behnke HD, Herrmann S (1978) Fine structure and development of laticifers in *Gnetum gnemon* L. Protoplasma 95: 371–384

Bell PR. Duckett JG, Myles D (1971) The occurrence of a multilayered structure in the motile spermatozoids of *Pteridium aquilinum*. J Ultrastruct Res 34: 181–189

Berlyn GP (1962) Developmental patterns in pine polyembryony. Am J Bot 49: 327–333

Berlyn GP (1967) The structure and germination in *Pinus lambertiana* Dougl. Yale Univ School Forestry Bull No. 71: 1–36

Berlyn GP, Miksche JP (1965) Growth of excised pine embryos and the role of the cotyledons during germination in vitro. Am J Bot 52: 730–736

Berlyn GP, Beck RC, Renfroe MH (1986) Tissue culture and the propagation and genetic improvement of conifers: Problems and possibilities. Tree Physiol 1: 227–240

Berridge EM, Sandy E (1907) Oogenesis and embryogeny in *Ephedra distachya*. New Phytol 6: 127–134

Bhatnagar SP, Singh MN (1984) Organogenesis in the cultured female gametophyte of *Ephedra foliata*. J Exp Bot 35: 268–278

Bhatnagar SP, Singh MN, Kapur N (1983) Preliminary investigations on organ differentiation in tissue cultures of *Cedrus deodara* and *Pinus roxburghii*. Indian J Exp Biol 21: 524–526

Bierhorst DW (1971) Morphology of Vascular Plants, pp 1–560. Macmillan, New York

Bino RJ, Dafni A, Meeuse ADJ (1984) Entomophily in the dioecious gymnosperm *Ephedra aphylla* Forsk. (= *E. alte* C.A. Mey.) with some notes on *E. campylopoda* C.A. Mey. 1. Aspects of the entomophilous syndrome. Proc Koninklijke Nederlandse Akad Wetenschappen Ser C 87: 1–13

Biswas Chhaya (1994) Gymnosperms: Morphogenetic studies. pp 25–36 In: Johri BM (ed) Botany in India: History and Progress. Vol 2. pp 1–480. Oxford IBH, New Delhi.

Bliss MC (1921) The vessel in seed plants. Bot Gaz 71: 314–326

Bock W (1954) *Primaraucaria*, a new araucarian genus from the Virginia Triassic. J Paleontol 28: 32–42

Bock W (1969) The American Triassic flora and global distribution, pp 406. Geol Cent Res Ser 3–4 North Wales

Bogorad L (1950) Factors associated with the synthesis of cholorophyll in the dark in seedlings of *Pinus jeffreyi*. Bot Gaz 111: 221–241

Bonga JM (1974) In vitro culture of microsporophylls and megagametophyte tissue of *Pinus*. In Vitro 9: 270–272

Bonga JM (1977) Applications of tissue culture in forestry, pp 93–108. In: Reinert J, Bajaj YPS (eds) Applied and Fundamental Aspects of Plant Cell, Tissue and Organ Culture. Springer, Berlin Heidelberg New York

Bonga JM, Fowler DP (1970) Growth and differentiation in gametophytes of *Pinus resinosa* cultures in vitro. Can J Bot 48: 2205–2207

Bonga JM, McInnis AH (1975) Stimulation of callus development from immature pollen of *Pinus resinosa* by centrifugation. Plant Sci Lett 4: 199–203

Bonnot EJ (1967) Le plan d'organisation fondamental de la spermatids de *Bryum capillare* (L) Hedw. CR Acad Sci Paris 265: 958–961

Borchert R (1967) The cultivation of tissues of *Cupressus lusitanica* in vitro. Physiol Plant 20: 608–616

Borchert R (1968) Spontane Diploidisierung in Gewebekulturen des Megagametophyten von *Pinus lambertiana*. Z Pflanzenphysiol 59: 389–392

Bornman CH, Butler V, Jensen WA (1979a) *Welwitschia mirabilis* : Fine structure of the germinating seed 5. The quiescent and imbibed gametophyte interface. Z Pflanzenphysiol 92: 115–132

Bornman CH, Butler V, Jensen WA (1979b) *Welwitschia mirabilis:* Fine structure of the germinating seed 6. The quiescent embryo. Z Pflanzenphysiol 92: 355–365

Bornman CH, Butler V, Jensen WA (1979c) *Welwitschia mirabilis:* Fine structure of the germinating seed 8. Interface of the developing feeder. Z Pflanzenphysiol 92: 417–430

Bose MN, Pal PK, Harris TM (1984) The *Pentoxylon* plant. Phil Trans R Soc London B 310: 77–108

Bouck GB, Cronshaw J (1965) The fine structure of differentiating sieve elements. J Cell Biol 25: 79–96

Bouman F (1984) The ovule, pp 123–157. In: Johri BM (ed) Embryology of Angiosperms, pp 1–830. Springer Verlag Berlin

Bourgkard F, Favre JM (1988) Somatic embryos from callus of *Sequoia sempervirens.* Plant Cell Rep 7: 445–448

Bower FO (1881) On the germination and histology of seedlings of *Welwitschia mirabilis.* Quart J Microsc Sci 21: 15–30

Boyle P, Doyle J (1953) Development in *Podocarpus nivalis* in relation to other podocarps 1. Gametophytes and fertilization. Sci Proc R Dublin Soc 26: 179–205

Boyle P, Doyle J (1954) Development in *Podocarpus nivalis* in relation to other podocarps 2. Embryogeny in *Eupodocarpus.* Sci Proc R Dublin Soc 26: 289–312

Bracegirdle B, Miles PH (1973) An Atlas of Plant Structure. pp 1-104. Vol 2. Heinemann Educational Books, London

Branscheidt P (1930) Zur Physiologie der Pollenkeimung und ihrer experimentalen Beeinflussung. Planta 11: 368–456

Brennan M, Doyle J (1956) The gametophytes and embryogeny of *Athrotaxis.* Sci Proc R Dublin Soc 27: 193–257

Brongniart A (1828) Prodrome d'une Histoire de Vegetaux fossiles. Paris

Brough P. Taylor MH (1940) An investigation of the life cycle of *Macrozamia spiralis* Miq. Proc Linn Soc NSW 65: 494–524

Brown RW (1943) Some prehistoric trees of the United States. J Forestry 41: 861–868

Brown CL, Gifford Jr EM (1958) The relation of the cotyledons to root development of pine embryos grown in vitro. Plant Physiol 33: 57–64

Brown CL, Sommer HE (1977) Bud and root differentiation in conifer cultures. Tappi 60: 72–73

Bryan GS (1952) The cellular proembryo of *Zamia* and its cap cells. Am J Bot 39: 433–443

Bryan GS, Evans RI (1956) Chromatin behaviour in the development and maturation of the egg nucleus of *Zamia umbrosa.* Am J Bot 43: 640–646

Buchholz JT (1929) The embryogeny of conifers. Proc 4th Intl Congr Plant Sci (New York) 1: 359–392

Buchholz JT (1936) Embryogeny of species of *Podocarpus* of the sub-genus *Stachycarpus.* Bot Gaz 98: 135–146

Buchholz JT (1941) Embryogeny of the Podocarpaceae. Bot Gaz 103: 1–37

Buchholz JT, Gray NE (1948) Taxonomic revision of *Podocarpus* 1. The sections of the genus and their subdivisions with special reference to leaf anatomy. J Arnold Arbor 29: 49–63

Bulard C (1952) Culture aseptique d'embryons de *Ginkgo biloba.* Rôle des cotyledons dans l'absorption du sucre et la croissance de la tige. C R Acad Sci Paris 2350: 739–741

Bulard C, LePage-Degivry MT (1968) Quelques précisions sur les conditions d'obtention d'une inhibition de la croissance épicotylaire chez *Ginkgo biloba* L. sous l'éffet de l'acide gibbérellique. C R Acad Sci Paris 2660: 356–359

Burgerstein A (1900) Ueber das Verhalten der Gymnospermenkeimlinge im Lichte und im Dunkeln. Ber Dtsch Bot Gesell. 18: 168–184

Burke JG (1975) Human use of the California nutmeg tree, *Torreya californica*, and of other members of the genus. Econ Bot 29: 127–138

Burlingame LL (1913) The morphology of *Araucaria brasiliensis* 1. The staminate cone and male gametophyte. Bot Gaz 55: 97–114

Burlingame LL (1914) The morphology of *Araucaria brasiliensis* 2. The ovulate cone and female gametophyte. Bot Gaz 57: 490–508

Burlingame LL (1915) The morphology of *Araucaria brasiliensis* 3. Fertilization, the embryo and the seed. Bot Gaz 59: 1–39

Butenko RG, Yakovleva SM (1962) Controlled organogenesis and regeneration of a whole plant in a culture of non-differentiated plant tissue. Izv Akad Nauk SSSR Biol Ser 2: 230–241

Butler V, Bornman CH, Jensen WA (1979a) *Welwitschia mirabilis* : Fine structure of the germinating seed 4. The imbibed gametophyte. Z Pflanzenphysiol 92: 95–113

Butler V, Bornman CH, Jensen WA (1979b) *Welwitschia mirabilis* : Fine structure of the germinating seed 7. The imbibed embryo. Z Pflanzenphysiol 92: 407–415

Butler V, Bornman CH, Jensen WA (1979c) *Welwitschia mirabilis* : Fine structure of the germinating seed 9. Inner cells of the developing feeder. Z Pflanzenphysiol 93: 45–61

Button J, Bornman CH, Carter M (1971) *Welwitschia mirabilis* : Embryo and Free-Cell Culture. J Expt Bot 23: 922–924

Calder MG (1953) A coniferous petrified forest in Patagonia. Bull Br Mus (Natu Hist) Geol 2: 97–137

Caldwell OW (1907) *Microcycas calocoma*. Bot Gaz 44: 118–141

Camefort H (1959) Sur la nature cytoplasmique des inclusions dites "vitellines" de l'oosphére du *Pinus laricio* Poir (var. *austrica*): Étude en microscopie électronique. C R Acad Sci Paris 248: 1568–1570

Camefort H (1960) Evolution de l'organisation du cytoplasme dans la cellule centrale et l'ossphére du *Pinus laricio* Poir. (var. *austriaca*). C R Acad Sci Paris 250: 3707–3709

Camefort H (1962) L'organisation du cytoplasme dans l'oosphére et la cellule central du *Pinus laricio* Poir. (var. *austrica*). Ann Sci Nat Bot 3: 269–291

Camefort H (1964) Observations sur la structure des chromosomes et des nucléoles de l'oosphere des Pins. C R Acad Sci Paris 259: 4335–4338

Camefort H (1965a) L'organisation du protoplasme dans la gaméte femelle, ou oosphére, du *Ginkgo biloba* L. J Microscop 4: 531–546

Camefort H (1965b) Une interprétation nouvelle de l'organisation du protoplasme de l'oosphere des pins; 407–436. En Travaux dédiés a Lucien Plantefol. Masson, Paris

Camefort H (1966) Etude en microscopie électronique de la dégénérescence du cytoplasme maternel dans les oosphéres embryonnés du *Pinus laricio* Poir. var *austriaca* (*P. nigra* Arn). C R Acad Sci Paris 263: 1443–1446

Camefort H (1967) Orgine et évolution structurale d'un cytoplasme propre á l' embryon, on néocytoplasme, chez les Pins. Ann Univ ARERS 5: 75–88

Camefort H (1968) Cytologie de la fécondation et de la proembryogénése chez quelques gymnospermes. Bull Soc Bot Fr 115: 137–160

Camefort H (1969) Fécondation et proembryogénése chez les Abiétacées (notion de néocytoplasme). Rev Cytol Biol vég 32: 253–271

Campbell R, Durzan D (1976) Vegetative propagation of *Picea glauca* by tissue culture. Can J For Res 6: 240–243

Carniel K (1966) Über die Körnchen-Schicht in den Pollensäcken von *Gnetum gnemon*. Östr bot Z 113: 368–374

Carothers IE (1907) Development of ovule and female gametophyte in *Ginkgo biloba*. Bot Gaz 43: 116–130

Carlquist S (1994) Wood and bark anatomy of *Gnetum gnemon* L. Bot J Linn Soc 116: 203–211

Carlquist S (1996) Wood, bark and stem anatomy of New World species of *Gnetum*. Bot J Linn Soc 120: 1–19

Carlquist S, Robinson AA (1995) Wood and bark anatomy of the African species of *Gnetum*; Origin of lateral meristems. Bot J Linn Soc 118: 123–137

Carmichael JS, Friedman WE (1995) Double fertilization in *Gnetum gnemon* : The Relationship between the Cell Cycle and Sexual Reproduction. The Plant Cell 7: 1975–1988

Carothers ZB, Kreitner GL (1967) Studies of spermatogenesis in the Hepaticae 1. Ultrastructure of the vierergruppe in *Marchantia*. J Cell Biol 33: 43–51

Carothers ZB, Kreitner GL (1968) Studies of spermatogenesis in the Hepaticae 2. Blepharoplast structure in the spermatid of *Marchantia*. J Cell Biol 36: 603–616

Chaloner WG, Lorch J (1960) An opposite-leaved conifer from the Jurassic of Israel. Palaeontology 2: 236–242

Chamberlain CJ (1906) The ovule and female gametophyte of *Dioon*. Bot Gaz 42: 321–358

Chamberlain CJ (1910) Fertilization and embryogeny in *Dioon edule*. Bot Gaz 50: 415–429

Chamberlain CJ (1912) Morphology of *Ceratozamia*. Bot Gaz 53: 1–19

Chamberlain CJ (1913) *Macrozamia moorei* : A connecting link between living and fossil cycads. Bot Gaz 55: 141–154

Chamberlain CJ (1916) *Stangeria paradoxa*. Bot Gaz 61: 353–372

Chamberlain CJ (1919) The Living Cycads. Chicago Univ Press, Chicago

Chamberlain CJ (1920) The living cycads and the phylogeny of seed plants. Am J Bot 7: 146–153

Chamberlain CJ (1935) Gymnosperms: Structure and Evolution, pp 1–484. Univ Press, Chicago

Chandler SF, Bateman C, Blomastedt C, Willyams D, Young R (1989) Forestry biotechnology at Calgene Pacific. Aust J Biotechnol 3: 281–284

Chen WM (1990) Chemical constituents and physiological activity of *Taxus* spp. Acta Pharmaceutica Sinica 25: 227–240

Cheng TY (1975) Adventitious bud formation in culture of Douglas-fir (*Pseudotsuga menzisii* [Mirb] Franco). Plant Sci Lett 5: 97–102

Cheng TY (1976) Vegetative propagation of western hemlock (*Tsuga heterophylla*) through tissue culture. Plant Cell Physiol 17: 1347–1350

Cheng TY, Voqui TH (1977) Regeneration of Douglas fir plantlets through tissue culture. Science 198: 306–307

Chesnoy L (1967) Nature et évolution des formations dites "asteroides" de la cellule centrale de l'archégone du *Juniperus communis* L. Étude en microscopie photonique et électronique. C R Acad Sci Paris 264D: 1016–1019

Chesnoy L 1969a) Sur la participation du gaméte mâle á la constitution du cytoplasme de l'embryon chez le *Biota orientalis* Endl. Rev Cytol Biol vég 32: 273–294

Chesnoy L (1969b) Sur l'origine du cytoplasme des embryons chez le *Biota orientalis* Endl (Cupressacées). C R Acad Sci Paris 268: 1921–1924

Chesnoy L (1971) Etude cytologique des gamétes, de la fécondation et de la proembryogenése chez le *Biota orientalis* Endl. Observation en microscopie photonique et électronique 1. Le gaméte femelle. Rev Cytol Biol vég 34: 257–304

Chesnoy L (1993) Les sécrétions dans la pollinisation des Gymnospermes. Acta Botanica Gallica 140: 145–156

Chesnoy L, Thomas MJ (1971) Electron microscopy studies on gametogenesis and fertilization in gymnosperms. Phytomorphology 21: 50–63

Ching TM (1970) Glyoxysomes in megagametophytes of germinating ponderosa pine seeds. Plant Physiol 46: 475–482

Choi JS, Friedman WE (1991) Development of the pollen tube of *Zamia furfuracea* (Zamiaceae) and its evolutionary implications. Am J Bot 78: 544–560

Chrysler MA (1926) Vascular tissue of *Microcycas calocoma*. Bot Gaz 82: 233–252

Cionini PG (1971) A DNA cytophotometric study on cell nuclei of the archegonial jacket in the female gametophyte of *Ginkgo biloba*. Caryologia 24: 493–500

Clapper RB (1954) Stimulation of pine oleoresin flow by fungus inoculation. Econ Bot 7: 269–284

Coker WC (1907) Fertilization and embryogeny in *Cephalotaxus fortunei*. Bot Gaz 43: 1–10

Coleman WK, Thorpe TA (1977) *In vitro* culture of Western red-cedar (*Thuja plicata* Donn) 1. Plantlet formation. Bot Gaz 138: 298–304

Constabel F (1965) Phenolics in the tissue cultures derived from *Juniperous communis* L. Studies in tannin synthesis, pp183–190 In: White PR, Grove AR (eds) Intl Conf Plant tissue Culture. McCutchan Publ, Berkely (Calif, USA)

Corti FE, Maugini E (1964) Passaggio di corpi figurati fra cellule del tappeto a cellula centrale nellárchegonio dei pini. Caryologia 17: 1–39

Coulter JM, Chamberlain CJ (1903) The embryogeny of *Zamia*. Bot Gaz 35: 184–194

Coulter JM, Chamberlain CJ (1917) Morphology of Gymnosperms, pp 1–466. 2nd edn. Chicago Univ Press, Chicago

Coulter JM, Chrysler MA (1904) Regeneration in *Zamia*. Bot Gaz 38: 130–139

Coulter JM, Land WJG (1905) Spermatogenesis and oogenesis in *Torreya taxifolia*. Bot Gaz 39: 161–178

Cragg GM, Schepartz SA, Suffness M, Grever MR (1993) The taxol supply crisis. New NCI policies for handling the large-scale production of novel natural product anticancer and anti-HIV agents. J Nat Prod 56: 1657–1668

Crepet WL (1972) Investigations of North American cycadeoids: Pollination mechanisms in *Cycadeoidea*. Am J Bot 59: 1048–1056

Crepet WL (1974) Investigations of North American cycadeoids: The reproductive biology of *Cycadeoidea*. Paleontographica B 148: 144–169

Crepet WL, Delevoryas T (1972) Investigations of North American cycadeoids: Early ovule ontogeny. Am J Bot 59: 209–215

Cridland AA (1964) *Amyelon* in American coal-balls. Palaeontology 7: 186–209

Cross GL (1939) The structure and development of the apical meristem in the shoots of *Taxodium distichum*. Bull Torrey Bot Club 66: 431–452

Cross GL (1941) Some histogenic features of the shoot of *Cryptomeria japonica*. Am J Bot 28: 573–582

Dallimore W, Jackson AB (1966) A Handbook of Coniferae, pp 1–729. 4th edn, E. Arnold, London

David A, David H (1979) Isolation and callus formation from cotyledon protoplasts of pine (*Pinus pinaster*). Z Pflanzenphysiol 94: 173–177

David H, Kausa I, David A (1978) In vitro studies of adventitious and axillary buds, shoot formation and plantlet regeneration in *Pinus pinaster* (Abstr.), p 30. In: Thorpe

TA (ed) Frontiers of Plant Tissue Culture 1978. Proc Fourth Intl Congr Plant Tissue and Cell Culture, Calgary (Canada)

David H, de Boucaud MT, Gaultier JM, David A (1986) Sustained division of protoplast-derived cells from primary leaves of *Pinus pinaster*, factors affecting growth and change in nuclear DNA content. Tree Physiol 1: 21–30

De Bary A (1984) Comparative anatomy of the vegetative organs of the Phanerogams and Ferns (translated by Bower FO, Scott DH), Oxford

De Cleene M, De Ley J (1976) The host range of crown gall. Bot Rev 42: 389–466

De Laubenfels DJ (1972) Flore de la Nouvelle-Calédonie et dépendances 4. Gymnospermes, p 168. Mus Natl Hist Natr

Delevoryas T (1953) A new male cordaitan fructification from the Kansas Carboniferous. Am J Bot 40: 144–150

Delevoryas T (1959) Investigations of North American cycadeoids: *Monanthesia*. Am J Bot 46: 657–666

Delevoryas T (1960) Investigations of North American cycadeoids: Trunks from Wyoming. Am J Bot 47: 778–786

Delevoryas T (1962) Morphology and evolution of fossil plants. New York London

Delevoryas T (1963) Investigations of North American cycadeoids: Cones of *Cycadeoidea*. Am J Bot 50: 45–52

Delevoryas T (1969) Glossopterid leaves from the Middle Jurassic of Oaxaca, Mexico. Science 165: 895–896

Delevoryas T (1971) Biotic provinces and the Jurassic-Cretaceons floral transition. Proc North American Paleontological Convention. Pt 1, pp 1660–1674

Delevoryas T, Hope RC (1971) A new Triassic cycad and its phyletic implications. Postilla 150: 1–14

De Luca P, Sabato S (1979) In vitro spermatogenesis of *Encephalartos* Lehn. Caryologia 32: 241–245

De Luca P, Sabato S (1980) Regeneration of coralloid roots on cycad megagametophytes. Plant Sci Lett 18: 27–31

De Luca P, Moretti A, Sabato S (1979) Regeneration in megagamemetophytes of cycads. Giorn Bot Ital 113: 129–143

Denne MP (1976) Effects of environmental change on wood production and wood structure in *Picea sitchensis* sedlings. Ann Bot (London) 40: 1017–1028

Deshpande BD, Bhatnagar P (1961) Apical meristems of *Ephedra foliata*. Bot Gaz 122: 279–284

De Silva BLT, Tambiah MS (1952) A contribution to the life history of *Cycas rumphii* Miq. Ceylon J Sci 12A: 1–22

De Sloover-Colinet A (1963) Chamber pollinique et gamétophyte mâle chez *Ginkgo biloba*. Cellule 64: 129–145

De Sloover JL (1963) Etudes sur les Cycadales-III Nucelli, gametophyte, famelle et embryon chez *Encephalartos poggei*. Cellule 64: 149–200

Dexheimer J (1969) Sur l' étude comparée des ultrastructures des grains de pollen des angiospermes et des préphanarogames. Rev Cytol Biol vég 32: 129–140

Dexheimer J (1970) Recherches cytophysiologiqus sur les grains de pollen. Rev Cytol Biol vég 33: 169–234

Dexheimer J (1973) Étude ultrastructurale du gametophyte femelle de *Ginkgo biloba* 1. Les cellules a reserves. Caryologia (Suppl) 25: 85–96

Dickinson HG (1970) The fine structure of a peritapetal membrane investing the microsporangium of *Pinus banksiana*. New Phytol 69: 1065–1068

Dickinson HG (1971) The role played by sporopollenin in the development of pollen in *Pinus banksiana*, pp 31–67. In: Brooks J, Grant PR, Muir MD, Van Gizel P, Shaw G (eds) Sporopollenin. Academic Press, London New York

Dickinson HG, Bell PR (1970a) The development of the sacci during pollen formation in *Pinus banksiana*. Grana 10: 101–108

Dickinson HG, Bell PR (1970b) Neocytoplasmic interaction at the nuclear envelope in post meiotic microspores of *Pinus banksiana*. J Ultrastr Res 33: 356–359

Dickinson HG, Bell PR (1972) The role of the tapetum in the formation of sporopollenin-containing structure during microsporogenesis in *Pinus banksiana*. Planta 107: 205–215

Dickinson HG, Bell PR (1976a) Development of taptum in *Pinus banksiana* preceeding sporogenesis. Ann Bot (London) 40: 103–113

Dickinson HG, Bell PR (1976b) The changes in the tapetum of *Pinus banksiana* accompanying formation and maturation of the pollen. Ann Bot (London) 40: 1101–1109

Dilcher DL (1969) *Podocarpus* from the Eocene of North America. Science 164: 299–301

Dodd MC, van Staden J (1981) Germination and viability studies on the seeds of *Podocarpus henkelii* Stapf. S African J Sci 77: 171–174

Dodd MC, van Staden J, Smith MT (1989) Seed development in *Podocarpus henkelii* Stapf. : An ultrastructural and biochemical study. Ann Bot (London) 64: 297–310

Dogra PD (1967) Seed sterility and disturbances in embryogeny in conifers with particular reference to seed testing and breeding in Pinaceae. Stud Forest Suecica 45: 1–87

Dogra PD (1992) Embryogeny of primitive gymnosperms *Ginkgo* and cycads-proembryo-basal plan and evolutionary trends. Phytomorphology 42: 157–184

Dorety HA (1908) The seedling of *Ceratozamia*. Bot Gaz 46: 203–220

Downie DG (1928) Male gametophyte of *Microcycas calocoma*. Bot Gaz 85: 437–450

Doyle J (1945) Developmental lines in pollination mechanism in the Coniferales. Sci Proc R Dublin Soc 24: 43–62

Doyle J (1954) Development in *Podocarpus nivalis* in relation to other podocarps 3. General conclusions. Sci Proc R Dublin Soc 26: 347–377

Doyle J, Brennan M (1971) Cleavage polyembryony in conifers and taxads : A survey 1. Podocarps, taxads and taxadioids. Sci Proc R Dublin Soc 4A: 57–88

Doyle J, Brennan M (1972) Cleavage polyembryony in conifers and taxads : A survey 2. Cupressaceae, Pinaceae and conclusions. Sci Proc R Dublin Soc 4A: 137–158

Doyle J, O'Leary M (1935) Pollination in *Pinus*. Sci Proc R Dublin Soc 21: 181–190

Duckett JG (1973) An ultrastructural study of the differentiation of the spermatozoid of *Equisetum*. J Cell Sci 12: 95–129

Duckett JG, Bell PR (1969) The occurrence of a multilayered structure in the sperm of a pteridophyte. Planta 89: 203–211

Duckett JG, Bell PR (1971) Studies on fertilization in archegoniate plants 1. Changes in the structure of spermatozoids of *Pteridium aquilinum* (L.) Kuhn during entry into the archegonium. Cytobiologie 4: 421–436

Duchartre P (1888) Notes sur l'envaciment de l'albumen d'un *Cycas*. Bull Soc Bot Fr 35: 243–251

Duhoux E (1980) Protoplast isolation of gymnosperm pollen. Z Pflanzenphysiol 99: 207–214

Dunstan DI (1988) Prospects and progress in conifer biotechnology. Can J For Res 18: 1497–1506

Dupler AW (1917) The gametophytes of *Taxus canadensis*. Bot Gaz 64: 115–136

Durzan DJ, Chalupa V (1968) Free sugars, amino acids, and soluble proteins in the embryo and female gametophyte of jack pine as related to climate and the seed source. Can J Bot 46: 417–428

Durzan DJ, Campbell RA (1974) Prospects for the mass production of improved stock of forest trees by cell and tissue culture. Can J For Res 4: 151–174

Durzan DJ, Gupta PK (1987) Somatic embryogenesis and polyembryogenesis in Douglasfir cell suspension cultures. Plant Sci 52: 229–235

Durzan DJ, Mia AJ, Ramaiah PK (1971) The metabolism and subcellular organization of the jack pine embryo (*Pinus banksiana*) during germination. Can J Bot 49: 927–938

Eames AJ (1913) The morphology of *Agathis australis*. Ann Bot (London) 27: 1–38

Eames AJ (1952) Relationships of the Ephedrales. Phytomorphology 2: 79–100

Eckey EW (1954) Vegetable Fats and Oils, ACS Monograph No 123. Reinhold, New York

Edgington SM (1991) Taxol out of the woods. Biotechnology 9: 933–937

Eicke R, Metzner-Kuster I (1961) Feinbauuntersuchungen an den Tracheiden von Cycadeen 1. *Cycas revoluta* und *Encephalartos* Spec. Ber Dtsch Bot Ges 74: 99–104

Ellis DD, Bilderback DE (1984) Multiple bud formation by cultured embryos of *Pinus ponderosa*. J Plant Physiol 115: 201–204

Endlicher S (1842) Mantissa botanica sistens genera plantarum supplementum secundeum, pp 1–114. Vienna

Endlicher SLF (1847) Synopsis Coniferarum. Sangalli

Engvild KC (1964) Growth and chlorophyll formation of dark-grown pine embryos on different media. Physiol Plantarum 17: 866–874

Esau K (1953) Plant Anatomy. Wiley, New York

Evert RF, Alfieri FJ (1965) Ontogeny and structure of coniferous sieve cells. Am J Bot 52: 1058–1066

Evert RF, Bormann CH, Butler V, Gilliland MG (1973a) Structure and development of sieve-cell protoplasts in leaf veins of *Welwitschia*. Protoplasma 76: 1–21

Evert RF, Bormann CH, Butler V, Gilliland MG (1973b) Structure and development of sieve areas in the leaf veins of *Welwitschia*. Protoplasma 76: 23–44

Fagerlind F (1941) Bau und Entwicklung der *Gnetum*—gametophyten. K Sv Vet-akad Handl 19: 1–55

Fagerlind F (1961) The initiation and early development of the sporangium in vascular plants. Svensk bot Tidskr 55: 229–312

Fagerlind F (1971) The initiation and primary development of the sporangial-forming organ systems in the genus *Ephedra* L. Cellule 68: 287–344

Favre-Duchartre M (1956) Contribution á l' étude de la reproduction chez *Ginkgo biloba*. Rev Cytol Biol vég 17: 1–218

Favre-Duchartre M (1957) Contribution á l' étude de la reproduction chez *Cephalotaxus drupacea*. Rev Cytol Biol vég 18: 305–343

Favre-Duchartre M (1958) *Ginkgo*, an oviparous plant. Phytomorphology 8: 377–390

Favre-Duchartre M (1960a) Contribution á l' étude de la reproduction sexuee chez l' *Araucaria araucana*. C R Acad Sci Paris 250: 4435–4437

Favre-Duchartre M (1960b) Contribution á l' étude des spermatozoïdes de *Taxus baccata*. Rev Cytol Biol vég 21: 329–337

Favre-Duchartre M (1967) Recherches cyto-biologiques sur lee rapprochment et le dévelopments des gamétophytes máles et femelles du *Torreya myristica*. C R Acad Sci Paris 264D: 574–576

Faye M, David A (1983) Isolation and culture of gymnosperm root protoplasts (*Pinus pinaster*). Physiol Plant 59: 359–362

Fett-Neto AG, DiCosmo F, Reynolds WF, Sakato K (1992) Cell culture of *Taxus* as a source of the antineoplastic drug taxol and related taxanes. Biotechnology 10: 1572–1575

Finer JJ, Kriebel HB, Becwar MR (1989) Initiation of embryogenic callus and suspension cultures of eastern white pine (*Pinus strobus* L). Plant Cell Rep 8: 203–206

Florin R (1931) Untersuchungen zur Stammesgeschichte der Coniferales und Cordaitales. K Sv Vet-akad Handl 10: 1–588

Florin R (1933a) Studien über die Cycadales des Mesozoikums nebst Erörterungen über die Spaltöffnungsapparate der Bennettitales. K Sv Vet-akad Handl 12: 1–134

Florin R (1933b) Über einige neue oder wenig bekannte asiatische *Ephedra* Arten der Sect Pseudobaccatae Stapf. K Sv Vet-akad Handl 12 : 1–44

Florin R (1938–45) Die koniferen des Oberkarbons und des unteren Perms. 1–8. Paleontographica 85: 1–729

Florin R (1939) The morphology of the female fructifications in cordaites and conifers of Palaeozoic age. Bot Notiser 36: 547–565

Florin R (1944) Die Koniferen des oberkarbons und des Unteren Perms. Palaeontographica B 85: 457–654

Florin R (1945) Die Koniferen des oberkarobons und des Unteren Perms. Palaeontographica B 85: 655–729

Florin R (1948) On the morphology and relationships of the Taxaceae. Bot Gaz 110: 31–39

Florin R (1949) The morphology of *Trichopitys heteromorpha* Saporta, a seed plant of Paleozoic age, and the evolution of the female flowers in the Ginkgoinae. Acta Horti Bergiani 15: 79–109

Florin R (1950a) On female reproductive organs in the Cordaitinae. Acta Horti Bergiani 15: 111–134

Florin R (1950b) Upper Carboniferous and Lower Permian Conifers. Bot Rev 16: 258–282

Florin R (1951) Evolution in Cordaites and Conifers. Acta Horti Bergiani 15: 285–388

Florin R (1954) The female reproductive structures of conifers and taxads. Bot Rev 29: 367–389

Florin R (1955) The systematics of the gymnosperms, pp 323–403. In: Kessel EL (ed) A Century of Progress in the Natural Sciences 1853–1953. California Acad Sci, San Francisco

Florin R (1958) On Jurassic taxads and conifers from North-Western Europe and Eastern Greenland. Acta Horti Bergiani 17: 257–402

Foster AS (1938) Structure and growth of the shoot apex in *Ginkgo biloba*. Bull Torrey Bot Club 65: 531–556

Foster AS, Gifford Jr EM (1959) Comparative Morphology of Vascular Plants, pp 1–555. WH Freeman, San Francisco

Foster RC, Marks GC (1966) The fine structure of the mycorrhizae of *Pinus radiata* D. Don. Aust J Biol Sci 19: 1027–1038

Francini E (1958) Ecologia comperata di *Pinus halepensis* ILL, *Pinus pinaster* SOL e *Pinus pinea* L. sulla base del comportamento del gametofito feminile. Acad Italiana Sci For 7: 107–172

Freidman WE (1987) Growth and development of the male gametophyte of *Ginkgo biloba* within the ovule (in vivo). Am J Bot 74: 1797–1815

Friedman WE (1990a) Double fertilization in *Ephedra*, a non-flowering seed plant: Its bearing on the origin of angiosperms. Science 247: 951–954

Friedman WE (1990b) Sexual reproduction in *Ephedra nevadensis* (Ephedraceae): Further evidence of double fertilization in a non-flowering seed plant. Am J Bot 77: 1582–1598

Friedman WE (1991) Double fertilization in *Ephedra trifurca*, a non-flowering seed plant: The relationship between fertilization events and the cell cycle. Protoplasma 165: 106–120

Friedman WE (1992a) Double fertilization in non flowering seed plants and its relevance to the origin of flowering plants. Intl Rev Cytol 140: 319–355

Friedman WE (1992b) Evidence of a pre-angiosperm origin of endosperm: Implications for the evolution of flowering plants. Science 255: 336–339

Friedman WE (1994) The evolution of embryogeny in seed plants and the developmental origin and early history of endosperm. Am J Bot 81: 1468–1486

Friedman WE (1995) Organismal duplication, inclusive fitness theory and altruism: Understanding the evolution of endosperm and the angiosperm reproductive syndrome. Proc Natl Acad Sci USA 92: 3913–3917

Friedman WE, Goliber TE (1986) Photosynthesis in the female gametophyte of *Ginkgo biloba*. Am J Bot 73: 1261–1266

Fritsch FE (1945) The Structure and Reproduction of the Algae. Vol 2, Cambridge Univ Press, Cambridge (England)

Gangulee HC, Kar AK (1982) College Botany, pp 1–1184. Vol 2, New Central Book Agency, Calcutta (India)

Gaussen H (1950-52) Les Gymnospermes actuelles et fossiles. Toulouse (France)

Gauthert RJ (1966) Factors affecting differentiation of plant tissues grown in vitro, pp 55–71. In: Beermann W (ed) Cell differentiation and morphogenesis. North Holland Publ, Amsterdam

Ghose SL (1924) A contribution to the morphology of *Agathis ovata* (Moore) Warb. J Indian Bot Soc 4: 79–86

Gifford EM (1943) The structure and development of the shoot apex of *Ephedra altissima* Desf. Bull Torrey Bot Club 70: 15–25

Gifford Jr EM, Lin J (1975) Light microscope and ultrastructural studies of the male gametophyte in *Ginkgo biloba*: The spermatogenous cell. Am J Bot 62: 974–981

Goo M, Negisi K (1952) Changes of reserve foods in the seeds of Kuromater (*Pinus thunbergii* Parl) during the course of germination. Bull Tokyo Univ Forests No. 43.

Good CW, Taylor TN (1970) On the structure of *Cordaites flicis* Benson from the Lower Pennsylvanian of North America. Palaeontology 13: 29–39

Gorska-Brylass A (1968) Callose in the cell walls of developing male gametophytes in gymnosperms. Acta Soc Bot Polon 37: 119–124

Gould RE (1971) *Lyssoxylon grigsbyi*, a cycad trunk from the Upper Triassic of Arizona and New Mexico. Am J Bot 58: 239–248

Gould RE (1975) A preliminary report on petrified axes of *Vertebraria* from the Permian of Eastern Australia, pp109–115. In: Campbell KSW (ed) Gondwana Geology: Papers presented at the third Gondwana Symposium, Canberra. Natl Univ Press, Canberra

Gould RE, Delevoryas T (1977) The biology of *Glossopteris* : Evidence from petrified seed-bearing and pollen-bearing organs. Alcheringa 1: 387–399

Grand'Eury FC (1877) Flore Carbonifere du Department de la Loire et du centre de la France. Men Acad Sci Paris 24: 1–624

Greenwood MS, Berlyn GP (1965) The regeneration of active root meristems in vitro by hypocotyl sections from dormant *Pinus lambertiana* embryos. Can J Bot 43: 173–175

Greguss P (1955) Identification of Living Gymnosperms on the Basis of Xylotomy, pp 1–263. Akad Kiado, Budapest

Gresshoff PM, Doy CH (1972) Development and differentiation of haploid *Lycopersicon esculentum* (tomato). Planta 107: 161–170

Grilli M (1963) Recherche al microscopio electtronic sulle cianoficee viventi nei tubereoli radicali di *Cycas revoluta*. Ann Microbiol Enzimol.

Griffith MM (1952) The structure and the growth of the shoot apex in *Araucaria*. Am J Bot 39: 253–263

Griffith MM (1971) Transfusion tissue in leaves of *Cephalotaxus*. Phytomorphology 21: 86–92

Gullvag BM (1966) The fine structure of some gymnosperm pollen walls. Grana Palynol 6: 435–475

Gunckel JE, Wetmore RH (1946) Studies of development in long shoots and short shoots of *Ginkgo biloba* L 1. The origin and pattern of development of the cortex, pith and procambium. Am J Bot 33: 285–295

Gunckel JE, Thimann KV, Wetmore RH (1949) Studies of development in long shoots and short shoots in *Ginkgo biloba* L 2. Growth habit, shoot expression and the mechanism of its control. Am J Bot 36: 309–316

Gupta PK (1988) Advances in Biotechnology of Conifers. Curr Sci 57: 629–637

Gupta PK, Durzan DJ (1986a) Somatic proembryogenesis from callus of mature sugar pine embryos. Biotechnology 4: 643–645

Gupta PK, Durzan DJ (1986b) Isolation and cell regeneration of protoplasts from sugar pine (*Pinus lambertiana*). Plant Cell Rep 5: 346–348

Gupta PK, Druzan DJ (1987a) Biotechnolgy of somatic polyembryogenesis and plantlet regeneration in loblolly pine. Biotechnology 5: 147–151

Gupta PK, Durzan DJ (1987b) Somatic embryos from protoplasts of loblolly pine proembryonal cells. Biotechnology 5: 710–712

Gupta PK, Durzan DJ, Finkle BJ (1987) Somatic polyembryogenesis is embryogenic cell masses of *Picea abies* (Norway spruce) and *Pinus taeda* (loblolly pine) after thawing from liquid nitrogen, Can J For Res 17: 1130–1134

Haines RJ (1983) Embryo development and anatomy in *Araucaria* Juss. Aust J Bot 31: 125–140

Haines RJ, de Fossard RA (1977) Propagation of hoop pine (*Araucaria cunninghamii* Ait). Acta Hort 78: 297–302

Haines RJ, Prakash N (1980) proembryo development and suspensor elongation in *Araucaria* Juss. Aust J Bot 28: 511–522

Hakansson A (1956) Seed development of *Picea abies* and *Pinus sylvestris*. Medd Stat Skigsforskinst 46: 1–23

Hakman I, Von Arnold S (1985) Plantlet regeneration through somatic embryogenesis in *Picea abies* (Norway spruce). J-Plant Physiol 121: 149–158

Hakman I, Fowke LC (1987) Somatic embryogenesis in *Picea glauca* (white spruce) and *Picea mariana* (black spruce). Can J Bot 65: 656–659

Hakman I, Fowke LC, Arnold SV, Eriksson T (1985) The development of somatic embryos in tissue cultures initiated from immature embryos of *Picea abies* (Norway Spruce). Plant Sci 38: 53–59

Hakman I, Rennie P, Fowke LC (1987) A light and electron microscope study of *Picea glauca* (white spruce) somatic embryos. Protoplasma 240: 100–109

Halliday J, Pate JS (1976) Symbiotic nitrogen fixation by blue green algae in the cycad, *Macrozamia reidlei;* physiological characteristics and ecological significance. Aust J Plant Physiol 3: 349–358

Harris TM (1932) The fossil flora of Scoresby Sound, East Greenland 2. Meddelelser om Grønland 85: 1–133

Harris TM (1933) A new member of the Caytoniales. New Phytol 32: 97–114

Harris TM (1935) The fossil flora of Scoresby Sound, East Greenland 4, Ginkgoales, Coniferales, Lycopodiales, and isolated fructifications. Meddelelser om Grønland 112: 1–176

Harris TM (1937) The fossil flora of Scorsby Sound, East Greenland 5. Meddelelser om Grønland 112: 1–112

Harris TM (1941) Cones of extinct Cycadales from the Jurassic rocks of Yorkshire. Phil Trans R Soc London B 231: 75–98

Harris TM (1951) The relationships of the Caytoniales. Phytomorphology 1: 29–39

Harris TM (1961) The fossil cycads. Palaeontology 4: 313-323

Harris TM (1964) The Yorkshire Jurassic Flora 2. Caytoniales, Cycadales and Pteridosperms. Brit Mus Natu Hist, London

Harris TM (1974) The Yorkshire Jurassic Flora 4. Ginkgoales and Czekanowskiales. Brit Mus Nat Hist London

Harris TM (1976) The mesozoic gymnosperms. Rev Palaeobot Palynol 21: 119–134

Hasegawa M, Higuchi T, Ishikawa H (1960) Formation of lignin in tissue cultue of *Pinus strobus*. Plant Cell Physiol 1: 173–182

Hatano K (1957) *a*- Keto acids in pine seeds. Bull Tokyo Univ Forests No 53.

Hatano K, Asakawa S (1964) Physiological processes in forest seeds during maturation, storage and germination. Intl Rev For Res 1: 279–323

Hatch AB (1937) The physical basis of mycotrophy in *Pinus*. Black Rock Forest Bull No. 6: 1–168

Hatch AB, Doak KD (1933) Mycorrhizal and other features of the root system of *Pinus*. J Arnold Arbor 14: 85–99

Haupt AW (1941) Oogenesis and fertilization in *Pinus lambertiana* and *P. monophylla*. Bot Gaz 102: 482–498

Heitz E (1960) Über die Geisselstruktur sowie die *Driergruppe* in den Leber- und Laubmoose, pp 934–937. In: Houwinck AL, Spit BJ (eds). Proc Eur Reg Conf Electron Microscopy 2: Delft : Ned Ver Electronenmicroscop

Hemmerley TE (1970) Economic uses of Eastern Red Cedar. Econ Bot 24: 39–41

Hilditch TP (1956) The Chemical Constituents of Natural Fats. London

Hirase S (1895) Étude sur le *Ginkgo biloba*. Bot Mag (Tokyo) 9: 240

Hodcent E (1965) Les enveloppes polliniques chez *Araucaria columnaris*. Extrait Bull Soc Bot Fr 112: 121–127

Hodcent E (1971) Precisions cytologiques et chronologiques concernant les gametogeneses male et female chez L' *Araucaria araucana*. Ann Univ ARERS 9: 22–26

Hooker JC (1863) On *Welwitschia* : A new genus of Gnetaceae. Trans Linn Soc London 24: 1–48

Horgan K, Holland L (1989) Rooting micropropagated shoots from mature radiata pine. Can J For Res 19: 1309–1315

Howes FN (1948) Nuts: Their Production and Everyday Uses. London

Hu HH, Cheng WC (1948) On the new family Metasequoiaceae and *Metasequoia glyptostroboides*, a living species of *Metasequoia* found in Szechuan and Hupen. Bull Fan Mem Inst Biol (N.S) 1: 153–161

Hu Yu-shi, Yao Bi-jun (1981) Transfusion tissue in gymnosperm leaves. Bot J Linn Soc 83: 263–272

Ikeno S (1896) Das Spermatozoid von *Cycas revoluta*. Bot Mag (Tokyo) 10: 367–368

Ikeno S (1898) Untersuchungen über die Entwicklung der Geschlechtsorgane und den Vorgang der Befruchtung bei *Cycas revoluta*. J Coll Sci Imp Univ Tokyo 12: 151–214

Isenberg IH (1956) Paper making fibers. Econ Bot 10: 176–193

Isikawa H (1974) In vitro formation of adventitious buds and roots on the hypocotyl of *Cryptomeria japonica*. Bot Mag (Tokyo) 87: 73–77

Jäger L (1899) Beiträge zur Kenntnis der Endospermbildung und zur Embryologie von *Taxus baccata* L. Flora 86: 241–288

Jain SM, Newton RJ, Soltes EJ (1988) Enhancement of somatic embryogenesis in Norway spruce (*Picea abies* L). Theor Appl Genet 76: 501–506

Jain SM, Dong N, Newton RJ (1989) Somatic embryogenesis in slash pine (*Pinus elliottii*) from immature embryos cultured in vitro. Plant Sci 65: 233–241

Jane FW (1970) The Structure of Wood, 2nd edn. Adam & Charles Black, London

Jeffrey EC, Torrey RE (1916) *Ginkgo* and the microsporangial mechanisms of the seed plants. Bot Gaz 62: 281–292

Johnson MA (1950) Growth and development of the shoot of *Gnetum gnemon* L. 1. The shoot apex and pith. Bull Torrey Bot Club 77: 354–367

Johnson LAS (1959) The families of cycads and the Zamiaceae of Australia. Proc Linn Soc NSW 34: 64–117

Johnson RW, Riding RT (1981) Structure and ontogeny of the stomatal complex in *Pinus strobus* L. and *P. banksiana* Lamb. Am J Bot 68: 260–268

Johri BM (1936) The life history of *Butomopsis lanceolata* Kunth. Proc Indian Acad Sci B 4: 139–162

Johri BM (1992) Haustorial role of pollen tubes. Ann Bot (London) 70: 471–475

Johri BM, Ambegaokar KB (1984) Embryology: Then and now. pp 1–47. In: Johri BM (ed) Embryology of Angiosperms. pp 1–830. Springer Verlag, Berlin Heidelberg New York Tokyo

Kang QS, Hou SS (1993) Recent advances in the research on the natural antitumor drug taxol. Natural Prod Res Dev 5: 61–68

Karnosky DF (1981) Potential for forest tree improvement via tissue culture. BioScience 31: 114–120

Kato M, Inoue T, Nagamitsu T (1995) Pollination biology of *Gnetum* (Gnetaceae) in a lowland mixed Dipterocarp forset in Sarawak. Am J Bot 82: 862–868

Katsuta M (1961) The synthesis of reserve proteins in ripening pine seeds. J Jap Forestry Soc 43: 157–162

Kaur D (1958) Embryology of *Cephalotaxus drupacea* var *pedunculata*. Proc 45th Indian Sci Congr (Madras), pp 291–292

Kaur D, Konar RN, Bhatnagar SP (1981) Male cone development in *Agathis robusta*. Phytomorphology 31: 104–112

Kausik SB (1974) Ontogeny of the stomata in *Gnetum ula* Brongn. Bot J Linn Soc London 68: 143–151

Kausik SB, Bhattacharya SS (1977) Comparative foliar anatomy of selected gymnosperms: Leaf structure in relation to leaf form in Coniferales and Taxales. Phytomorphology 27: 146–160

Keng H (1969) Aspects of morphology of *Amentotaxus formosana* with a note on the taxonomic position of the genus. J Arnold Arbor 50: 432–446

Khan R (1943) Contributions to the morphology of *Ephedra foliata* Boiss 2. Fertilization and embryogeny. Proc Natl Acad Sci India (Allahabad) 13: 375–375

Khoshoo TN (1962) Cytological evolution in the gymnosperms: Karyotype, pp 119–135. In : Maheshwari P, Johri BM, Vasil IK (eds) Proc Summer School Botany, Darjeeling, Min Sci Res Cultural Affairs, Govt India, New Delhi

Khoshoo TN, Ahuja MR (1963) The chromosomes and relationships of *Welwitschia mirabilis*. Chromosoma 14: 522–533

Kirby E, Cheng T (1979) Colony formation from protoplasts derived from Douglas fir cotyledons. Plant Sci Lett 14: 145–154

Klimaszewska (1989) Plantlet development from immature zygotic embryos of hybrid larch through somatic embryogenesis. Plant Sci 63: 95–103

Kokubugata G, Kondo K (1996) Differential fluorescent-banding patterns in chromosomes of four species of *Cycas* (Cycadaceae). Bot J Linn Soc 120: 51–55

Konar RN (1958) A quantitative survey of some nitrogenous substances and fats in the developing embryos and gametophytes of *Pinus roxburghii* Sarg. Phytomorphology 8: 174–176

Konar RN (1960) The morphology and embryology of *Pinus roxburghii* Sarg. with a comparision with *Pinus wallichiana* Jack. Phytomorphology 10: 305–319

Konar RN (1963a) Anatomical studies on Indian pines with special reference to *Pinus roxburghii* Sarg. Phytomorphology 13: 388–402

Konar RN (1963b) Studies on submerged callus culture of *Pinus gerardiana* Wall. Phytomorphology 13: 165–169

Konar RN (1963c) In vitro studies on *Pinus roxburghii* Sarg, pp 224–229. In : Maheshwari P. Rangaswamy NS (eds) Symp Plant Tissue and Organ Culture. Int Soc Plant Morphologists, Univ Delhi

Konar RN (1963d) A haploid tissue from the pollen of *Ephedra foliata* Boiss. Phytomorphology 13: 170–174

Konar RN, Ramchandani Sundri (1958) The morphology and embryology of *Pinus wallichiana* Jack. Phytomorphology 8: 328–346

Konar RN, Banerjee SK (1963) The morphology and embryology of *Cupressus funebris* Endl. Phytomorphology 13: 321–338

Konar RN, Oberoi YP (1965) In vitro development of embryoids on the cotyledons of *Biota orientalis*. Phytomorphology 15: 137–140

Konar RN, Oberoi YP (1969a) Anatomical studies on *Podocarpus gracilior*. Phytomorphology 19: 122–133

Konar RN, Oberoi YP (1969b) Studies on the morphology and embryology of *Podocarpus gracilior* Pilger. Beitr Biol Pflanzen 45: 329–376

Konar RN, Singh MN (1979) Production of plantlets from the female gametophytes of *Ephedra foliata* Boiss. Z Pflanzenphysiol 95: 87–90

Konar RN, Singh MN (1980) Induction of shoot buds from tissue cultures of *Pinus wallichiana*. Z Pflanzenphysiol 99: 173–177

Konar RN, Moitra A (1980) Ultrastructure, cyto- and histochemistry of female gametophyte of gymnosperms. Gamete Res 3: 67–97

Krassilov VA (1974) *Podocarpus* from the Upper Cretaceous of eastern Asia and its bearing on the theory of conifer evolution. Palaeontology 17: 365–370

Krogstrup P (1986) Embryo-like structures from cotyledons and ripe embryos of Norway spruce (*Picea abies*). Can J For Res 16: 664–668

Krogstrup P, Eriksen EN, Moller JD, Roulund H (1988) Somatic embryogenesis in Sitka spruce (*Picea sitchensis*) (Bong. Carr). Plant Cell Rep 7: 594–597

Kuhlwein H (1937) Zur Physiologie der Pollenkeimung insbesonders der Frage nach dem Befruchtungsverzug bei Gymnospermen. Beih Bot Zbl 57: 37–104

Kubitzki K (1990) Gnetatae. pp 378–391. In: Kranmer KV, Green PS (eds) The families and genera of vascular plants 1. Springer, Berlin

Lacey WS, van Dijk DE, Gordon-Gray KD (1974) New Permian *Glossopteris* flora from Natal. African J Sci 70: 131–141

Lacey WS, van Dijk DE, Gordon-Gray KD (1975) Fossil plants from the Upper Permian in the Mooi River district of Natal, South Africa. Annals Natal Museum 22: 349–420

Lainé E, David A (1990) Somatic embryogenesis in immature embryos and protoplasts of *Pinus caribaea*. Plant Sci 69: 215–224

Land WJG (1904) Spermatogenesis and oogenesis in *Ephedra trifurca*. Bot Gaz 38: 1–18

Land WJG (1907) Fertilization and embryogeny in *Ephedra trifurca*. Bot Gaz 44: 273–292

LaRue CD (1936) The growth of plant embryos in culture. Bull Torrey Bot Club 63: 365–382

LaRue CD (1948) Regeneration in the megagametophyte of *Zamia floridana*. Bull Torrey Bot Club 75: 597–603

LaRue CD (1954) Studies on growth and regeneration in gametophytes and sporophylls of gymnosperms. Brookhaven Natl Lab Symp 6: 187–208

Laudi G, Bonatti PM (1973) Ultrastructure of chloroplasts of some Chlamydospermae (*Ephedra twediana, Gnetum montana, Welwitschia mirabilis*). Caryologia 26: 107–114

Lawson AA (1907) The gametophytes, fertilization and embryo of *Cephalotaxus drupacea.* Ann Bot (London) 21: 1–23

Lawson AA (1926) A contribution to the life history of *Bowenia.* Trans R Soc Edinb 54: 357–394

Lee CL (1955) Fertilization in *Ginkgo biloba.* Bot Gaz 117: 79–100

Le Goc MJ (1917) Effect of foreign pollination on *Cycas rumphii.* Ann Bot Gard Peradiniya 6: 187–194

Lehmann-Baerts M (1967) Étude sur les Gnétales - 12. Ovule, gamétophyte femelle et embryogenése chez *Ephedra distachya* L. Cellule 67: 53–87

Lelu MA, Boulay M, Arnaud Y (1987) Obtention de cals embryogénes á partier de cotylédons de *Picea abies* (L). Karst. prélevés sur de jeunes plantes âgées de 3 á 7 jours aprés germination. C R Acad Sci Ser III Sci Vie 305: 105–109

Lemoigne Y (1967) Paléoflore á Cupressales dans le Trias-Rhétien du Contentin. C R Acad Sci Paris 264: 715–718

LePage-Degivry MT (1968) Mise en évidence d'une dormance associée a une immaturité de l'embryon chez *Taxus baccata* L. C R Acad Sci Paris 266D: 1028–1030

Lepage-Degivry MT (1970) Acide abscissique et dormance chez les embryons de *Taxus baccata* L. C R Acad Sci Paris 271D: 482–484

LePage-Degivry MT (1973a) Etude en culture in vitro de la dormance embryonnaire chez *Taxus baccata* L. Biol Plant 15: 264–269

LePage-Degivry MT (1973b) Intervention d'un inhibiteur lié dans la dormance embryonnaire de *Taxus baccata* L. C R Acad Sci Paris 277D: 177–180

LePage-Degivry MT (1973c) Influence de l'acide abscissique sur le dévelopement des embryons de *Taxus baccata* L. cultivés in vitro. Z Pflanzenphysiol 70: 406–413

LePage-Degivry MT, Garello G (1973) La dormance embryonnaire chez les embryons de *Taxus baccata* L. : Influence de la composition du milieu liquide sur l'induction de la germination. Physiol Plant 29: 204–207

Li T (1934) The development of *Ginkgo* embryos in vitro. Peiping Nat Tsing Hua Univ Sci Rep B2: 41–52

Li T, Shen T (1934) The effect of 'pantothenic' acid on the growth of the radicle of *Ginkgo* embryos in artificial media. Peiping Nat Tsing Hua Univ Sci Rep B2: 53–60

Lindley J (1853) The Vegetable Kingdom. London

Litvay JD, Johnson MA, Verma D, Einspahr D, Weyrauch K (1981) Conifer suspension culture medium development using analytical data from developing seeds. Tech Pap Ser Inst Paper Chem 115: 1–17

Lloyd DG, Wells MS (1992) Reproductive biology of a primitive angiosperm, *Pseudowintera colorata* (Winteraceae), and the evolution of pollination systems in the Anthophyta. Plant Syst Evol 181: 77–95

Long AG (1944) On the prothallus of *Lagenostoma ovoides* Will. Ann Bot (London) 8: 105–117

Long AG (1963) Some specimens of *Lyginopteris papilio* Kidston associated with stems of *Pitys.* Trans R Soc Edinburgh 65: 211–224

Long AG (1979) Observations on the Lower Carboniferous genus *Pitus* Witham. Trans R Soc Edinburgh 70: 111–127

Loo SW, Wang FH (1943) The culture of young conifer embryos in vitro. Science 98: 544

Looby WJ, Doyle J (1939) The ovule, gametophytes and proembryo in *Saxegothaea.* Sci Proc R Dublin Soc 22: 95–117

Looby WJ, Doyle J (1944a) The gametophytes of *Podocarpus andinus*. Sci Proc R Dublin Soc 23: 227–237

Looby WJ, Doyle J (1944b) Fertilization and early embryogeny in *Podocarpus andinus*. Sci Proc R Dublin Soc 23: 257–270

Lopez-Perez MJ, Giminez-Solves A, Calonge FD, Santos-Ruiz AC (1974) Evidence of glyoxysomes in germinating pine seeds. Plant Sci Lett 2: 377–386

Lotsy JP (1899) Contribution to the life history of the genus *Gnetum* 1. The grosser morphology of reproduction in *Gnetum gnemon*. Ann Jard Bot Buitenzorg 1: 46–114

Loze JC (1965) Édude de l'ontogenése de appareil reproducteur femelle de l'if *Taxus baccata* . Rev Cytol Biol vég 28: 211–256

Lu CY, Thorpe TA (1987) Somatic embryogenesis and plantlet regeneration in cultured immature embryos of *Picea glauca*. J Plant Physiol 128: 297–302

Lyon HL (1904) The embryogeny of *Ginkgo*. Minn Bot Studies 3: 275–290

MacDuffie RC (1921) Vessels of the Gnetalean type in angiosperms. Bot Gaz 71: 438–445

Maheshwari P (1935) Contributions to the morphology of *Ephedra foliata* Boiss 1. The development of male and female gametophytes. Proc Indian Acad Sci 1: 586–606

Maheshwari P (1950) An introduction to the Embryology of Angiosperms. pp 1–453. McGraw-Hill, New York

Maheshwari P, Singh H (1960) Economic importance of conifers. J Univ Gauhati 11 (Sci): 1–28

Maheshwari P, Vasil Vimla (1961a) *Gnetum*. Bot Monograph No. 1 : 1–142. Council Sci Ind Res (CSIR), New Delhi

Maheshwari P, Vasil Vimla (1961b) The stomata of *Gnetum*. Ann Bot (London) 25: 313–319

Maheshwari P, Sanwal Madhulata (1963) The archegonium in gymnosperms : A review. Memoirs Indian Bot Soc No 4 : 104–119

Maheshwari P, Singh H (1967) The female gametophyte of gymnosperms. Biol Rev 42: 88–130

Maheshwari P, Biswas Chhaya (1970) *Cedrus*. Bot Monograph No. 5 : 1–115. Council Sci Ind Res (CSIR), New Delhi

Maheshwari P, Konar RN (1971) *Pinus*. Bot Monograph No. 7 : 1–130. Council Sci Ind Res (CSIR), New Delhi

Mantell CL (1950) The natural hard resins–their botany, sources and utilization. Econ Bot 4: 203–242

Manton I (1964) The possible significance of some details of flagellar bases in plants. J L R Microsc Soc 82: 279–285

Mapes G (1982) Morphology and anatomy of ovulate *Lebachia* cones. Bot Soc Am Mis Publ 162:61

Mapes MO, Zaerr JB (1981) The effect of the female gametophyte on the growth of cultured Douglas fir embryos. Ann Bot (London) 48: 577–583

Marsden MPF, Steeres TA (1955) On the primary vascular system and the nodal anatomy of *Ephedra*. J Arnold Arbor 36: 241–258

Martens P (1959) Études sur les Gnétales 3. Structure et ontogenése du cône et de la fleur femelles de *Welwitschia mirabilis*. Cellule 60: 171–286

Martens P (1963) Études sur les Gnétales 6. Recherches sur *Welwitschia mirabilis* 3. L'ovule et le sac embryonnaire. Cellule 63: 309–329

Martens P (1971) Les Gnétophytes. Borntraeger, Berlin Stuttgart

Martens P (1974) Étude sur les Gnétales 14. Recharches sur *Welwitschis mirabilis* 6. La Cupule Trachéidale et le probléme de la vascularisation ovulaire chez les spermatophytes. Cellule 70: 261–292

Martens P (1975) Études sur les Gnétales 15. Recherches sur *Welwitschia mirabilis* 7. Histologie et histogenése de la fleur femelle. Cellule 71: 103–144

Martens P (1977a) *Welwitschia mirabilis* and Neoteny. Am J Bot 64: 916–920

Martens P (1977b) Recherches sur *Welwitschia mirabilis* 9. Vascularisation et histogenése de la plantule. Cellule 72: 131–194

Martens P, Waterkeyn L (1963a) Les prétendus bourgeons cotylédonaires et les trois verticilles foliaires du *Welwitschia*. C R Acad Sci Paris 257: 2048–2052

Martens P, Waterkeyn L (1963b) Stem apex of *Welwitschia*. Phytomorphology 13: 359–363

Martens P, Waterkeyn L (1964a) Organization and functioning of the shoot apical meristem of *Welwitschia* (Abstr), p 423. Proc 10th Intl Bot Congr, Edinburgh Demonstr

Martens P, Waterkeyn L (1964b) Recherches sur *Welwitschia mirabilis* 4. Germination et plantules. Structure, fonctionnement et productions du méristéme caulinaire apical. (Études sur les Gnetales 7). Cellule 65: 5–64

Martens P, Waterkeyn L (1974) Étude sur les Gnétales 13. Recherches sur *Welwitschia mirabilis* 5. Évolution ovulaire et embryogenése. Cellule 70: 163–258

Martin PC (1950) A morphological comparison of *Biota* and *Thuja*. Proc Penn Acad Sci 24: 65–112

Maugini E, Fiordi C (1970) Passagio di materiale dalle cellule del tappeto alla cellula centrale dell 'archegonio e al proembrione di *Ginkyo biloba* L. Caryologia 23: 415–440

McWilliam JR (1958) The role of the micropyle in the pollination of *Pinus*. Bot Gaz 120: 109–117

Mehra PN (1938) The germination of pollen grains in artificial cultures in *Ephedra foliata* Boiss and *Ephedra gerardiana* Wall. Proc Indian Acad Sci B 8: 218–230

Mehra PN (1947) Method of formation of diploid pollen and development of the male gametophyte in *Ephedra saxatilis* Royle. J Indian Bot Soc (Iyengar Comm Vol 1946) : 121–132

Mehra PN (1949) Effect of sulfanilamide on mitotic division in pollen grains of *Ephedra*. Bot Gaz 111: 53–63

Mehra PN (1950) Inequality in size of the male nuclei in the genus *Ephedra*. Ann Bot (London) 14: 331–339

Mehra PN (1988) Indian Conifers : Gnetophytes and Phylogeny of Gymnosperms, pp 1–264. Pramodh P Kapur, Raj Bandhu Ind Complex, New Delhi

Mehra PN (1991) Branched haustorial pollen tubes of *Podocarpus gracilior* Pilger. Proc Indian Natl Sci Acad B 57: 171–174

Mehra PN, Malhotra RK (1947) Stages in the embryogeny of *Cupressus sempervirens* with particular reference to the occurrence of multiple male cells in the male gametophyte. Proc Natl Acad Sci India (Allahabad) 17 Pt B: 129–153

Meyen SV (1984) Basic features of gymnosperm: Systematics and phylogeny as evidenced by the fossil record. Bot Rev 50: 1–111

Meyen SV (1986) Gymnosperm systematics and phylogeny: A reply to comentaries by CB Beck, CN Miller, and GW Rothwell. Bot Rev 52: 300–320

Mia AJ, Durzan DJ (1974) Cytochemical and subcellular organization of the shoot apical meristem of dry and germinating jack pine embryos. Can J For Res 4: 39–54

Miki BLA, Reich TJ, Iyer VN (1987) Microinjection: An experimental tool for studying and modifying plant cells, pp 249–265. In: Hohn T, Schell J (eds) Plant DNA infectious agents. Springer-Verlag, New York

Millay MA, Taylor TN (1974) Morphological studies of Paleozoic saccate pollen. Palaeontographica B 147: 75–99

Millay MA, Taylor TN (1976) Evolutionary trends in fossil gymnosperm pollen. Rev Palaeobot Palynol 21: 65–91

Miller Jr CN (1977) Mesozoic conifers. Bot Rev 43: 217–280

Miller Jr CN (1982) Current status of Paleozoic and Mesozoic conifers. Rev Palaeobot Palynol 37: 99–114

Miller Jr CN (1985) A critical review of SV Meyen's "Basic features of gymnosperm systematics and phylogeny as evidenced by fossil record". Bot Rev 51: 295–318

Minocha SC (1980) Callus and adventitious shoot formation in excised embryos of white pine (*Pinus strobus*). Can J Bot 58: 366–370

Miyake K (1906) Über die spermatozoiden von *Cyvas revoluta*. Ber Deutsch Bot Gesell 24: 78–83

Mizukami I, Gall J (1966) Centriole replication 2. Sperm formation in the fern *Marsilea*, and the cycad *Zamia*. J Cell Biol 29: 97–111

Mohammad GH, Vidaver WE (1988) Root production and plantlet development in tissue-cultured conifers. Plant Cell, Tissue and Organ Culture 14: 137–160

Mohammad GH, Patel ER, Vidaver WE (1989) The control of adventitious root production in tissue-cultured Douglas-fir. Can J For Res 19: 1322–1329

Moitra A, Bhatnagar SP (1982) Review Article : Ultrastructure, cytochemical, and histochemical studies on pollen and male gamete development in gymnosperms. Gamete Res 5: 71–112

Moussel B (1972) Étude en microscopie électronique des cols archégoniaux de l'*Ephedra distachya* L. C R Acad Sci Paris 274: 3372–3374

Moussel B (1974) Étude histologique, cytochimique et ultrastructurale des réserves du gametophyte femelle de l' *Ephedra distachya* L an cours de l' oogénése et de la proembryogénése. Rev Gen Bot 81: 315–334

Moussel B (1978) Double fertilization in the genus *Ephedra*. Phytomorphology 28: 336–345

Moussel B, Moussel C (1973) Évolution rhythmique de l'appareil de golgi et du plasmalemma en liason avec la croissance de la paroi enveloppant le prothalleen femelle cenocytique de l'*Ephedra distachya* L. Caryologia 25 Suppl: 97–108

Muhammad AF, Sattler R (1982) Vessel structure of *Gnetum* and the origin of angiosperms. Am J Bot 69: 1004–1021

Munting AJ, Willemse TM (1987) External influences on development of vascular cambium and its derivatives. Phytomorphology 37: 261–273

Muramanis L, Evert RF (1966) Some aspects of sieve cell ultrastructure in *Pinus strobus* L. Am J Bot 53: 1065–1078

Murashige T, Skoog F (1962) A revised medium for rapid growth and bioassays with tobacco tissue cultures. Physiol Plant 15: 473–497

Nagmani R, Bonga JM (1985) Embryogenesis in subcultured callus of *Larix decidua*. Can J For Res 15: 1088–1091

Nagmani R, Becwar MR, Wann SR (1987) Single cell origin and development of somatic embryos in *Picea abies* (L.) Karst (Norway spruce) and *P. glauca* (Moench) voss (white spruce). Plant Cell Rep 6: 157–159

Namboodiri KK, Beck CB (1968) A comparative study of the primary vascular system of conifers 3. Stelar evolution in gymnosperms. Am J Bot 55: 464–472

Nathorst AG (1902) Beitrage zur Kenntnis einiger mesozoischen Cycadophyten. Kongl Svenska Vetensk-Akad Handl 36: 1–28

Nathorst AG (1909) Über die Gattung *Nilssonia* Brong. Kongl Svenska Vetensk-Akad Handl 43

Nautiyal DD, Singh S, Pant DD (1976) Epidermal structure and ontogeny of stomata in *Gnetum gnemon*, *G. montanum* and *G. ula*. Phytomorphology 26: 282–296

Negi Vimla, Madhulata (1957) Male gametophyte and megasporogenesis in *Gnetum*. Phytomorphology 7: 230–236

Niklas KJ, Norstog K (1984) Aerodynamics and pollen grain depositional patterns on cycad megastrobili : Implications on the reproduction of three cycad genera (*Cycas*, *Dioon* and *Zamia*). Bot Gaz 145: 92–104

Noel ARA, van Staden J (1975) Seed-coat structure and germination in *Podocarpus henkelii*. Z pflanzen 77: 174–186

Norstog K (1965a) Some observations on the gametophytes of *Zamia integrifolia*. Phytomorphology 15: 46–49

Norstog K (1965b) Induction of apogamy in megagametophytes of *Zamia integrifolia*. Am J Bot 52: 993–999

Norstog K (1967a) Fine structure of the spermatozoids of *Zamia* with special reference to the flagellar apparatus. Am J Bot 54: 831–840

Norstog K (1967b) Some characteristics of growth and development in cycad tissue cultures. 5th Conf Cycad Toxicity. Natl Inst Health, Bethesda (USA) 4: 1–8

Norstog K (1968) Fine structue of the spermatozoids of *Zamia* : Observations on the microtubule systems and related structures. Phytomorphology 18: 350–356

Norstog K (1972) Role of archegonial neck cells of *Zamia* and other cycads. Phytomorphology 22: 125–130

Norstog K (1974) Fine structure of the spermatozoids of *Zamia* : The vierergruppe. Am J Bot 61: 449–456

Norstog K (1975) The motility of cycad spermatozoids in relation to structure and function, pp 135–142. In: Duckett JG, Racey PA (eds) The Biolgy of the Male Gamete. Biol J Linn Soc 7 (Suppl 1)

Norstog K (1977) The spermatozoids of *Zamia chigua* Seem. Bot Gaz 138: 409–412

Norstog K (1982) Experimenal embryology of gymnosperms, pp 25–51. In: Johri BM (ed) Experimental Embryology of Vascular Plants. Springer-Verlag, Berlin

Norstog K (1990) Spermatozoids of *Microcycas calocoma* : Ultrastructure. Bot Gaz 151: 275–284

Norstog K, Rhamstine E (1967) Isolation and culture of haploid and diploid cycad tissues. Phytomorphology 17: 374–381

Nyman B (1966) Studies on the fat metabolism of light and dark germinated seeds of scots pine *Pinus sylvestris* L. Physiol Plant 19: 63–75

Oliver FW, Scott DH (1904) On the structure of the Palaeozoic seed *Lagenostoma lomaxi*, with a statement of the evidence upon which it is referred to *Lyginodendron*. Phil Trans R Soc London B 197: 193–247

Osborn TGB (1960) Some observations on the life history of *Podocarpus falcatus*. Aust J Bot 8: 243–255

Pacini E, Franchi GG, Hesse M (1985) The tapetum: Its form, function, and possible phylogeny in Embryophyta. Pl Syst Evol 149: 155–185

Paliwal GS (1992) Phloem in Pinophyta. In : Venkatachala BS, Dilcher DL, Maheshwari HK (eds) Essays in Evolutionary Plant Biology (A Birbal Sahni Birth Centenary Tribute). Palaeobot 41: 114–127

Paliwal GS, Behnke HD (1973) Light microscopic study of the organization of phloem in the stem of *Gnetum gnemon*. Phytomorphology 23: 183–193

Palmer E, Pitman N (1972) Trees of Southern Africa. Vol 1. Balkema, Capetown

Pankow H (1962) Histogenetische studien an den Blüten einiger Phanerogamen. Bot Stud (Jena) 13: 1–106

Pant DD (1957) The classification of gymnospermous plants. Palaeobot 6: 65–70

Pant DD (1973) *Cycas* and Cycadales, pp 1–255. Central Book Depot Allahabad (India)

Pant DD (1977) The plant of *Glossopteris*. J Indian Bot Soc 56: 1–23

Pant DD (1982) The lower Gondwana gymnosperms and their relationships. Rev Palaeobot Palynol 37: 55–70

Pant DD (1986) Reproductive biology of the Glossopterids and their affinities. L'Evolution des Gymnospermes Approche biologique et paléobiologique. Resumes des communications. 1 'Universite' Sci Techniques Languador Montpellier du 23 au 25 Septembre 1986

Pant DD, Mehra B (1963) Development of stomata in leaves of three species of *Cycas* and *Ginkgo biloba* L. J Linn Soc London 58: 491–496

Pant DD, Singh R (1990) Preliminary observations on insect-plant relationships in Allahabad plants of *Cycas*. Palms and Cycads 28: 10–14

Paolillo Jr DJ (1965) On the androcyte of *Polytrichum* with special reference to the *Driergruppe* and limosphere (Nebenkern). Can J Bot 43: 669–676

Paolillo Jr DJ. Kreitner GL, Reighard JA (1968) Spermatogenesis in *Polytrichum juniperinum* 1. The origin of the apical body and the elongation of the nucleus. Planta 78: 226–247

Parthasarathy MV (1975) Ultrastructure of sieve cells in cycads (Abstr). 12th Intl Bot Congress

Patel KR, Shekhawat NS, Berlyn GP, Thorpe TA (1984) Isolation and culture of protoplasts from cotyledons of *Pinus coulteri* D Don. Plant Cell, Tissue and Organ Culture 3: 85–90

Pearson HHW (1912) On the microsporangium and microspore of *Gnetum*, with some notes on the structure of the inflorescence. Ann Bot (London) 26: 603–620

Pearson HHW (1914) Notes on the morphology of certain structures concerned in reproduction in the genus *Gnetum*. Trans Linn Soc London (Bot) II 8: 311–330

Pearson HHW (1929) Gnetales, Cambridge Univ Press, London

Pennell RI, Bell PR (1985) Microsporogenesis in *Taxus baccata* L.: The development of the archesporium. Ann Bot (London) 56: 415–427

Pennell RI, Bell PR (1986) Microsporogenesis in *Taxus baccata* L.: The formation of the tetrad and development of the microspores. Ann Bot (London) 57: 545–555

Pennell RI, Bell PR (1987) Megasporogenesis and the subsequent cell lineage within the ovule of *Taxus baccata* L. Ann Bot (London) 59: 693–704

Pettitt JM (1966) A new interpretation of the structure of the megaspore membrane in some gymnosperm ovules. J Linn Soc (Bot) 59: 253–263

Pettitt JM (1977) The megaspore wall in gymnosperms : Ultrastructure in some zooidogamous forms. Proc R Soc London B 195: 497–515

Pigg KB, Stockey RA, Taylor TN (1986) Studies of Palaeozoic seed ferns: Additional studies of *Microspermopteris aphyllum* Baxter. Bot Gaz 147: 126–136

Pijl, Van Der L (1953) On the flower biology of some plants from Java with general remarks on fly-traps (species of *Annona, Artocarpus, Typhonium, Gnetum, Arisaema* and *Abroma*). Ann Bogor 1: 77–99

Pilger R (1926) Gymnospermae. In : Engler A, Prantl K, Die natürlichen Pflanzenfamilien. 13: 121–407. Engelmann, Leipzing

Pilger R, Melchior H (1954) Gymnospermae. In: Engler A, Syllabus der Pflanzenfamilien. 1: 1: 312–344. Borntraeger, Berlin

Pillai SK (1963) Zonal structure and seasonal variations in the shoot apex of *Podocarpus gracilior*. Proc Indian Acad Sci B 57: 58–67

Pillai A (1966) Root apical organization in gymnosperms. Root apex of *Ephedra foliata* with a sugestion on the possible evolutionary trend of root apical structures in gymnosperms. Planta 70: 26–33

Plumstead EP (1952) Description of two new genera and six new species of fructifications borne on *Glossopteris* leaves. Trans Geol Soc S Afric?. 55: 281–328

Plumstead EP (1956) Bisexual fructifications borne on *Glossopteris* leaves from S Africa. Palaeontographica B 100: 1–25

Potonié H (1899) Lehrbuch der Pflanzenpalaeontologie, Berlin

Radforth NW (1936) The development in vitro of the proembryo of *Ginkgo*. Trans R Can Inst 21; 87–94

Radforth NW, Pegoraro L (1955) Assessment of early differentiation in *Pinus* proembryos transplanted to in vitro conditions. Trans R Soc Can 49: 69–82

Radforth EW, Bonga JM (1960) Differentiation induced as season advances in the embryo-gametophyte complex of *Pinus nigra* var *austriaca* using indole acetic acid. Nature 185:332

Radforth NW, Trip P, Bonga JM (1958) Polarity in the early embryogeny of *Ginkgo biloba* L. Trans R Soc Can 52: 55–58

Rao LN (1961) Life history of *Cycas circinalis* L. Part 1. Microsporogenesis, male and female gametophytes and spermatogenesis. J Indian Bot Soc 40: 599–619

Rao LN (1963) Life history of *Cycas circinalis* L. Part 2. Fertilization, embryogeny and germination of the seed. J Indian Bot Soc 42: 319–332

Rao AR (1981) The affinities of the Pentoxyleae. Palaeobotanist 28–29: 207–209

Rao NM, Mehta AR (1969) Callus tissue from the pollen of *Thuja orientalis* L. Indian J Exp Biol 7: 132

Raunsgaard PK, Fries EM (1986) Ultrastructure of *Caytonanthus* pollen, evolutionary significance. L'evolution des Gymnospermes. Approche biologique et paléobiologique. Resumes des communications. 1'Université Sci Techniques Languador Montpellier du 23 au 25 September 1986

Raymer MC (1947) Behaviour of Corsican pine stock following different nursery treatments. Forestry 21: 204–216

Razmologov VP (1973) Tissue culture from the generative cell of the pollen grain of *Cupressus* spp. Bull Torrey Bot Club 100: 18–22

Reilly K, Washer J (1977) Vegetative propagation of radiata pine by tissue culture: Plantlet formation from embryonic tissue. NZ J For Sci 7: 199–206

Reinert J, White PR (1956) The cultivation in vitro of tumor tissues and normal tissus of *Picea glauca*. Physiol Plant 9: 177–189

Reinke J (1872) Parasitische *Anabaena* in wurzeln der cycadeen. Gottingen Nachrichten 57: 107

Renault B (1879) Structure comparée de quelques tiges de la flore Carbonifére. Mus Hist Natu Paris Nouv Archives Ser 2, 2: 213–326

Renzoni GC, Viegi L, Stefani A, Onnis A (1990) Different in vitro germination responses in *Pinus pinea* pollen from two localities with different levels of pollution. Ann Bot Fennici 27: 85–90

Riding RT, Gifford Jr EM (1973) Histochemical changes occurring at the seedling shoot apex of *Pinus radiata*. Can J Bot 51: 501–512

Risser PG, White PR (1964) Nutritional requirements of spruce tumor cells in vitro. Physiol Plant 17: 620–635

Rodin RJ (1953a) Distribution of *Welwitschia mirabilis*. Am J Bot 40: 280–285

Rodin RJ (1953b) Seedling morphology of *Welwitschia*. Am J Bot 40: 371–378

Rodin RJ (1958) Leaf anatomy of *Welwitschia* 2. A study of mature leaves. Am J Bot 45: 96–103

Rodin RJ (1963) Anatomy of the reproductive bracts in *Welwitschia*. Am J Bot 50: 641–648

Rodin RJ, Kapil RN (1969) Comparative anatomy of the seed coats of *Gnetum* and their probable evolution. Am J Bot 56: 420–431

Rohr R (1973a) Production de cals par les gametophytes mâles de *Taxus baccata*

L. cultives sur un milieu artificiel : Étude en microscopie photonique et électronique. Caryologia 25 (Suppl) : 177–189

Rohr R (1973b) Ultrastructure des spermatozoides de *Taxus baccata* L. obtenus a partir de cultures aseptiques de microspores sur un milieu arrtificiel. C R Acad Sci Paris D 277: 1869–1871

Rohr R (1977) Evolution en culture in vitro des prothalles femelles âgées chez le *Ginkgo biloba* L. Z Pflanzenphysiol 85: 61–69

Rothwell GW (1982) New interpretations of the earliest conifers. Rev Palaeobot Palynol 37: 7–28

Rothwell GW (1985) The role of comparative morphology and anatomy in interpreting the systematics of fossil gymnosperms. Bot Rev 51: 319–327

Rothwell GW (1986) Morphology and Reproductive Biology in the origin of Cordaites and Conifers. L'evolution des Gymnospermes Approche biologique et paleobiologique. Resumés Des Communications l'université Sci Techniques du Languedoc. Montpellier du 23 au 25 September 1986

Roy Chowdhury Chhaya (1962) The embryogeny of conifers : A review. Phytomorphology 12: 313–338

Sacher JA (1956) Observations on pine embryos grown in vitro. Bot Gaz 117: 206–214

Sahni B (1920) On certain archaic features in the seed of *Taxus baccata*, with remarks on the antiquity of the Taxineae. Ann Bot (London) 34: 117–132

Sahni B (1948) The Pentoxyleae : A new group of Jurassic gymnosperms from the Rajmahal Hills of India. Bot Gaz 110: 47–80

Sahni B, Johri BM (1936) Pollen grains in the stylar canal and in the ovary of an angiosperm. Curr Sci 4: 587–589

Sanwal Madhulata (1962) Morphology and embryology of *Gnetum gnemon* L. Phytomorphology 12: 243–264

Saxton WT (1910) The development of the embryo of *Encephalartos*. Bot Gaz 49: 13–18

Saxton WT (1930) The root nodules of Podocarpaceae. S African J Sci 27: 323–325

Saxton WT (1934) Notes on conifers 8. The morphology of *Austrotaxus spicata* Compton. Ann Bot (London) 48: 411–427

Saxton WT (1936) Notes on conifers 10. Some normal and abnormal structures in *Taxus baccata*. Ann Bot (London) 50: 519–522

Schenk A (1867) Fossil flora der Grenzschichten des keupers und Lias Frankens. Wiesbaden

Schmidt A (1924) Über die chlorophyllbildung in Koniferenembryo. Bot Arch 5: 260–285

Schnarf K (1937) Anatomie der Gymnospermen-Samen. Borntrager, Berlin

Schopf JM (1976) Morphologic interpretations of fertile structures in *Glossopteris* gymnosperms. Rev Palaeobot Palynol 21: 25–64

Schuller A, Reuther G, Geier T (1989) Somatic embryogenesis from seed explants of *Abies alba*. Plant Cell, Tissue and Organ Culture 17: 53–58

Schuster J (1932) Cycadaceae. In : Engler A (ed) Das Pflanzenreich, 4 (1). Leipzig

Schweitzer HJ (1963) Der weibliche zapfen von *Pseudovoltzia liebeana* und seine Bedeutung für die Phylogenie der Koniferen. Palaeontographica B 113: 1–29

Scott DH (1909) Studies in Fossil Botany, 2nd edn. Vol 1 London

Scott DH (1923) Studies in Fossil Botany Vol 2: 1–446. A and C Black London

Sedgwick PJ (1924) Life history of *Encephalartos*, Bot Gaz 77: 300–310

Seeliger I (1954) Studien am Sprossvegetationskegel von *Ephedra fragilis* var *campylopoda* (CA Mey) Stapf. Flora 141: 114–162

Sharma BD (1994) Gymnosperms : Morphology, Systematics, Reproductive Biology, pp 1–23. In: Johri BM (ed) Botany in India : History and Progress. Vol 2: pp 1–480. Oxford IBH, New Delhi

Sharma V (1975) Development and morphology of sclereids in some species of *Gnetum*. Curr Sci 44: 714–716

Shiu-Ying Hu (1969) *Ephedra* (Ma-Huang) in the New Chinese Materia Medica. Econ Bot 23: 346–351

Simola LK (1974) The ultrastructure of dry and germinating seeds of *Pinus sylvestris*. Acta Bot Fenn 103:1–31

Simola LK, Santanen A (1990) Improvement of nutrient medium for growth and embryogenesis of megagametophyte and embryo callus lines of *Picea abies*. Physiol Plant 80: 27–35

Singh H (1961) The life history and systematic position of *Cephalotaxus drupaca* Sieb et Zucc. Phytomorphology 11: 153–197

Singh H (1978) Embryology of Gymnosperms, pp 1–302. Geb Bornträger, Berlin

Singh H, Maheshwari Kamala (1962) A contribution to the embryology of *Ephedra gerardiana* Wall. Phytomorphology 12: 361–372

Singh H, Oberoi YP (1962) Life history of *Biota orientalis*. Phytomorphology 12: 373–393

Singh H, Chatterjee Jyotsna (1963) A contribution to the life history of *Cryptomeria japonica* D. Don. Phytomorphology 13: 429–445

Singh MN, Konar RN, Bhatnagar SP (1981) Haploid plantlet formation from female gametophyte of *Ephedra foliata* Boiss. in vitro. Ann Bot (London) 48: 215–220

Sinnott EW (1913) The morphology of the reproductive structures in Podocarpineae. Ann Bot (London) 27: 39–82

Smith DR, Horgan KJ, Aitken-Christie J (1982) Micropropagation of *Pinus radiata* for afforestation, pp 723–724. In: Fujiwara A (ed) Plant Tissue and Cell Culture. Maruzen, Tokyo

Sokolowa C (1890) Naissance de l'endosperme dans le sac embryonnaire de quelque gymnospermes. Byull Mosk Obshch Ispyt Prir 4: 446–497

Sommer HE, Brown CL (1974) Plantlet formation in pine tissue cultures. Am J Bot 61: 11

Sommer HE, Brown CL, Kormanik PP (1975) Differentiation of plantlets in longleaf pine (*Pinus palustris* Mill) tissue cultures in vitro. Bot Gaz 136: 196–200

Sporne KR (1965) The Morphology of Gymnosperms, pp 1–216. Hutchinson, London

Srivastava BP (1946) Silicified plant remains from the Rajmahal Series of India. Proc Natl Acad Sci India (Allahabad) 15: 185–211

Srivastava LM (1963a) Cambium and vascular derivatives of *Ginkgo biloba*. J Arnold Arbor 44: 165–192

Srivastava LM (1963b) Seconadary phloem in Pinaceae. Univ Calif Publ Bot 36: 1–142

Srivastava LM (1970) The seconadary phloem of *Austrobaileya scandens*. Can J Bot 48: 341–359

Srivastava LM, O'Brien TP (1966) On the ultrastructure of cambium and its vascular derivatives. 2. Secondary phloem of *Pinus strobus*. Protoplasma 61: 277–293

Staff IA, Ahern CP (1993) Symbiosis in cycads with special reference to *Macrozamia communis*, pp 200–210. In: Stevenson DW, Norstog J (eds) The biology, structure and systematics of the Cycadales. Proc CYCAD 90, 2nd Intl Conf on Cycad Biology. Palm and Cycad Societies Australia, Queensland, Australia

Sterling C (1948a) Gametophyte development in *Taxus cuspidata*. Bull Torrey Bot Club 75: 147–165

Sterling C (1948b) Proembryo and early embyogeny in *Taxus cuspidata*. Bull Torrey Bot Club 75: 469–485

Sterling C (1949a) Embryonic differentiation in *Taxus cuspidata.* Bull Torrey Bot Club 76: 116–133

Sterling C (1949b) Preliminary attempts in larch embryo culture. Bot Gaz 111: 90–94

Sterling C (1958) Dormant apical bud of *Agathis lanceolata.* Bot Gaz 120: 49–53

Stevenson DW (1981) Observations on root and stem contraction in cycads (Cycadales) with special reference to *Zamia pumila* L. Bot J Linn Soc 81: 275–281

Stevenson DW (1990a) Morphology and Systematics of the Cycadales. Mem NY Bot Gdn 57: 8–55

Stevenson DW (1990b) *Chigua,* a new genus in the Zamiaceae with comments on its biographic significance. Mem N Y Bot Gdn 57: 169–172

Stewart WN (1976) Plystely, primary xylem, and the Peteropsida. Birbal Sahni Inst Palaeobot, Lucknow, India

Stewart WN (1981) The Progymnospermopsida : The construction of a concept. Can J Bot 59: 1539–1542

Stewart WN (1983) Palaeobotany and the evolution of plants, pp 1–405. University Press, Cambridge (England)

Stewart KD, Gifford Jr EM (1967) Ultrastructure of the developing megaspore mother cell of *Ginkgo biloba.* Am J Bot 54: 375–383

Stiles W (1912) The Pdocarpaceae. Ann Bot (London) 26: 443–514

Stockey RA (1982) The Araucariaceae: An evolutionary perspective. Rev Palaeobot Palynol 37: 133–154

Stockey RA, Taylor TN (1978a) Scanning electron microscopy of epidermal patterns and cuticular structure in genus *Araucaria.* Scanning Electron Micoscopy 2: 223–228

Stockey RA, Taylor TN (1978b) Cuticular features and epidemal patterns in the genus *Araucaria* De Jussieu. Bot Gaz 139: 490–498

Stockey RA, Ko Helen (1986) Cuticle micromorphology of *Araucaria* De Jussieu. Bot Gaz 147: 508–548

Stopes MC, Fujii K (1906) The nutritive relationship of the surrounding tissues to the archegonia of gymnosperms. Beih Bot Zbl 20: 1–24

Strasburger E (1871) Die Bestäubung der Gymnospermen. Jena Z Med Naturw 6: 249–262

Strasburger E (1872) Die Coniferen und die Gnetaceen. H Dabis Jena

Strasburger E (1891) Über den Bau und die verreichtungen der Leitungsbahnen in den Pflanzen. Histol Beitr 3. Jena

Strasburger E (1930) Textbook of Botany (Engl Tansl By Lang WH). London

Sugihara Y (1947) The male gametes of *Cephalotaxus drupacea* Siebold et Zucc. Bot Mag Tokyo 6: 45–46

Sugihara Y (1969) On the embryo of *Cryptomeria japonica.* Phytomorphology 19: 110–111

Surange KR (1966) The present position of the genus *Glossopteris,* pp 316–327. Proc Autumn School Botany, Mahabaleshwar (India)

Surange KR, Maheshwari HK (1970) Some male and female fructifications of Glossopteridales from India. Plaeontographica B 129: 178–192

Surange KR, Chandra S (1971) *Dankania indica* gen et sp. nov.: A glossopteridean fructification from the Lower Gondwana of India. Palaeobotanist 20: 264–268

Surange KR, Chandra S (1972) Fructifications of Glossopteidae from India. Palaeobot 21: 1–17

Surange KR, Chandra S (1975) Morphology of the gymnospermous fructifications of the *Glossopteris* flora. Palaeontographica B 149: 153–180

Surange KR, Chandra S (1976) Morphology and affinities of *Glossopteris.* Palaeobot 25: 509–524

Swamy BGL (1948) Contribution to the life history of a *Cycas* from Mysore (India). Am J Bot 35: 77–88

Swamy BGL (1973) Contributions to the monograph on *Gnetum* 1. Fertilization and proembryo. Phytomorphology 23: 176–182

Swamy BGL (1974) Contributions to the monograph on *Gnetum* 3. Some observations on microsporogenesis and germinations of pollen. Phytomorphology 24: 206–210

Takaso T (1984) Structural changes in the apex of the female strobilus and the initiation of the female reproductive organ (ovule) in *Ephedra distachya* L. and *E. equisetina* Bge. Acta Bot Neerl 33: 257–266

Takaso T (1985) A developmental study of the integument in Gymnosperms 3. *Ephedra distachya* L. and *E. equisetina* Bge. Acta Bot Neerl 34: 33–48

Takaso T, Bouman F (1986) Ovule and seed ontogeny in *Gnetum gnemon* L. Bot Mag Tokyo 99: 241–266

Takeda H (1913) Development of the stoma in *Gnetum gnemon*. Ann Bot (London) 27: 365–366

Takhtajan AL (1956) Telomophyta. Academiae Scientiarum, Moscow

Takhtajan AL (1969) Flowering Plants: Origin and Dispersal, pp 1–310. Oliver and Boyd, Edinburgh

Takhtajan AL (1980) Outline of the classification of flowering plants (Magnoliophyta). Bot Rev 46: 225–359

Tang W (1987a) Heat production in cycad cones. Bot Gaz 148: 165–174

Tang (1987b) Insect pollination in the cycad *Zamia pumila* (Zamiaceae). Am J Bot 74: 90–99

Tarchi AMF (1969) Callose wall in sporogenesis of *Biota orientalis* Endl. preliminary observations. G Bot Ital 103: 515–535

Tautorus TE, Attree SM, Fowke LC, Dunstan DI (1990) Somatic embryogenesis from immature and mature zygotic embryos, and embryo regeneation from protoplasts in black spruce (*Picea mariana* Mill.) Plant Sci 67: 115–124

Taylor TN, Millay MA (1979) Pollination biology and reproduction in early seed plants. Rev Palaeobot Palynol 27: 329–355

Taylor TN, Millay MA (1981) Morphological variability of Pennsylvanian lyginopterid seed ferns. Rev Palaeobot Palynol 32: 27–62

Tepfer SS, Greyson RI, Craig WR, Hindman JL (1963) In vitro culture floral buds of *Aquilegia*. Am J Bot 50: 1035–1045

Terrazas Teresa (1991) Origin and activity of successive cambia in *Cycas* (Cycadales) Am J Bot 78: 1335–1344

Thieret JW (1958) Economic botany of the cycads. Econ Bot 12: 3–41

Thomas HH (1925) The Caytoniales : A new group of angiospermous plants from the Jurassic rocks of Yorkshire. Phil Trans R Soc London B 213: 299–363

Thomas HH (1955) Mesozoic pteridosperms, Phytomorphology 5: 177–185

Thomas HH, Bancroft N (1913) On the cuticle of some recent and fossil cycadean fronds. Trans Linn Soc London Ser 2 Bot 48: 155–204

Thomas MJ (1970) Premiérs recherches sur les besoins nutritifs des embryons isolés du *Pinus sylvestris* L. : Embryons differenciés. C R Acad Sci Paris D 270: 2648–2651

Thomas BA, Spicer RA (1986) The evolution and palaeobiology of land plants. Croom Helm, London

Thompson WP (1916) The morphology and affinities of *Gnetum*. Am J Bot 3: 135–184

Tiagi YD (1966) A contribution to the morphology and vascular anatomy of *Ephedra foliata* Boiss. Proc Natl Acad Sci 36: 417–436

Torrey JG (1966) The initiation of organized development in plants. Advances in morphogenesis. 5: 39–81

Tourte Y, Hurel-Py G (1967) Ontogénie et ultrastructure de l'appareil cinétique des spermatozoides du *Pteridium aquilinium* L. C R Acad Sci Paris Sér D 265: 1289–1292

Townrow JA (1967) On *Voltziopsis* : A southern conifer of Lower Triassic age. Pap Proc R Soc Tasmania 101: 173–188

Trease GE, Evans WC (1983) Pharmacognosy, pp 1–812. 12th Edn. Bailliere Tindall

Tremblay FM (1990) Somatic embryogenesis and plantlet regeneration from embryo isolated from stored seeds of *Picea glauca*. Can J Bot 68: 236–242

Treub M (1884) Recherches sur les Cycadees 3. Embryogenie du *Cycas circinalis*. Ann Jard Bot Buitenzorg 4: 1–11

Trevisan L (1980) Ultrastructural notes and considerations of *Ephedripites*, *Eucommiidies* and *Monosulcites* pollen grains from Lower Cetaceous sediments of Southern Tuscany (Italy). Pollen Spores 22: 85–132

Troup RS (1921) The Silviculture of Indian Trees. Vol 3. Oxford University Press, Oxford (UK)

Tulecke W (1953) A tissue derived from the pollen of *Ginkgo biloba*. Science 117: 599–600

Tulecke W (1957) The pollen of *Ginkgo biloba* : In vitro culture and tissue formation. Am J Bot 44: 602–608

Tulecke W (1959) The pollen cultures of CD LaRue : A tissue from the pollen of *Taxus*. Bull Torrey Bot Club 86: 283–289

Tulecke W (1960) Arginine-requiring strains of tissue obtained from *Ginkgo* pollen. Plant Physiol 35: 19–24

Tulecke W (1964) A haploid tissue culture from the female gametophyte of *Ginkgo biloba* L. Nature 203: 94–95

Tulecke W (1967) Studies on tissue cultures derived from *Ginkgo biloba* L, Phytomorphology 17: 381–386

Tulecke W, Nickell LG (1960) Methods, problems, and results of growing plant cells under submerged conditions. Trans N Y Acad Sci Ser 3 22: 196–206

Tulecke W, Sehgal Nanda (1963) Cell proliferation from the pollen of *Torreya nucifera*. Contr Boyce Thompson Inst (Yonkers, USA) 23: 33–46

Tulecke W, Taggart R, Colavito L (1965) Continuous cultures of higher plant cells in liquid media. Contr Boyce Thompson Inst (Yonkers USA) 23: 33–46

Vasil Vimla (1959) Morphology and embyology of *Gnetum ula* Brongn. Phytomorphology 9: 167–215

Vasil Vimla (1963) In vitro culture of embryos of *Gnetum ula* Brongn, pp 278–280. In: Maheshwari P, Rangaswamy NS (eds) Symp Plant Tissue and Organ Culture. Intl Soc Plant Mophologists, Univ Delhi

Vasil IK, Aldrich HC (1970) A histochemical and ultrastructural study of the ontogeny and differentiation of pollen in *Podocarpus macrophyllous* D. Don, Protoplasma 71: 1–37

Vasiliyeva GV (1969) A contribution to the comparative anatomy of leaves of the *Araucaria* De Jussieu. Bot Zh SSSR 54: 448–459

Verhagen SA, Wann SR (1989) Norway spruce somatic embryogenesis : High frequency initiation from light cultured mature embryos. Plant Cell, Tissue and Organ Culture 16: 103–111

Vishnu–Mittre (1952) A male flower of the Pentoxyleae with remarks on the stucture of the female cones of the group. Palaeobot 2: 75–84

Vishnu–Mittre (1957) Studies on the fossil flora of Nipania (Rajmahal series), India– Pentoxyleae. Palaeobot 6: 31–46

Von Aderkas P, Bonga JM (1988) Formation of haploid embryoids of *Larix decidua*. Am J Bot 75: 690–700

Von Aderkas P, Bonga JM, Nagmani R (1987) Promotion of embryogenesis in cultured megagametophytes of *Larix decidua*. Can J For Res 17: 1293–1296

Von Aderkas P, Klimaszewska K, Bonga JM, (1990) Diploid and haploid embryogenesis in *Larix leptolepis, L. decidua* and their reciprocal hybrids. Can J For Res 20: 9–14

Von Arnold S (1987) Improved efficiency of somatic embryogenesis in mature embryos of *Picea abies* (L.) Karst. J Plant Physiol 128: 233–234

Von Arnold S, Hakman I (1986) Effects of sucrose on initiation of embryogenic callus from mature zygotic embryos of *Picea abies* (L.) Karst. (Norway spruce). J Plant Physiol 122: 251–265

Von Arnold S, Woodward S (1988) Organogenesis and embryogenesis in mature zygotic embryos of *Picea sitchensis*. Tree Physiol 4: 291–300

Von Willart DJ (1985) *Welwitschia mirabilis-* New aspects in the biology of an old plant. Adv Bot Res Vol 2 : 157–191

Von Breitenbach F (1965) The indigenous trees of Southern Africa. Govt. Printer, Pretoria (South Africa)

Vovides AP, Olivares M (1996) Karyotype polymorphism in the cycad *Zamia loddigesii* (Zamiaceae) of the Yucatan Pninsula, Mexico. Bot J Linn Soc 120: 77–83

Walton J (1928) On the structure of a Paleozoic cone-scale and the evidence it furnishes of the primitive nature of the double conescale in the conifers. Mem Proc Manchester Lit Philos Soc 73: 1–6

Wani MC, Taylor HL, Wall ME, Coggon P, McPhail AT (1971) Plant antitumour agents VI. The isolation and structure of taxol, a novel antileukemic and antitumour agent from *Taxus brevifolia*. J Am Chem Soc 93: 2325–2327

Waterkeyn L (1954) Études sur les Gnétales 1. Le strobile femelle, l'ovule et la grain de *Gnetum africanum* Welw. Cellule 56: 105–146

Waterkeyn L (1959) Études sur les Gnétales 2. Le strobile mále, la microsporogenése et le gamétohyte mále de *Gnetum africanum* Welw. Cellule 60: 1–78

Webb KJ, Street HE (1977) Morphogenesis in vitro of *Pinus* and *Picea*. Acta Hort 78: 259–269

Webber HJ (1901) Spermatogenesis and fecundation of *Zamia*. USDA Bur Plant Ind Bull No. 2: pp 1–92

Wershing HF, Bailey IW (1942) Seedling as experimental material in the study of "redwood" in conifers. J For 40: 411–414

Wettstein R (1935) Handbuch der systematischen Botanik. Deuticke, Leipzig Vienna

White CT (1947) Notes on two species of *Araucaria* in New Guinea and a proposed new section of the genus. J Arnold Arbor 28: 259–260

White PR (1967) Some aspects of differentiation in cells of *Picea glauca* cultivated in vitro. Am J Bot 54:334–353

White PR, Risser PG (1964) Some basic parameters in the cultivation of spruce tissues. Physiol Plant 17: 600–619

Whitmore TC (1980a) A monograph of *Agathis*. Plant Syst Evol 135: 41–49

Whitmore TC (1980b) Utilization, potential and conservation of *Agathis*: A genus of tropical Asian conifers. Econ Bot 34: 1–12

Wieland GR (1906) American Fossil Cycads, pp 1–295. Carnegi Institution, Washington

Wilde MH (1975) A new interpretation of microsporangiate cones in Cephalotaxaceae and Taxaceae. Phytomorphology 25: 434–450

Wilde MH, Eames AJ (1952) The ovule and "seed" of *Araucaria bidwilli* with discussion of the taxonomy of the genus, 2. Taxonomy. Ann Bot (London) 16: 27–47

Wilson EH (1926) *Thuja orientalis* Linn. J Arnold Arbor 7: 71–74

Willemse MTM (1968) Development of micro- and megagametophyte of *Pinus sylvestis* L. An electron microscopic investigation. Acta Bot Neerl 17: 330–331

Willemse MTM (1971a) Morphological and quantitative changes in the population of cell oganelleles during microsporogenesis of *Pinus sylvestis* L. 1. Morphological changes from zygotene until prometaphase 1. Acta Bot Neerl 20 : 261–274

Willemse MTM (1971b) Morphological and quantitative changes in the population of cell oganelleles during microsporogenesis of *Pinus sylvestris* L. 2. Morphological changes from prometaphase 1 until the tetrad stage. Acta Bot Neerl 20 : 411–427

Willemse MTM (1971c) Morphological and quantitative changes in the population of cell organelles during microsporogenesis of *Pinus sylvestris* L. 3. Mophological changes during the tetrad stage and in the young microspore. A quantitative approach to the changes in the population of cell organelleles. Acta Bot Neerl 20 : 498–523

Willemse MTM, (1971d) Morphological changes in the tapetal cell during microsporogenesis of *Pinus sylvestris* L. Acta Bot Neerl 20: 611–623

Willemse MTM, Linskens HF (1969) Développment du microgametophyte chez le *Pinus sylvestris* entre la meiose et la fécondation. Rev Cytol Biol vég 32: 121–128

Williamson WC (1887) On the organization of the fossil plants of the Coal measures 13 Phil Trans R Soc London B 178: 289–304

Willis JC (1966) A Dictionary of Flowering Plants and Ferns, pp 1–1245. Cambridge Univ Press, Cambridge (England)

Wolniak MS (1976) Organelle distribution and apportionment during meiosis in the microsporocyte of *Ginkgo biloba* L. Am J Bot 63: 251–258

Wooding FBP (1966) The development of the sieve cells of *Pinus pinea*. Planta 69: 230–243

Wooding FBP (1968) Fine structure of callus phloem in *Pinus pinea*. Planta 83: 99–110

Wooding FBP, Northcote DJ (1965) Association of the endoplasmic reticulum and the plastids in *Acer* and *Pinus*. Am J Bot 52: 526–531

Zhong J-J (1995) Recent advances in cell cultures of *Taxus* spp. for production of the natural anticancer drug taxol. Plant Tissue Culture and Biotechnology 1: 75–80

Ziegler H (1959) Über die Zusammensetzung des "Bestaubungstropfens" und den Mechanismus seiner Sekretion. Planta 52: 587–599

Plant Index
(*Extinct)